The Nigerian Rice Economy

IFPRI

This book is published by the University of Pennsylvania Press (UPP) on behalf of the International Food Policy Research Institute (IFPRI) as part of a joint-publication series. Books in the series present research on food security and economic development with the aim of reducing poverty and eliminating hunger and malnutrition in developing nations. They are the product of peer-reviewed IFPRI research and are selected by mutual agreement between the parties for publication under the joint IFPRI-UPP imprint.

The Nigerian Rice Economy

Policy Options for Transforming Production, Marketing, and Trade

Edited by Kwabena Gyimah-Brempong, Michael Johnson, and Hiroyuki Takeshima

Published for the International Food Policy Research Institute

PENN

University of Pennsylvania Press
Philadelphia

Published by
University of Pennsylvania Press
Philadelphia, Pennsylvania 19104-4112
www.upenn.edu/pennpress

Library of Congress Cataloging-in-Publication Data is available.

ISBN 978-0-8122-4895-1

Printed in the United States of America on acid-free paper
10 9 8 7 6 5 4 3 2 1

Contents

Tables and Figures

Tables

Figures

Acronyms and Abbreviations

ABS	average budget share(s)
ADP	Agricultural Development Project
AfRGM	African rice gall midge
ATA	Agricultural Transformation Agenda
C&F	cost and freight
CGE	computable general equilibrium
COMTRADE	United Nations Commodity Trade Statistics Database
CPI	consumer price index
EA	enumeration area
EMM	economywide multimarket
FAO	Food and Agriculture Organization of the United Nations
FAOSTAT	statistical database of the Food and Agriculture Organization of the United Nations
FMARD	Federal Ministry of Agriculture and Rural Development
FOB	free on board
GDP	gross domestic product
GES	Growth Enhancement Support program
IFPRI	International Food Policy Research Institute
IITA	International Institute of Tropical Agriculture
IMF	International Monetary Fund
IRRI	International Rice Research Institute
LES	Linear Expenditure System

LGA	local government area
LSMS–ISA	Living Standards Measurement Study–Integrated Surveys on Agriculture
MBS	marginal budget share(s)
MRM(s)	modern rice mill(s)
MT	metric tons
NAERLS	National Agricultural Extension and Research Liaison Services
NARI(s)	national agricultural research institute(s)
NBS	National Bureau of Statistics
NCRI	National Cereals Research Institute
NEEDS	National Economic Empowerment and Development Strategy
NERICA	New Rice for Africa
NRA	Nominal Rate of Assistance
NSS	National Seed Service
O&M	operations and maintenance
OFN	Operation Feed the Nation
ppm	parts per million
PPP	purchasing power parity
PrOpCom	Promoting Pro-poor Opportunities in service and Commodity Markets
R&D	research and development
RMM	rice milling model
RYMV	rice yellow mottle virus
S&T	science and technology
SAPs	structural adjustment programs
SCPZ	staple crop processing zone
SSA	Africa south of the Sahara
USDA	United States Department of Agriculture
WARDA	West Africa Rice Development Association (now Africa Rice Center)
WDI	World Development Indicators of the World Bank

Foreword

In recent decades, Africa south of the Sahara has become increasingly reliant on rice imports. This is the result of globalization, urbanization, and diet change, as well as slow transformation and modernization of domestic agricultural sectors in many parts of the region. Because rice is an important source of diet and income, the increase in importation has become a symbol of the food security challenges facing many African countries.

Nigeria is the most populous country in Africa, with the largest economy. It also has the largest rice production area in Africa, comparable in size and diversity to many Asian countries. However, despite ever-growing demand and a long history of government efforts, its rice sector remains vastly underdeveloped, and the country is one of the largest rice importers in the world.

The Nigerian Rice Economy assesses the policy challenges and opportunities for transforming and expanding Nigeria's rice economy. The authors discuss the rice economy's evolution, structure, and agroecological constraints, as well as policy issues related to consumption, production, milling, and trade, each substantiated by rigorous quantitative analyses. The book provides in-depth insights into focused areas, including the effects of rice price policy on distribution, production technology constraints, optimal milling sector structure, trade policy effectiveness, and economywide implications of key rice sector interventions. The authors also suggest options to improve the rice sector.

Achieving rice self-sufficiency in the short term, as envisaged by the Nigerian government, may be too costly in terms of required resources and social costs. Price and trade policies alone may be ineffective in inducing private-sector responses, as these policies are often stymied by limited supply responses, cross-border leakages, limited market integration, and low substitutability of imported rice with local rice. High rice prices can rather significantly hurt the majority of low-income consumers. But policymakers can transform the rice sector, and there is room for improved policy. Production technologies and market infrastructure improvement are at the center of an effective rice development strategy that includes reducing production costs, raising sector competitiveness, and enhancing the biophysical potential of Nigerian rice.

Shenggen Fan
Director General, IFPRI

Acknowledgments

This book is the result of the collaboration and active support of many individuals on two continents—Africa and North America—either in reviewing previous drafts, in sharing input during the presentation of preliminary findings at various venues, or in encouraging the completion of the book. For this, we are grateful.

In particular, we wish to thank various Nigerian officials, including the minister of agriculture and rural development, Dr. Akinwumi Adesina; Dr. Olumuyiwa Osiname, the chair of the Rice Value Chain Development Team of the Federal Ministry of Agricultural and Rural Development (FMARD); and Dr. Hussaini, director of the Planning, Research and Statistics Division of FMARD, for providing important insights into Nigerian conditions. We also greatly appreciate the staff from Agricultural Development Projects, Fadama Development Offices, the National Bureau of Statistics, and the Nigeria Customs Office for providing various secondary information and facilitating authors' fieldwork.

We are indebted to the intellectual support and guidance from various other senior researchers at the International Food Policy Research Institute (IFPRI), including Jawoo Koo, Bart Minten, Tewodaj Mogues, and Ephraim Nkonya. We are grateful to Gershon Feder, the chair of the IFPRI Publication Review Committee, for providing invaluable suggestions and constructive criticisms along various stages of the writing of this book. We are also indebted to Marco Wopereis (Africa Rice Center) and Randy Barker for their guidance in identifying experts in Nigerian agronomy, as well as for sharing important insights into general rice-sector issues. We are also indebted to two anonymous reviewers, as their invaluable inputs greatly improved the manuscript.

We are particularly grateful to the excellent research support of Angga Pradesha and Amarachi Utah at IFPRI headquarters in Washington, DC, and of Ian Masias and Hyacinth Edeh, both at IFPRI's Nigeria Strategy Support Program (NSSP) Abuja office. We are also grateful for the efficient administrative and logistical support provided to us by staff in Abuja at both IFPRI and its host institution, the International Fertilizer Development Institute. They made sure our travel plans and appointments occurred in a timely, safe, and secure fashion.

The feedback from participants at various venues in Africa proved especially valuable. Among these are an IFPRI-NSSP conference in Abuja, Nigeria (2013); a seminar at the International Institute of Tropical Agriculture campus in Ibadan, Nigeria (2013); a seminar at the Institute of Food Security in Makurdi, Nigeria (2013); a presentation at the Third Africa Rice Congress in Yaoundé (2013); and the IFPRI Dissemination Workshop in Accra, Ghana (2015).

The completion of the book would not have been possible without the financial support of a number of donors. Among these, core support was provided by the Nigerian mission of the United States Agency for International Development (USAID/Nigeria). Other financial support was indirectly provided by the Policies, Institutions and Markets research program of CGIAR and the Japanese government through its bilateral funding to IFPRI. A number of individuals among the donor community in Nigeria also deserve mention, as they provided both technical and strategic support along the way. These include Howard Batson and Alefia Merchant of USAID/Nigeria, and Atsuko Toda of the Abuja office of the International Fund for Agricultural Development.

Finally, we would like to thank the director general of IFPRI, Dr. Shenggen Fan, for providing continuous encouragement toward the completion of this book.

The views expressed in this book do not necessarily reflect the views of IFPRI, the National Cereals Research Institute, or the donors. The authors are solely responsible for all errors.

Chapter 1

RICE IN THE NIGERIAN ECONOMY AND AGRICULTURAL POLICIES

Kwabena Gyimah-Brempong, Michael Johnson, and Hiroyuki Takeshima

Over the past few decades, rice has become one of the leading food staples in Nigeria, surpassing cassava in food expenditures. Throughout this period, consumption has increased faster than production, resulting in a growing dependency on imports. By 2014, about half of the rice consumed in Nigeria was imported. As the most populous country in Africa south of the Sahara (SSA), Nigeria has quickly become the leading importer of rice on the continent and, more recently, in the world.

This growing dependence on rice imports is a major concern of Nigeria's government, and since the early 1980s numerous programs have been implemented to encourage domestic rice production and achieve rice self-sufficiency (or at least to reduce the growth in imports). In particular, rice featured prominently in the Agricultural Transformation Agenda (ATA), which had guided the Federal Ministry of Agriculture and Rural Development (FMARD) in Nigeria under the administration of President Goodluck Jonathan as the central agenda of the country's agricultural policy.[1] The ATA included major investments and programs related to rice production, processing, and marketing. Trade policies (import tariffs and even import bans) have also been used in an attempt to slow the growth in imports, with import tariffs on milled rice increasing to 110 percent beginning in 2013.

In spite of these policies, the Nigerian rice sector has yet to be transformed into a more productive one that can compete with foreign imports. This situation is not unique to Nigeria and applies to the rest of SSA, where the sector's slow growth has puzzled many international donors and research communities (Otsuka and Larson 2012). As one of the largest producers and consumers of rice in SSA, Nigeria has been at the center of this puzzle.

The principal objective of this book is to review and assess the potential for Nigeria to transform its domestic rice sector to become competitive with

1 This book was written when Nigeria was under President Jonathan's administration. Throughout the book, rice policies or the policy framework mentioned are those that had been implemented during this administration, unless otherwise specified.

imports. We assess the policy alternatives for bringing about this transformation and also briefly discuss the opportunity costs of achieving such competitiveness. In particular, three key strategies that have recently been adopted by the government are examined in more detail with regard to their potential long-run welfare implications for transforming the rice economy and making domestic brands competitive with imports:

1. Introducing public-sector interventions to stimulate paddy production through the dissemination and adoption of better seeds and other modern inputs.

2. Improving the postharvest processing and milling sectors to promote premium and high-quality local brands of rice.

3. Introducing import tariffs to help protect the domestic rice sector.

This chapter assists readers to gain a better understanding of how Nigeria arrived at its recent policy framework for agriculture and the rice sector in particular. The chapter first describes the evolution of rice imports and the growing imbalance between production and consumption. It then presents a brief overview of Nigeria's economy and recent history, the basic structure of the agricultural sector, and a summary of recent key rice policies. Finally, the chapter describes the key set of questions asked in this book and how each chapter addresses them from different perspectives on the rice sector in Nigeria.

Nigeria's Rice Trade in a Global Context

Nigeria has become the world's biggest importer of rice within the last ten years. As Table 1.1 shows, Nigeria's share of global rice imports has risen from 7 percent in the early years of the 21st century to 8.2 percent over the most recent five years for which data are available (2008–2012). Among the top rice importers, Nigeria is followed closely by the Philippines, Iran, Indonesia, and the European Union. The bulk of rice imports to Nigeria come from Thailand, Vietnam, and India, who together supply about 60 percent of the rice traded in global markets.

The reliance on global rice markets raises serious concerns for policy with regard to ensuring food security and maintaining a healthy balance of the country's foreign-exchange reserves. This is especially true when faced with a dramatic rise in prices, as occurred during the recent 2008 food crisis. For rice in particular, prices rose by about 255 percent between 2007 and 2008, even higher than in the last major food crisis in 1974, when they increased by 200 percent (Headey and Fan 2008).

TABLE 1.1 Top ten major rice importers and exporters in the world, 2000–2012 (percent)

Top 10 major importers (share of global imports)				Top 10 major exporters (share of global exports)			
2000–2004		2008–2012		2000–2004		2008–2012	
Indonesia	8.8	Nigeria	8.2	Thailand	30.1	Thailand	27.3
Nigeria	7.0	Philippines	6.7	Vietnam	14.0	Vietnam	19.3
EU	5.3	Iran	5.6	India	13.6	India	13.7
Philippines	5.1	Indonesia	4.8	United States	11.9	Pakistan	10.3
Saudi Arabia	5.0	EU	4.8	China	7.9	United States	10.1
Iraq	4.4	Iraq	4.0	Pakistan	7.7	Uruguay	2.6
Iran	4.2	Saudi Arabia	3.9	Uruguay	2.7	Brazil	2.4
Brazil	3.5	Malaysia	3.6	Egypt	2.4	Cambodia	2.1
Senegal	3.5	Côte d'Ivoire	3.4	Burma	1.8	Burma	2.1
South Africa	3.1	Senegal	3.0	Australia	1.4	China	1.9

Source: United States Department of Agriculture international database (USDA 2013).

As in 1974, the experience of the 2008 food crisis led many governments of net importing countries in the developing world to reduce their vulnerability to price shocks by striving for self-sufficiency in rice production. Nigeria is no exception. In the aftermath of the most recent crisis and rising consumer preference for imported rice, the Nigerian government has set a goal of making the country self-sufficient in rice production. The perception among Nigerian policymakers is that the increasing trend of rice imports is fiscally and politically unsustainable, as it threatens the country's food security by displacing local production, draining scarce foreign-exchange reserves, and making the country a hostage to any volatility of supply in global markets (Adesina 2012).

Despite this, domestic demand for rice has continued to grow at an even faster pace in Nigeria and elsewhere in Africa since the global food crisis of 2008. As Table 1.2 shows, rice consumption in Nigeria increased by about 8.4 percent per year in the most recent period (2010 to 2012) compared to 5.3 percent for the rest of SSA. Lower global rice prices, increased household incomes, and continuing growth in urban populations may explain this most recent upturn in the trend in rice imports.

Another factor explaining the increased demand has been a preference for imported rice among urban consumers due to its higher quality with respect to swelling capacity, taste, and grain shape and cleanliness (Bamidele, Abayomi, and Esther 2010; Lançon et al. 2003b). Local rice, on the other hand, is often broken, not polished, and contains stones and other debris. Finally, the

TABLE 1.2 Average annual volume and growth rates of milled rice supply in Nigeria and the rest of West Africa and Africa south of the Sahara (SSA), 1980–2013

	Volume, annual average (million metric tons)				Growth rates, annual average (%)			
	1980–1989	1990–1999	2000–2009	2010–2012	1980–1989	1990–1999	2000–2009	2010–2012
Consumption								
Nigeria	1.2	2.3	3.6	5.0	3.1	2.8	4.6	8.4
Rest of West Africa	1.8	2.5	4.6	7.4	3.0	4.3	5.9	10.1
Rest of SSA	3.7	4.6	7.2	9.7	2.4	2.2	4.2	5.3
Production								
Nigeria	0.7	1.8	2.1	2.6	10.9	0.7	3.4	2.0
Rest of West Africa	1.5	1.9	2.8	5.0	3.4	2.6	5.4	5.4
Rest of SSA	2.0	2.5	3.3	4.7	2.3	1.6	4.6	–1.1
Imports								
Nigeria	0.6	0.4	1.6	2.6	2.3	15.0	3.9	19.8
Rest of West Africa	0.8	1.1	2.3	3.3	2.6	4.3	3.8	13.7
Rest of SSA	1.3	1.7	3.3	4.5	2.5	5.1	4.3	13.6

Source: United States Department of Agriculture international database (USDA 2013).

removal of input subsidies, price supports, and protective import barriers in the aftermath of structural adjustment programs of the 1980s and 1990s also played a key role by exposing the lack of competitiveness of local rice production in terms of technologies, costs, and milling efficiencies relative to imports (Moseley, Carney, and Becker 2010).

As the demand has accelerated, no country in the region has been able to match this through domestic production. In Nigeria, imports have increased the fastest, at 20 percent per year on average during the last three years for which data are available, 2010–2012 (Table 1.2).[2] Local production has not grown as fast. In Nigeria, local rice production grew by only about 2 percent per year over the last three years. This is much lower than the average 5.4 percent in the rest of the countries in West Africa.[3]

2 While it is possible that some imports to Nigeria may be reexported to neighboring countries such as Niger, Chad, and Cameroon, this is difficult to confirm given scanty data on informal cross-border trade. But we expect these shares to be small given the sheer size of the Nigerian market relative to these other countries. We note that there is a difference between market demand and consumption, as the latter includes both market demand and subsistence production for own-consumption.

3 In this book, West Africa comprises Benin, Burkina Faso, Côte d'Ivoire, Gambia, Ghana, Guinea, Guinea-Bissau, Liberia, Mali, Mauritania, Niger, Nigeria, Senegal, Sierra Leone, and Togo.

In an effort to reverse the current trend of rising imports, the Nigerian government recently introduced a number of key policies and investment strategies. These include the provision of improved seeds, subsidized fertilizer, mechanization services, and incentives for private-sector investments in irrigation. At the macro level, rice import tariffs were increased in order to discourage imports and encourage domestic production. Improvements in paddy production, rice processing, and marketing have also been encouraged with the support of public-sector reforms and investments.

The recent policy reforms have included deregulating seed and fertilizer markets and establishing private-sector marketing corporations to help coordinate supply and demand and set grades and standards for many agricultural commodities. Physical investments have also been made to establish staple crop processing zones (SCPZs) that have been intended to encourage the clustering of food-processing industries, including rice milling, in proximity to raw materials and end markets.

A Brief Overview of Nigeria's Economy and History

The Nigerian economy is the largest in Africa, having recently overtaken South Africa after rebasing its gross domestic product (GDP) in 2014. It makes up two-thirds of all economic activity in West Africa and one-fifth in SSA. In 2012, Nigeria's GDP was $180.9 billion in constant 2005 US dollars (Table 1.3). The country also has the largest population in the region. For every two people living in West Africa or for every five people in SSA, one is a Nigerian.

In recent years, the country has continued to experience steady and positive economic growth and is becoming one of the fastest growing economies in the region. As Table 1.3 shows, GDP growth rates have averaged about 7 percent or more per year, and per capita growth has been about 4 percent over the last decade, faster overall than the average per capita GDP growth for West Africa or SSA (Table 1.4).

The irony for Nigeria is that despite the positive economic growth, it remains—due to its sheer size and high poverty rate—home to most of the poor and hungry living in West Africa. It has one of the highest incidences of poverty, with 62 percent of the population living on less than $1.25 (purchasing power parity [PPP]) a day in 2010. In contrast, Mali, which has a lower per capita GDP, reported a poverty rate of 50.4 percent in 2010. In Ghana, the poverty rate was less than half the rate in Nigeria during the same year (Table 1.4).

In addition to poverty, the prevalence of hunger is also high. The proportion of underweight children of less than five years of age in Nigeria fell only

TABLE 1.3 Selected socioeconomic indicators for Nigeria, 1995–2012

						Annual growth rates (%)			
	1995	2000	2005	2010	2012	1995–2000	2000–2005	2005–2010	2010–2012
Select economic indicators									
GDP (constant 2005 billion US$)	57.8	67.9	112.2	159.0	180.9	2.9	11.8	7.0	6.7
GDP per capita (constant 2005 US$)	533.4	552.2	804.2	995.7	1,071.5	0.4	9.0	4.2	3.7
Agriculture GDP (constant 2005 billion US$)	15.1	18.5	36.4	49.8	54.7	4.3	15.8	6.5	4.8
Share of agriculture in GDP (%)	26.1	27.3	32.4	31.3	30.3	1.3	3.6	–0.5	–1.7
Share of manufactures in GDP (%)	3.5	3.0	2.8	3.0	3.0	–3.2	–2.6	1.5	0.8
Share of industry in GDP (%)[a]	47.9	47.4	41.3	35.4	32.8	–1.3	–3.4	–5.9	–3.8
Share of services in GDP (%)	22.5	22.2	23.5	30.3	33.9	0.2	0.3	5.4	5.7
Imports of goods and services (% of GDP)	42.2	32.0	31.0	29.9	35.6	–0.4	–0.2	0.6	9.2
Petroleum exports (constant 2005 billion US$)	13.0	25.9	48.1	67.9	84.8	6.9	16.1	5.0	11.8
Official exchange rate (naira/$)	21.9	101.7	131.3	150.3	156.8	40.9	5.5	3.1	2.1
Consumer price index (2005 = 100)	27.4	48.2	100.0	161.4	178.9	11.0	15.2	9.9	11.5
Electric power consumption (billion kWh)	15.9	14.7	23.5	26.1	27.0	–1.3	10.9	–0.1	1.7
Select social indicators									
Population (millions)	108.4	122.9	139.6	159.7	168.8	2.5	2.6	2.7	2.8
Urban population (millions)	42.1	52.0	63.9	78.3	84.8	4.3	4.2	4.2	4.1
Urbanization rate (% of population)	38.8	42.4	45.8	49.0	50.2	1.7	1.6	1.4	1.2
Poverty headcount at $1.25 a day, PPP (%)	68.7	NA	61.8	62.0	NA	–0.8	–0.8	1.5	NA
Hunger Index (%)[b]	22.6	17.9	16.3	15.0	NA	–4.6	–1.9	–1.6	NA
Underweight children under five (%)[c]	35.1	24.7	26.5	24.2	NA	–6.8	1.4	–1.8	NA

Source: World Development Indicators of the World Bank (2014).

Notes: GDP = Gross domestic product; NA = not available; PPP = purchasing power parity. [a]About 90 percent of the value of industry is from the petroleum sector. [b]A composite index calculated by the International Food Policy Research Institute (IFPRI) using three measures: the prevalence of undernourished children, prevalence of adult undernutrition, and child mortality rate. See von Grebmer et al. (2013) for more details on this index. [c]For prevalence of undernourished children, this is 1993–1997, 1998–2002, 2003–2007, and 2008–2012.

TABLE 1.4 West Africa selected indicators, 2012

	Population		GDP per capita		Agricultural GDP	Poverty headcount ratio (at $1.25 a day, PPP)[a]		Prevalence of underweight in children under five years[b]	
	2012 (millions)	Growth, 2002–2012 (%)	2012 value (2005 US$)	Growth, 2002–2012 (%)	Growth, 2002–2012 (%)	2001 (%)	2010 (%)	2000 (%)	2010 (%)
Côte d'Ivoire	19.8	2.2	942.2	–0.1	1.8	23.3	23.8	18.2	15.4
Ghana	25.4	2.2	559.2	4.5	3.8	39.1	28.6	20.3	13.4
Mali	14.9	3.1	464.4	1.7	6.2	61.2	50.4	30.1	18.9
Nigeria	168.8	2.8	861.3	5.9	6.4	63.1	68.0	24.7	24.2
Senegal	13.7	3.0	771.6	1.1	4.0	44.2	29.6	20.3	14.4
West Africa	318.5	2.8	811.4	4.5	5.4	NA	NA	NA	NA
SSA	913.1	2.7	903.3	2.7	4.7	NA	NA	NA	NA

Source: World Bank (2014).

Note: GDP = gross domestic product; NA = not available; PPP = purchasing power parity; SSA = Africa south of the Sahara. [a](1) For 2001, the ratios for Côte d'Ivoire are for 2002, for Ghana 1998, and for Nigeria 2004; (2) For 2010, the ratios for Côte d'Ivoire are for 2008, for Ghana 2006, and for Senegal 2011. [b]From von Grebmer et al. (2013), various years: for 2000, average between 1998 and 2002; for 2010, average between 2008 and 2012.

marginally between 2000 and 2010, from 24.7 percent to 24.2, respectively. The comparable rates for Mali and Ghana dropped sharply during the same period (Table 1.4).

The structure of the Nigerian economy, its institutions and macroeconomy, and sociopolitical history can all partially explain the dichotomy between positive economic growth with no change in the incidences of poverty and hunger. Relatively low investment in and poor performance of agriculture may have led to such a dichotomy. The sector not only contributes the most to GDP, it employs over two-thirds of the working population in Nigeria, and therefore it is of critical importance for food security, rural incomes, and poverty reduction.

Research has shown that many African countries that rely heavily on agriculture for their GDP have experienced more equitable growth and poverty reduction whenever overall economic growth was agriculture led (Diao et al. 2010). In the case of Nigeria, a study using macroeconomic modeling underlines this finding by showing how potentially larger gains in incomes and poverty reduction are likely when fiscal policies are targeted at stimulating growth in the agricultural sector (Akanbi and Du Toit 2011). Until recently, however, Nigeria's economy relied heavily on petroleum production and exports to generate growth.

Based on the most recent estimates from the World Bank's World Development Indicators (WDI) database, the agricultural sector appears to have grown quite rapidly since the beginning of the 21st century. As shown in Table 1.4, the sector grew at an average annual rate of 6.4 percent between 2002 and 2012.[4] However, the growth has been driven mainly by an expansion in area planted to staple crops, as yields have changed little over the same period.[5]

Moreover, growth in the agricultural sector has not been a pro-poor one—it has had little effect on the welfare of a majority of the poor in Nigeria because the agricultural subsector(s) driving the growth may have weaker linkages with the households and locations most affected by poverty, such as those that are net purchasers of food (Diao et al. 2010). Additionally,

4 Recent estimates by the National Bureau of Statistics (NBS), which is the principal agency overseeing the national statistics of Nigeria, report a 4.5 and 5.1 percent growth in constant 1990 naira terms during the second and third quarters of 2013, respectively (Nigeria, NBS 2013b). Thus, the growth has somewhat slowed down in 2013.

5 For rice, for example, average yields have actually declined over time in Nigeria from a high of 2.1 metric tons (MT)/ha in the 1980s to 1.5 MT/ha after the turn of the century (based on FAO 2014).

insufficient investments have been made into basic infrastructure and social services (such as education and health) in rural areas in tandem with the growth in the agriculture sector (see Dim and Ezenekwe 2013). Evidently the many years of neglect of agriculture, basic infrastructure, institutions, and services in the past have taken a toll. The neglect is visible in the state of disrepair, unreliability, and inefficiencies of basic infrastructure that exists in many parts of the country.

A key factor that has challenged Nigeria's ability to manage its agricultural development agenda, including rice, is its status as a major petroleum exporter. Inevitably, therefore, the negative effects of the so-called Dutch Disease syndrome—a condition in which a boom in export earning does not translate into broad-based growth in the rest of the economy—come into play. Also referred to as a "resource curse," this affects countries that export a single resource commodity that leads to rising foreign-exchange earnings, either due to a price boom in global markets or a substantial increase in export volumes due to global demand.

The accompanying increase in export revenue and foreign-exchange reserves leads to an appreciation of the domestic currency and makes food imports (in addition to other goods and services) cheaper (Oyejide 1986). In the process, it undermines the competitiveness of domestic production of both tradeables and non-tradeables such as food staples. The situation can be made worse if increases in government revenue generated from the export earnings are not effectively transferred to other sectors of the economy. This is a reality that Nigeria has had to grapple with throughout its history since the discovery of large petroleum reserves.

Nigeria's resource curse occurred more as a result of poor governance and inherently weak institutions that existed in managing the petroleum sector and government revenues generated from it (Robinson, Torvik, and Verdier 2006; Sala-i-Martin and Subramanian 2013). Nigeria has struggled throughout its history in forming strong democratic institutions and transparent processes for governing its petroleum sector. The country is made up of several ethnic and religious groups, each with its own distinctive language, culture, and history. The country's diversity often makes it difficult to have a unified state; in this context, coordinated governance becomes more challenging.

Additionally, while the country's three-tier federal system of governance (i.e., a national, state, and local level) was intended to allow for greater autonomy and to avoid conflict across its various social groups and regions, it has instead encouraged rent-seeking behavior and politically motivated behavior over the control of government revenues and resources at the national level. This is because over 90 percent of the government's revenue is generated at the top tier from petroleum exports. As a result, local and state authorities have to rely heavily on the federal government for much of their capital expenditure needs. Without sufficient transparency and accountability in place, such a vertical fiscal imbalance introduces a higher risk for rent seeking and other politically motivated behavior in the allocation of revenues, especially during a petroleum boom (Gboyega and Shukla 2011; Diao et al. 2010; Robinson, Torvik, and Verdier 2006).

The combination of a large and diverse population, varying degrees of resource endowments, and generally poor transparency and accountability of governing institutions has contributed to a difficult and turbulent political history for Nigeria. Table A.1 in Appendix A provides a very brief chronology of this history. Fortunately, a more stable macroeconomic environment has been created over the past decade, and major reform efforts to improve public financial management, infrastructure and services, and transparency and accountability in the petroleum sector have been introduced (Gboyega and Shukla 2011). The efforts seem to be paying off, with signs of low inflation, a steady supply of foreign-exchange reserves, and stable exchange rates (Table 1.3). Demand for rice, however, has continued to grow, fueled by economic growth.

Importance of Agriculture and Evolution of Rice Policies

The discussion above shows that agriculture plays a dominant role in Nigeria's economy. The role of agriculture as a key source of employment, food security, and rural incomes is primarily due to the richly endowed and diverse agroecological landscape in Nigeria straddled by two of Africa's major rivers, the Niger and Benue (Figure 1.1). Freshwater resources are relatively abundant in Nigeria due to these two major rivers and other large bodies of water. Large swaths of land serve as river basins along these two rivers, locally referred to as *fadamas*, which are particularly suitable for rice production. Periodically, however, water access can be affected by droughts and/or floods (Kuku-Shittu et al. 2013).

Nigeria's agricultural landscape can be broadly broken down into three major agroecological zones: humid, subhumid, and semiarid. The most commonly grown agricultural crops in the humid zone of the south are tree crops (e.g., cocoa, oil palm, plantain, and rubber); root crops (yam, cassava, and

FIGURE 1.1 Political and physical map of Nigeria

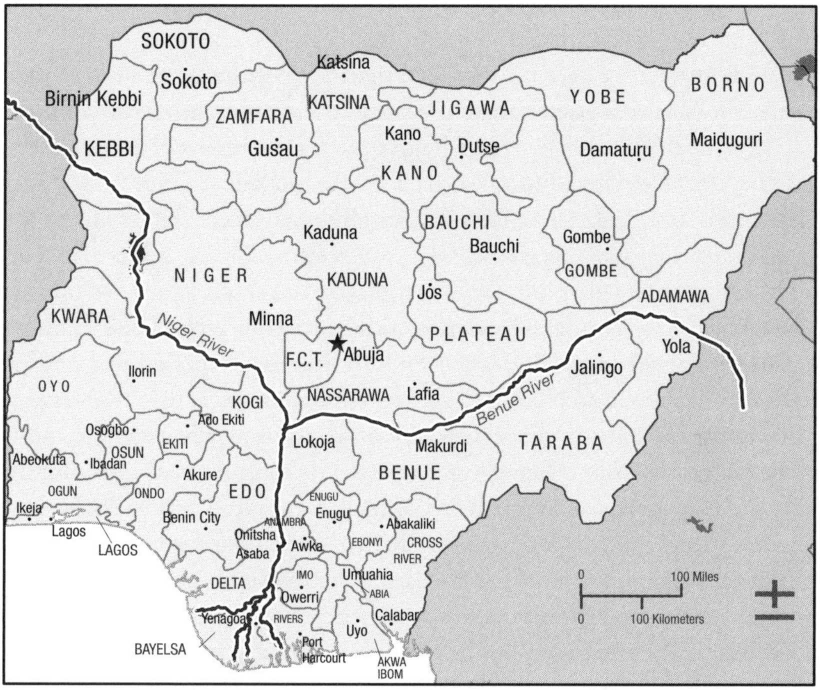

Source: Downloaded from the Commonwealth Secretariat (http://secretariat.thecommonwealth.org) on December 16, 2013.

Note: FCT = Federal Capital Territory. Nigeria is divided into six geopolitical zones: the North Central zone consists of Benue, FCT, Kogi, Kwara, Nasawara, Niger, and Plateau states; the North East zone consists of Adamawa, Bauchi, Borno, Gombe, Taraba, and Yobe states; the North West zone consists of Jigawa, Kaduna, Kano, Katsina, Kebbi, Sokoto, and Zamfara states; the South East zone consists of Abia, Anambra, Ebonyi, Enugu, and Imo states; the South South consists of Akwa Ibom, Bayelsa, Cross River, Delta, Edo, and Rivers states; and the South West zone consists of Ekiti, Lagos, Ogun, Ondo, Osun, and Oyo states.

cocoyam), and cereals (maize and some rice). As a rain forest area, timber is also an important commodity. The subhumid zone is just north of the humid forest zone in the south and covers much of the middle belt of Nigeria. It is characterized by more open forests and savanna grasslands. Crop production is significant and includes root crops (e.g., yam and cassava) and cereals (e.g., maize, rice, and sorghum). The semiarid zone occupies the northern part of Nigeria. Typical crops grown in this zone include rice, vegetables, millet, sorghum, cowpeas, groundnuts, and cotton. Livestock and gum arabic are also produced.

Food-crop production is the dominant activity in terms of value within agriculture. Cassava and yam contribute about a third of the total value of output, cereals 26 percent, and other food crops (such as plantains, melons, fruits, and vegetables) contribute another 26 percent.[6] Among cereals, rice has become a leading crop in value terms since the 1970s, surpassing other major cereals such as maize, millet, and sorghum.

The importance of agriculture—and rice in particular—has been recognized by the Nigerian government as central to food security in the country, especially since 1970. This is true whether the regime in power at the time was a military or civilian one. However, each successive regime rarely took a systematic approach in planning for agricultural development, preferring to focus instead on commodity-specific initiatives that were targeted at increasing output for food security purposes and to promote rural employment (Manyong et al. 2005). Figure 1.2 summarizes some of the initiatives introduced throughout the country's history and relative to trends in rice supply (production and rice imports) and demand (represented here by population growth).

Before the 1970s, as in the rest of Africa, the primary policy focus for agriculture was on the promotion of export crops, especially palm oil, groundnuts, cocoa, and cotton. Research institutes for these commodities were established soon after independence (Ehigiamusoe 2012). Nigeria was the world's leading exporter of palm oil, accounting for 45.2 percent of the world market in 1961.[7] Cocoa and groundnut exports were also significant, with 16.6 and 11.1 percent of global exports, respectively, in the same year. Export commodity boards inherited from the colonial era provided essential support for extension and the provision of modern inputs, in addition to crop procurement.

After the petroleum oil price surge in the middle of the 1970s, export shares of these crops began to erode. In 2012, Nigeria's share of world exports was 1.9 percent for palm oil, 8.7 percent for cocoa, and 7.4 percent for groundnuts. Prior to the windfall gains from petroleum exports, agricultural products accounted for up to 90 percent of export revenues. Nigeria was self-sufficient in food production, with a small surplus for exports to the West African subregion. However, the food production index in Nigeria decreased from 100 in 1960 to 88 in 1975 and to 67 in 1980. This was despite increased policy attention to food crops in the 1970s, especially in the aftermath of severe droughts in the Sahel from 1972 to 1974. That was the first period during which Nigeria imported a large quantity of rice (Ojo and Adebayo 2012).

6 Diao et al. (2010), based on national- and state-level data on agricultural production, agricultural yield, and market prices, which all come from FMARD.

7 Statistical database of the Food and Agriculture Organization of the United Nations (http://www.faostat.fao.org/), accessed December 21, 2013.

FIGURE 1.2 Trends in production, imports, population, and evolution of rice policies and agricultural investments

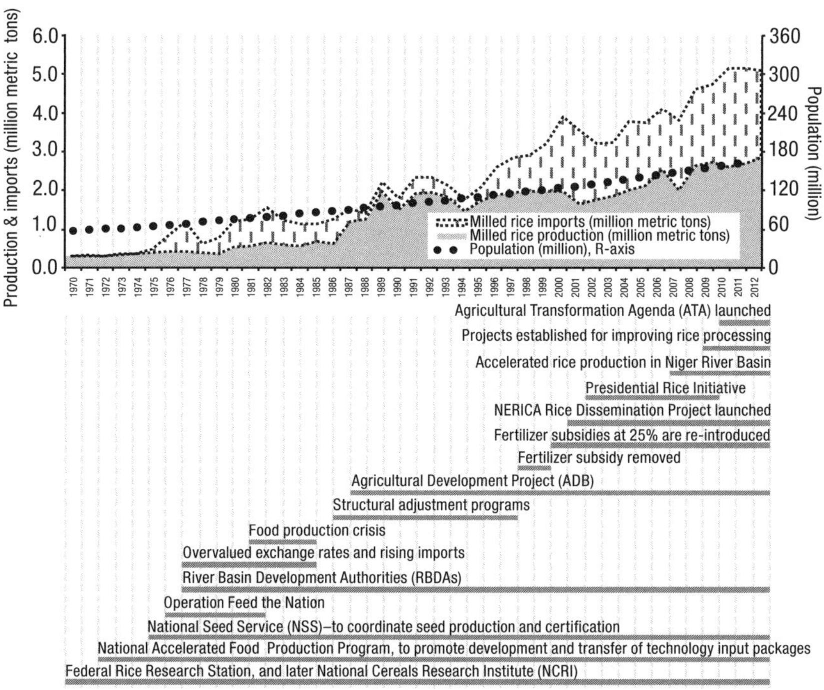

Source: United States Department of Agriculture international database (USDA–ERS 2012) for production and imports; FAOSTAT (FAO 2014) for population; and various documents for import tariffs and investment strategies (e.g., Emodi and Madukwe 2008; Daramola 2005; Akpokodje, Lançon, and Erenstein 2001).

The shift toward food crops in the 1970s included the establishment of the National Cereals Research Institute (NCRI) to oversee the research on major cereal crops, regional marketing boards for food crops, the National Seed Service (NSS) to coordinate seed production and certification, and the launching of the National Food Production Program. The marketing boards were set up to expand farmers' access to modern inputs, guarantee crop procurement, and stabilize prices through buffer stock schemes (Ehigiamusoe 2012).[8]

A number of grand plans to revolutionize the agricultural sector followed, including Operation Feed the Nation (OFN) under the Obasanjo regime in 1976, later renamed the Green Revolution Initiative under Shagari's civilian

8 These were later abolished as part of the structural adjustment programs in 1985.

government from 1979 to 1983 (Ojo and Adebayo 2012). The share of government expenditure going to agriculture and rural development was always very small (about 5 percent) relative to other sectors (Akande 2005). Nevertheless, a number of large investments were implemented, including the establishment of the River Basin Development Authorities in 1977 to improve irrigation development and water control and establishment of the Abakaliki Rice Project in 1978 for rice production and processing. Both are still in operation today.

Much of the focus of investments was on large-scale irrigation and mechanization—involving a mix of state-run and private-sector enterprises—rather than smallholder production systems. A land use decree in 1978 helped to ensure land ownership for large-scale operations (Akande 2005). However, none of the state-run enterprises proved profitable. The few private-sector enterprises simply benefited from government subsidies and market interventions but contributed little to total output. As a result, an agricultural revolution as envisioned under the OFN and Green Revolution Initiative did not take place.

Rice imports were relatively unregulated throughout the 1970s. Import tariffs ranged between 10 and 67 percent. By the latter part of the 1970s, imports of rice and other food commodities had begun to rise sharply due to rising global petroleum prices and higher export earnings, while domestic food production had begun to decline. Rice import growth was stimulated by structural changes, including an overvalued exchange rate, inflation and wage increases, and accelerated rural-to-urban migration.

A crisis emerged in the early 1980s, as petroleum prices declined and the country faced food shortages as a result of low levels of domestic food production, balance of payments problems, and fast-depleting foreign-exchange reserves. The government tried to curb imports through more restrictive import licensing requirements. This was followed by a complete ban on rice imports in 1986, complemented by structural adjustment programs described below (Ojo and Adebayo 2012; Akpokodje, Lançon, and Erenstein 2001). The ban on rice imports was not lifted until 1995.

Following the macroeconomic and food crisis in the early 1980s, as in many other countries in the region, Nigeria introduced structural adjustment programs (SAPs) in 1985. The government shifted its focus to increasing the productivity and output of food-crop production within smallholder agriculture through the provision of extension services and modern inputs through the Agricultural Development Project (ADP), which is the institution established in the 1970s to oversee agricultural and rural development projects in each state.[9] As part of these efforts, smallholder rice production schemes were introduced in the *fadama* areas to promote the adoption of simple, low-cost, improved irrigation technology through credit arrangements in order to boost production (Akpokodje, Lançon, and Erenstein 2001; Nkonya et al. 2008).[10]

After the end of military rule in 1999, which marked the beginning of the current Fourth Republic (Table A.1 in Appendix A), newly elected president Obasanjo launched a National Economic Empowerment and Development Strategy (NEEDS), with broad economic development goals such as poverty eradication, employment generation, and wealth creation (Iwuchukwu and Igbokwe 2012). For agriculture, NEEDS included targets of 6 percent annual growth in agricultural GDP, increased earnings from agricultural exports, and a 95 percent self-sufficiency ratio in food production. A number of commodity-specific presidential initiatives were also developed with these goals in mind, including ones for rice, cassava, vegetable oil, tree crops, livestock, and aquaculture products.

The aims of the Presidential Initiative on Rice were to achieve national self-sufficiency in rice production by 2005 and the ability to export by 2007 (Phillip et al. 2009). To pursue these goals, the initiative focused on improving access to technology packages at a 50 percent subsidized rate (rice box, or R-Box, technologies), which emphasized minimum tillage and key inputs such as seed and agrochemicals. To ensure higher-quality domestic rice brands, the government also granted concessions to large-scale milling companies to import brown rice with a 50 percent import tariff rather than the 100 percent on milled rice whenever there was insufficient local paddy rice to supply the mill (Daramola 2005). During this time, the NCRI and the former West Africa Rice Development Association (currently known as the Africa Rice Center) also assisted with the multiplication of breeder and foundation seeds, while the NSS worked with seed companies and ADPs to ensure production and dissemination of quality-certified seeds. Additionally, the construction of rice mills was proposed in ten locations (Adejumo-Ayibiowu 2010).

Other rice projects were also launched soon after the turn of the century. One major breakthrough was the release and dissemination of Multinational New Rice for Africa upland varieties. By 2005, the second phase of the National Fadama Development Project funded by the World Bank was introduced and

9 The ADPs are still active today in various states across the country.

10 This was officially referred to as the first National Fadama Development Project (or Fadama I), funded by the World Bank, and it lasted until 1999. A second phase was introduced in 2005, and currently the third phase is being implemented.

referred to as Fadama II. It involved 26 states and proved more successful in boosting production and incomes than the first phase by addressing many of the shortcomings of Fadama I. In particular, there was a shift away from a top-down and supply-driven public-sector approach to a community-driven approach (Nkonya et al. 2008). Such achievements are critical for future development efforts considering that only a small portion of the country's total *fadama* potential has been tapped (Djurfeldt, Aryeetey, and Isinika 2011).

As rice imports continued to rise throughout the post-2000 period, tariffs were raised from about 50 percent in the latter half of the 1990s to 85 percent in 2001 and to 110 percent by 2005. Following the removal of the tariff in 2008 in response to high world prices in that year, tariffs were reimposed and increased to 110 percent again by 2013. Despite the domestic price incentives created by these tariffs, there has been no significant supply response at the national level. Public investments to improve yields, such as irrigation systems in river basins, development and dissemination of high-yielding seed varieties, and improved access to fertilizer have been insufficient. Similarly, efforts to improve product quality through the promotion of modern milling technologies have generally been unsuccessful, despite being critical for the competitive position of domestically produced rice (Demont et al. 2013).

Recent Rice Policy Framework

The general objective of the Nigerian government's policy framework has recently been to transform the agricultural and rural sector in order to promote overall economic growth, poverty reduction, increased rural employment and incomes, and national food security. Under the Agricultural Transformation Agenda (ATA), the framework called for a number of policy, institutional, and financing instruments to drive sector growth through the improvement of agricultural performance all along a value chain, from on-farm production to value-added processing and marketing activities.

Key among the initiatives was the Growth Enhancement Support (GES) program, which provided targeted input subsidies for fertilizer and seeds made available through electronic vouchers in partnership with private-sector input dealers. Other investment programs included the establishment of SCPZs to promote industrial clusters for food-processing activities; the setting up of marketing corporations to strengthen ties between agro-industry, markets, and farmers; the establishment of innovative agricultural financing mechanisms that enable risk sharing between government and the banking industry; and coinvestments through state governments and private-sector partnerships.

For the rice sector specifically, as part of the ATA strategy, FMARD put forward a plan whose goal was to transform the rice sector and achieve self-sufficiency. Referred to as the Rice Value Chain Transformation Plan (FMARD 2012), the principal goal was to improve productivity and output significantly. The government expected to accomplish this by doing the following:

- Increasing the competitiveness of local rice all along the value chain. At the production level, this involved greater mechanization, intensification of paddy rice production in rainfed lowland and irrigated rice ecologies, increased use of fertilizer, and improved rice varieties. At the processing and marketing level, farmers and millers were encouraged to create formally registered groups to facilitate access to improved technologies, markets, and mills.
- Expanding accessibility to inputs such as fertilizers and agrochemicals (through initiatives such as the GES program).
- Promoting modern practices and management along the commodity chain to raise the quality of rice (e.g., use of quality seeds, threshing, drying, and milling).
- Creating an enabling policy and market environment to encourage growth in the sector.
- Introducing innovative agricultural financing mechanisms to improve the accessibility to credit all along the value chain.

These policies were to be implemented while protection for the domestic rice sector was provided through continuing application of import tariffs.

Objectives and Approach of the Book

The principal objective of this book is to assess the policy challenges and opportunities for transforming and expanding the Nigerian rice economy. Transformation is critical for achieving sustainable growth, especially given the role that rice plays in food security and poverty reduction. Despite a long history of rice production, the Nigerian government's intervention in the rice sector, and the rapidly growing demand for rice, the sector has not been transformed from its premodern and low-productivity status. It still has a relatively low level of private-sector participation and faces costly market inefficiencies. While the recent government's strategy, including the ATA under President Jonathan's administration, has been promising, has had certain innovative elements, and has appeared to be backed by a strong political will, it remains unclear whether it will succeed in transforming the rice economy.

The ability of the rice economy to grow and be transformed into a more vibrant and competitive one requires a medium-to-long-term perspective, considering that many of the underlying constraints facing the sector are basic and fundamental to the overall development of the agricultural sector and are therefore less likely to be altered by short-term policies. Understanding the kinds of challenges and the various policy options needed both in the short and long run requires a holistic approach that analyzes each of the principal components of the rice economy: demand, production, processing, and trade.

In the process, a number of relevant policy questions can be directly addressed: How is rice consumption growth in Nigeria associated with structural changes in the economy, such as income growth and urbanization? Is there biophysical and economic potential for domestic rice production to transform and expand over the short to medium term? Is there potential to improve the quality and competitiveness of local rice in domestic markets? What is the most efficient way to take advantage of this potential for both production and processing? Are there differential abilities and efficiencies among existing mill types (small to large) for improving quality and meeting demand? Are there lessons that can be drawn from elsewhere in West Africa and Asia? What are some of the constraints in historically protective rice trade policies in Nigeria? What are the welfare implications and opportunity costs of alternative strategies for growing and transforming the rice economy?

The book has been organized around these key questions. Chapter 2 reviews and discusses rice consumption and demand in Nigeria. This is followed in Chapter 3 by a descriptive overview of the current state of rice production systems and economics in Nigeria. Chapter 4 provides an assessment of the potential to induce a supply response in paddy production given these current conditions, as well as indicating what will be needed to grow and transform the sector in the future. These chapters are intended to set the context of the policy challenges Nigeria faces in order to stimulate growth in local paddy production. In Chapters 5 and 6, we shift attention to the postharvest rice milling and marketing sector by examining its current structure, conduct, and performance, as well as associated policy challenges and opportunities for modernizing the sector.

Chapter 7 analyzes rice import policies in terms of their effectiveness and tradeoffs with regard to affecting the volume of imports, domestic prices, and tariff revenues in the Nigerian context. Finally, in Chapter 8, we combine the analyses and conclusions from Chapters 2 through 7 on demand, production, processing, and trade to discuss the economywide implications of policy options raised in each of these chapters for transforming and increasing the competitiveness of the domestic rice sector. Chapter 9 summarizes the key findings and presents conclusions on the challenges and opportunities facing policymakers in transforming the rice economy in Nigeria.

Our analysis seeks to answer three major questions regarding policies that influence the Nigerian rice economy. The first is how to go about increasing productivity: one approach is to promote the use of fertilizer, improved seeds, mechanization, and irrigation. The second focuses on how to go about improving the efficiency of postharvest processing and marketing activities, such as through the modernization of the rice milling industry. And the third is how to more effectively manage trade policies, such as import tariffs, as well as price policies in general, in order to support the transformation process.

Aside from its core objective of informing development strategy by assessing the policy challenges and opportunities for transforming and growing the Nigerian rice economy, the book is also intended to be of practical use for policy analysts, students, and researchers alike, especially those who are interested in the various econometric and modeling approaches used. These include the application of econometric tools to estimate demand elasticities in Chapter 2; the application of simple farm budgeting tools and econometric approaches for estimating supply response elasticities in Chapter 4; development and use of mathematical programming methods to assess the optimal mix of large-scale and medium- or small-scale millers and location of rice milling operations in Chapter 6; the use of simple price and import tariff analysis approaches in Chapter 7; and finally, the use of an economywide multimarket model in Chapter 8 for assessing the tradeoffs of various policy alternatives on the overall rice sector and national economy.

Chapter 2

EVOLUTION OF RICE CONSUMPTION AND DEMAND PATTERNS

Kwabena Gyimah-Brempong and Oluyemisi Kuku-Shittu

The ultimate objective of food policy generally and rice policy specifically for Nigeria is to ensure food security and improved household welfare for the majority of Nigerians. As already highlighted in Chapter 1, this has particular relevance for Nigeria given the government's concerns about the country's growing dependency on rice imports, as evident in the introduction of policies to encourage domestic rice production and achieve rice self-sufficiency. However, such policies can negatively affect household welfare, especially if prices rise sharply as a result of the policies and a majority of Nigerian consumers happen to consider rice a major staple. Therefore, to understand how rice policy might impact the lives of the average Nigerian, it is necessary to understand the importance of rice in the Nigerian diet as well as the demand behavior for rice by the average Nigerian household, which is the objective of this chapter.

The chapter begins by highlighting how rice has become a dominant food staple in the Nigerian diet over a relatively short period of time. This is accomplished by examining trends in rice consumption, both on aggregate and per capita terms, over the postcolonial period, including regional and urban/rural differences. This is followed by a discussion on the possible explanations or determinants of the growth in demand, including a look at local trends in rice prices.

Following this, the chapter empirically estimates the behavioral patterns of demand for rice in Nigeria by rural and urban populations and income status. The analysis uses a linear expenditure system demand model and the most recent data from the 2011 Living Standards Measurement Study–Integrated Surveys on Agriculture (LSMS–ISA) (Nigeria, NBS and World Bank 2011), which is the nationally representative household data collected by the World Bank and the NBS. The results of the analysis provide parameter estimates of average budget shares (ABS) and marginal budget shares (MBS), income, and own- and cross-price elasticities, which can be invaluable for informing future rice policies. A summary and discussion of the key findings from the analysis and their policy implications conclude the chapter.

Trends in Rice Consumption

Nigeria has a rich history of rice production and consumption, as indigenous rice species (local rice) have been grown and consumed in the country for over 300 years (Akinbile 2007). However, the dominance of rice as a major staple in the Nigerian diet is a fairly recent phenomenon, as can be seen in Figure 2.1. Aggregate rice consumption grew from a few thousand metric tons at the time of independence in 1960 to over 5.2 million tons in 2012.[1]

It is interesting to note that the dramatic increase in rice consumption began in the late 1970s. This coincided with—and may have been caused by—the astronomical increase in world prices of crude oil at that time, which gave Nigeria large amounts of foreign currency reserves. This in turn led to the appreciation of the naira, making it cheaper to import food (rice) rather than to produce it at home.[2] Figure 2.1 makes it clear that the growth in consumption was not limited to rice but included other major cereals and food crops, such as yams and cassava, although the increase in rice consumption preceded those of yam and cassava by a decade.

The rapid growth of the Nigerian population as well as increased income has been partly responsible for the upsurge in demand for all food commodities, including rice. However, there are two major reasons to pay special attention to rice. First, its position has changed from virtually not being eaten in the 1960s to becoming a major staple crop today. Based on the amount of kilo calories of food staples available per person and per day in Nigeria, the two charts in Figure 2.1 show that the supply of milled rice from domestic production and imports has increased at a rapid rate, and the crop has almost caught up with cassava and yams as a principal staple. Indeed, measured by the share of the household budget devoted to different food commodities, as we will show later in this chapter, rice has now become the leading food staple in Nigeria.

The second reason is that of all the major staple crops shown in Figure 2.1, rice is the only staple for which imports are increasingly being used to meet domestic demand. As Figure 1.2 in Chapter 1 shows, local rice production has not kept up with a growing domestic demand. Consequently, imports have made up for the growing difference between consumption and local production.

The exponential growth in rice consumption that has occurred in Nigeria is also being replicated in other parts of Africa, particularly in West Africa (Figure 2.2), and in other parts of the world. Per capita rice consumption in Nigeria has yet to catch up with the average for West Africa, mainly because of

---

1 Metric tons (hereafter "tons") are used throughout the book.

2 This is a classic case of the Dutch Disease arising from natural resource discovery and exploitation. This is further discussed in Chapter 5 on rice trade in Nigeria.

FIGURE 2.1 Availability of rice and other major staples, 1961–2011

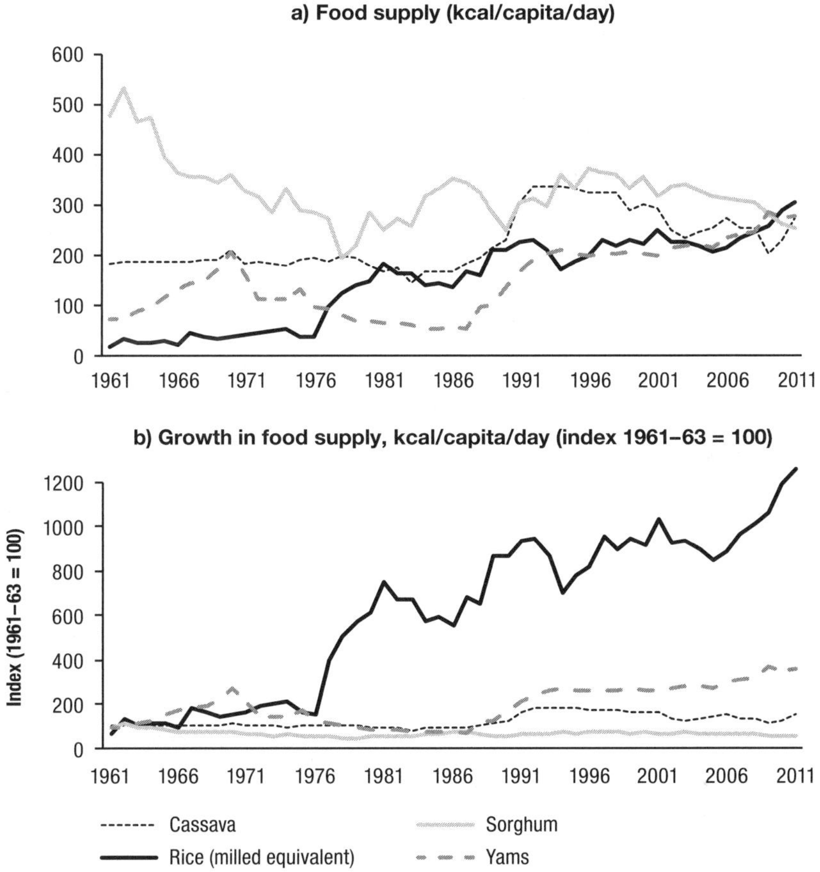

Source: FAO (2014).

Note: While the horizontal axes of 2.1a and 2.1b are measured at the same scale, the vertical axes of the two figures are different. The vertical axis of Figure 2.1a is measured in absolute numbers, but the vertical axis in Figure 2.1b is measured as an index, so changes should be measured as relative to the base period (1961–1963).

the large gap between Nigeria and major rice-consuming countries such as Liberia, Sierra Leone, and Senegal. Setting aside these three major rice consumers, however, Nigeria's consumption is impressive: it has risen above the average for the rest of Africa and is increasing at about the same rate as the rest of West Africa.

It is also instructive to note that since the late 1970s, trends in per capita rice consumption in Africa generally and West Africa in particular have mimicked the trend in per capita consumption in Nigeria. While rice consumption in Nigeria and West Africa in general is well below levels in Asia and major

FIGURE 2.2 Per capita rice consumption: (a) Nigeria and Africa; (b) Nigeria and Asia

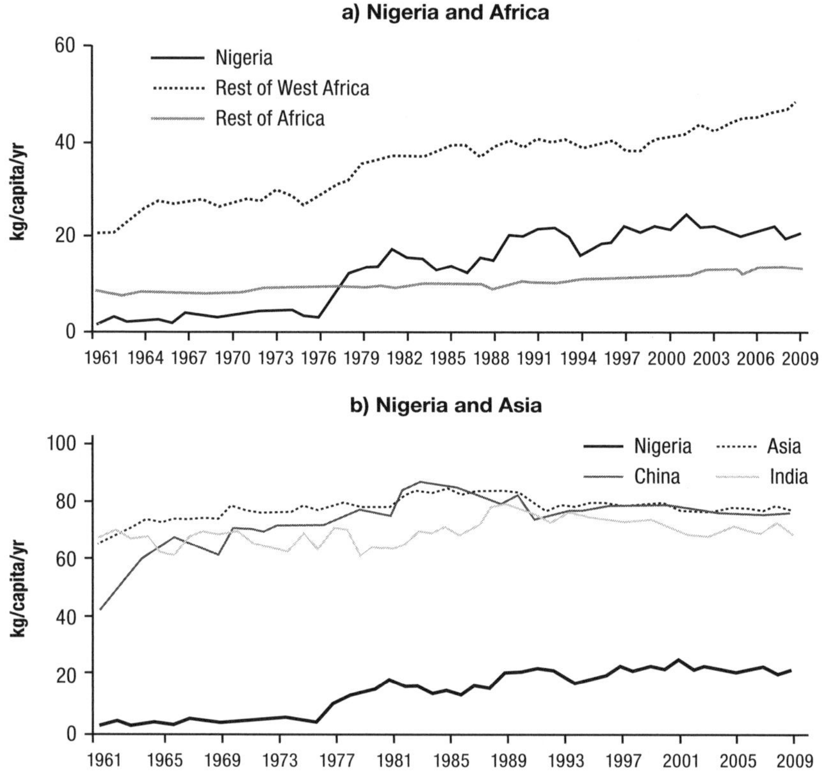

Source: FAO (2014).

rice-consuming countries such as China and India, where consumption is about 60–80 kilograms (kg) per capita per year, both per capita and aggregate consumption in Nigeria are rising faster than in other regions of the world.

It is also interesting to note in Figure 2.1 that per capita availability of rice has grown considerably in the past several decades and has become comparable to other major staples in Nigeria such as yams, cassava, and sorghum. The growth suggests that Nigerian consumers may be substituting rice for cassava and possibly other cereals in consumption. If current trends continue, it is possible that rice will surpass cassava as the leading staple in both volume and value in less than a decade.

Current Rice Consumption

Table 2.1 shows the ranking of food consumption in kilograms per capita and per capita expenditure. Rice has become one of the most dominant staples in the

TABLE 2.1 Ranking of per capita consumption and expenditure

Commodity	Annual consumption (kg/capita)	Rank by consumption quantity per capita	Annual expenditure (naira/capita)	Rank by expenditure per capita
All rice	32.1	2	3,951	1
Local rice	17.6		1,893	
Imported rice	14.5		2,058	
Maize	24.1	4	1,164	4
Sorghum	28.3	3	960	5
Millet	19.8	5	786	6
Cassava	33.1	1	2,374	3
Yam	18.2	6	2,824	2

Source: Authors' calculation from Nigeria, NBS and World Bank (2011).

FIGURE 2.3 Aggregate consumption of rice and other major staples in 2011 (million tons)

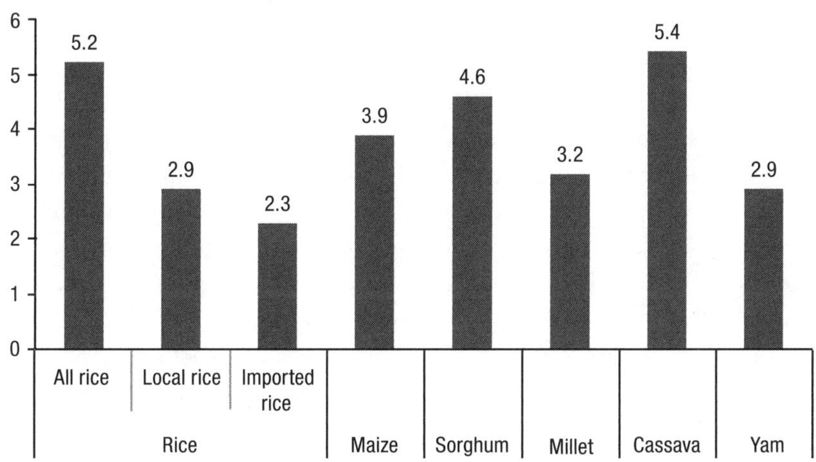

Source: Authors' calculation from Nigeria, NBS and World Bank (2011).

Nigerian diet. It now ranks first among all staple food items in terms of expenditures and is second only to cassava in terms of quantities consumed. Figure 2.3 shows the aggregate quantities of major food items consumed in Nigeria. By volume, the per capita consumption of rice is 3 percent less than that of cassava; however, by value, per capita consumption of rice exceeds that of cassava by as much as 67 percent, indicating how dominant rice has become in the Nigerian food budget.

Rice not only dominates other cereals, it also dominates the average Nigerian budget. Table 2.2 shows that 5.2 million tons of rice were consumed in Nigeria

TABLE 2.2 Rice consumption and production in 2011

Commodity	Total consumption (million tons)[1]	Total consumption share (%)	Total production (million tons)[2]	Total production share (%)
All rice	5.20		5.16	
Local rice	2.85	54.8	2.71	52.5
Imported rice	2.35	45.2	2.45	47.5

Source: (1) Authors' calculation from Nigeria, NBS and World Bank (2011); (2) United States Department of Agriculture International Database (USDA 2013).

in 2011 (the last year for which we have complete data). Of this amount, only 2.85 million tons were produced locally, while 2.35 million tons were imported. Table 2.3 shows average budget shares of a large number of food and nonfood items in the average Nigerian household's budget. The average Nigerian household spent over 6 percent of its total income on rice consumption in 2011. This budget share was the highest among food staples in this time period, and the pattern is similar for both urban and rural households.

Compared to other staples, rice is also widely consumed across the population: 84 percent of households reported consuming rice at home (Table 2.4).[3] There are, however, minor spatial differences in the pattern of rice consumption: while 91 percent of urban households in Nigeria consume rice, a slightly lower percentage—80 percent—of rural households do so. Consumption of imported and domestic rice also differs by location. While 72 percent of rice consumed by urban households is imported, only 33 percent of that consumed by rural households is imported. Table 2.3 shows that with the exception of nonfood items and dining out, rice dominates all other food items in the Nigerian household budget. The only exception is rural households, where fruits and vegetables make up a slightly larger share of households' food budget than rice.

RICE CONSUMPTION BY SECTOR AND ZONE

As noted earlier, there are differences in rice consumption across demographic and regional groupings in Nigeria. Table 2.4 shows that rural households consume more local rice than imported rice, while urban households consume more imported than local rice. Table 2.5 presents more disaggregated information on the structure of rice consumption by geopolitical zones in Nigeria.[4]

3 It is important to note that the data on which this analysis is based comes from LSMS–ISA (Nigeria, NBS and World Bank 2011), which surveyed households on their consumption patterns over the last seven days prior to being surveyed. While it is possible that there are households that consume rice regularly but had not done so in the last seven days and vice versa, we nevertheless assume this is a reasonable snapshot of the average proportion of households that consume rice in Nigeria.

4 For the geopolitical zones in Nigeria, see the footnote of Figure 1.1.

TABLE 2.3 Average household budget share across commodities

	Average budget share (%)		
Commodity	Urban	Rural	National
All rice	5.7	7.0	6.6
Local rice	1.8	4.7	3.8
Imported rice	3.8	2.4	2.8
Maize	1.2	2.7	2.2
Wheat	1.7	1.5	1.6
Sorghum/millet	1.3	5.7	4.3
Other grains	0.1	0.1	0.1
Cassava	3.2	4.8	4.3
Yam	4.6	5.3	5.0
Other roots	1.3	1.6	1.5
Pulses	3.0	3.9	3.6
Oils and fats	4.7	6.7	6.1
Fruits/vegetables	5.2	7.1	6.5
Milk	1.1	1.0	1.0
Poultry	0.9	1.1	1.0
Other meat	5.3	6.1	5.9
Fish	5.5	5.9	5.8
Other food	3.1	3.8	3.5
Dining out	12.8	9.3	10.4
Nonfood	39.5	26.4	30.5
Total	100	100	100*

Source: Authors' calculation from Nigeria, NBS and World Bank (2011).
Note: *Because of rounding, the percentages may not always add up to exactly 100.

TABLE 2.4 Share of households consuming rice by region

Region	Share of households consuming rice (%)	Share of households consuming imported rice (%)
Urban	90.8	71.8
Rural	79.9	33.1
National	83.8	48.2

Source: Authors' calculations based on Nigeria, NBS and World Bank (2011).

Generally, in each zone rice ranks very high in terms of expenditure as well as consumption volume compared to other main crops (Table B.1 in Appendix B). Per capita consumption ranges from a low of 20.9 kg per year in the South South to a high of about 39.6 kg per year in the North West, where rice is more important in the consumption baskets of households. While rice is

TABLE 2.5 Consumption patterns by region

Group	Urban			Rural		
	Expenditure (million naira/year)	Quantity (thousand tons/year)	(kg/capita)	Expenditure (million naira/year)	Quantity (thousand tons/year)	(kg/capita)
North Central						
All rice	26,660	191	28.6	54,605	495	28.8
Local rice	10,758	96	14.4	37,126	389	22.6
Imported rice	15,902	95	14.2	17,479	106	6.2
North East						
All rice	19,689	138	39.7	61,477	543	29.3
Local rice	12,580	99	28.5	51,237	482	26.0
Imported rice	7,109	39	11.2	10,240	61	3.3
North West						
All rice	62,462	514	53.0	142,428	1,260	39.5
Local rice	43,401	389	40.1	124,084	1,144	35.9
Imported rice	19,061	125	12.9	18,344	116	3.6
South East						
All rice	13,022	192	36.2	29,446	363	26.3
Local rice	2,225	25	4.7	12,059	111	8.1
Imported rice	10,797	167	31.5	17,387	252	18.2
South South						
All rice	41,772	234	24.4	53,440	314	20.9
Local rice	2,117	19	2.0	7,780	78	5.2
Imported rice	39,655	215	22.4	45,660	236	15.7
South West						
All rice	101,059	774	32.8	30,575	184	26.4
Local rice	1,171	9	0.4	1,339	10	1.5
Imported rice	99,888	765	32.4	29,236	174	24.9

Source: Authors' calculations based on Nigeria, NBS and World Bank (2011).

not always the dominant food item in terms of per capita consumption in all regions, it always ranks very highly regardless of geopolitical zone.

Table 2.5 also shows that the type of rice consumed differs across zones of Nigeria as well as between rural and urban areas. There is a preference for local rice in the northern zones, where it accounts for over 80 percent of the rice consumed, compared to the southern zones, where only about 12 percent is local rice. While imported rice and local rice are consumed in roughly equal proportions in urban areas of the North Central zone, the consumption of local rice dominates that of imported rice in rural areas of the North Central and both rural and urban areas in the North East and North West. The opposite applies in the southern zones (South West, South South, and South East), where imported rice dominates, even in the rural areas, where very little local rice is consumed. These patterns of consumption may hold for several reasons that are discussed in greater detail below.

RICE CONSUMPTION PREFERENCES BY REGION

There are several varieties of local rice grown and consumed in different regions of Nigeria, and they are often prized for specialized purposes. Generally, local varieties of rice grown in Nigeria tend to be short to medium grained, absorb water upon cooking, and can be made into a paste for local dishes—qualities that are not possessed by imported rice. Examples of some of the local rice varieties cultivated and consumed in the three northern regions include *galaware* (African wild rice), *yar-yarmidi*, *yar-kera*, *yar-kilaki*, *yar-das*, *dukusa*, *yar-banki*, *yar-mubi*, *yar-tundun wada*, WITA 4, and *yar-mubi*.[5] Of these varieties, *yar-tundun wada* is of long grain with brightness, while others are of short grains, bright, and well filled.

They all possess high-quality processing characteristics and organoleptic (length, taste, and swelling) properties. These properties give local varieties an edge over imported rice in the preparation of several local dishes frequently consumed in the region, such as *tuwo*, *sinasir*, and *waina/masa* (PrOpCom 2007a, 2007b, 2007c). Imported rice cannot be used to prepare these local dishes because it does not have the aforementioned characteristics.

In the southern zones, there are also several varieties of local rice produced and consumed. However, the varieties of rice consumed in the South South and South East zones are very similar. Local rice in the Igbo-dominated South East is known as *osikapa*. The characteristics of this rice variety are that the grains are usually of white, yellow-red, and variegated type. Okeke et al. (2008) described the local rice varieties as being "cream colored and tasting better than some of the 'exotic' imported rice." He also likened the *osikapa* rice to *fufu* (a popular West African staple made of cassava, maize, or wheat flour).

The South East is also often associated with the consumption of Abakaliki rice—named after the local rice mill in Ebonyi State (for the location of each state, see Figure 1.1).[6] While the generic name is Abakaliki rice, it is in fact a mixture of several rice varieties. Oko, Ubi, and Efisue (2012) present a comprehensive analysis of some of the Abakaliki rice varieties. They identified the superior qualities of local rice varieties in Ebonyi State of the South East as having "high amylopectin content, tender texture, and length expansion

5 See http://www.propcommaikarfi.org/wp-content/uploads/2013/08/12-Kano-Rice-Baseline-Report-4-07.pdf for a more comprehensive breakdown of the different northern rice varieties.

6 See http://www.nigerianorientnews.com/?p=1400.

on cooking without affecting the width."[7] This variety of rice is usually consumed boiled and eaten with stew.

The local rice in the South West has become synonymous with *ofada* rice, a popular variety that is discussed in more detail in Box 2.1 as it is an interesting case study of a popular niche market for local rice. However, *ofada* is not

BOX 2.1 Case study of *ofada* rice

Ofada rice is a variety found in the South West and grown almost exclusively in Ogun State. It takes its name from Ofada, the town in Ogun State where it was first cultivated. It is characterized by a very strong aroma, as well as the tendency to swell or increase in size, resulting in the claim that "a little quantity fills one up." It is generally eaten with a vegetable-based stew that is sometimes made with locust bean.

The grains are short but large, aromatic if brown and non-aromatic if white, and perceived to be more nutritious than imported rice (PrOpCom 2007c). *Ofada* rice is not an indigenous African rice; it has existed in Ogun State only since the 1940s, when according to rumor it was smuggled into Nigeria by a soldier who returned from Asia and planted this variety of rice in his hometown of Ofada.

In recent years, *ofada* rice has developed from being viewed as an inferior variety to a premium, highly desired rice variety. There is no verifiable data on acreage under cultivation or quantities consumed, but from anecdotal evidence the demand for *ofada* rice has been growing exponentially since the early 2000s, particularly in the South West, where it has become indigenous.

It appears that this increased awareness of *ofada* rice coincided with an intervention by the Department for International Development of the United Kingdom, known as Promoting Pro-poor Opportunities in Service and Commodity Markets, which funded a commodity and service market for *ofada* rice. The initiative was carried out in conjunction with many local agencies, including the Agricultural Development Project (ADP) and the rice farmers' associations, to increase knowledge about the production process and public awareness of *ofada* rice.

Figure B2.1 shows the trends in real prices of *ofada* rice and imported rice in Abeokuta between the last quarter of 2010 and December 2012. Imported rice prices were very stable over the period. *Ofada* rice prices, on the other hand, have been consistently higher than the real price of imported rice. *Ofada* rice has increasingly been viewed as premium rice, superior to

7 Oko, Ubi, and Efisue (2012), pp. 11–17.

other local varieties, often served at high-level parties, and served as a delicacy at very exclusive restaurants.

There have been many questions regarding this meteoric ascension of *ofada* rice to the top of the local rice food chain. It was initially assumed that it was due to improvement in the processing of this particular brand of local rice. However, field visits have shown that this is not strictly true. While processors are responding to consumer demands for cleaner local rice by packaging *ofada* in more attractive ways, the rice is being processed not with specialized machines but using manual labor.

FIGURE B2.1 Real prices, in naira, of imported and *ofada* rice in Abeokuta, Ogun State (October 2010–December 2012)

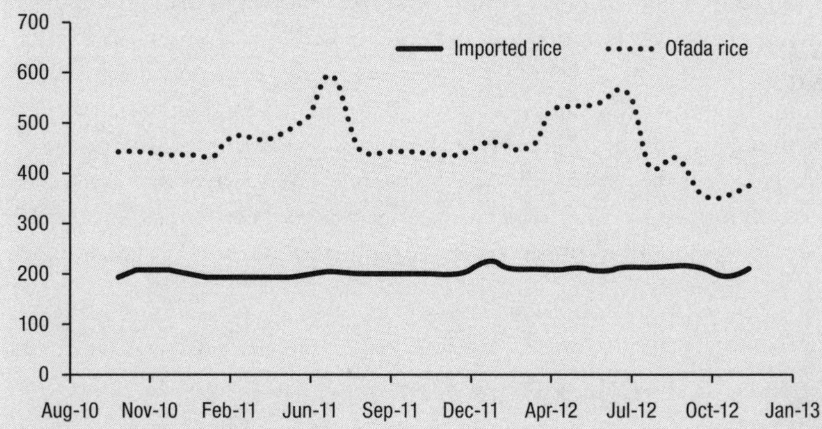

Source: National Bureau of Statistics data.

Ofada rice has a very special aroma, which is achieved through a fermentation process. The aroma has now become its distinguishing feature. It has become a staple at high-class parties. The rice has become branded by being served in leaves with a special sauce. In addition, the high price is regarded by some as a status symbol. *Ofada* rice has also been marketed as having many health benefits. It is unpolished rice, so it still retains good nutritional value.

The key message that emerges from the *ofada* rice case study is that Nigerians are willing to pay premium prices for local rice (even higher than for imported rice) if it meets their quality and taste standards. These types of rice can compete effectively with rice imports. One characteristic of *ofada* is the fact that it has been branded and marketed as unique. Perhaps the lesson for policymakers is that besides improving quality, branding and marketing Nigerian rice through information campaigns might be very effective. Perhaps creating more niche-market rice is another option for policymakers.

the only local rice variety grown and consumed in the South West. A large variety of New Rice for Africa (NERICA) species are cultivated in the South West, in addition to many indigenous species (e.g., *igbimo* in Ekiti State), but *ofada* remains the most popular and best known. *Ofada* is not a single variety and therefore unique; however, the way it is processed, packaged, and prepared for cooking gives it its unique characteristics (PrOpCom 2007c), as well as the fact that it is generally eaten on special occasions.

CONSUMPTION FROM OWN-PRODUCTION

According to the LSMS–ISA (2011) data, an overwhelming proportion of rice consumers buy their rice on the market; however, a significant number and proportion of consumers produce their own rice for consumption. Own-production and consumption accounts for about 20 percent of all local rice consumed in Nigeria and about 10 percent of all rice consumed in the nation. Nationally, about 13 million individuals reported producing the rice they consumed in 2011. About 82 percent of these own-production consumers are in the northern zones, with the North East and the North West having the largest share of individuals who produce the rice they consume. On the other hand, own-production consumers are negligible in the southern zones, especially in the South West.

Factors Affecting Rice Demand in Nigeria

Several factors, including rapid urbanization, accelerated population growth, increased per capita income, and changes in family occupational structures and lifestyles have contributed to increased demand for rice (Akpokodje, Lançon, and Erenstein 2001; Akande 2002; UNEP 2005; DeMont et al. 2013). Urbanization is a major factor in rice demand because of the lifestyle changes it engenders, requiring foods that are convenient and quicker to prepare, and rice meets these conditions very satisfactorily. It is clean and easier and less time consuming to prepare than traditional Nigerian staples such as cassava or yam.

While most consumers combine imported and local rice in their diets, urban households generally have a preference for imported rice (Table 2.6). The characteristics that have endeared many of these consumers to imported rice include higher quality—defined as a higher swelling capacity, better taste, and preferred grain shape—as well as cleanliness, as imported rice tends to be polished, nonbroken, and free from stones and other debris (Bamidele, Abayomi, and Esther 2010; Lançon et al. 2003a). This type of

TABLE 2.6 Household budget shares and income elasticities of demand

Commodity	(1) Average budget share (%)	(2) Marginal budget share (%)	(3) Expenditure elasticity	(4) Market price (naira)	(5) Average household expenditure (naira)	(6) Value of gamma[a] (naira)	(7) Share of gamma in expenditure (6)/(5) (%)
All rice	6.6	4.2	0.63				
Local rice	3.8	1.7	0.44	93	273	214	78.4
Imported rice	2.8	2.5	0.88	166	203	113	55.7
Dining out[b]	10.4	16.8	1.61	900	751	145	19.3

Source: Linear Expenditure System model results.

Note: [a] Gamma represents the part of expenditure for subsistence consumption of food. [b] The market price of dining out here represents the average cost of dining out for a family of four. The average total household expenditure on dining out is low relative to the price of dining out, suggesting that most households do not dine out.

rice is better suited to cooking foreign dishes, which are often favored by urban households.

As with imported rice, local rice is consumed for several reasons: it is cheaper, and it possesses organoleptic attributes that make it a vital component in certain local delicacies (Bamidele, Abayomi, and Esther 2010). These delicacies include *tuwo* in the north, where local rice is preferred for its ability to absorb water and to be pounded into paste, and *ofada* rice in the South West, which is preferred for its unique aroma. The problem with local rice is that it is often not properly processed, has a high percentage of broken grain, and usually includes foreign matter (e.g., stones). Most varieties of local rice produced in Nigeria are of short-grain types. These are usually viewed as inferior rice, and most consumers of local rice (particularly in urban areas) aspire to be able to afford imported rice, which is cleaner, requires less processing time, and can be utilized to prepare a wide variety of foreign dishes (e.g., *jollof* rice, fried rice) as well as local dishes.

Trends in Rice Prices

An important determinant of the quantity purchased of any good is its price. To understand the demand for rice, it is therefore necessary to discuss changes in the real prices of rice over time in Nigeria. Figure 2.4 presents trends in real prices of rice nationally for imported rice and several varieties of local rice between 2003 and 2011. The prices are real naira prices, with 2003 as the base year for a kilogram of rice. Generally, the real price of all rice—imported and different varieties of local rice—trended downward over the period, although the rate of decline was not smooth; prices tended to fluctuate from one year to the next even though the downward trend is clear. The domestic price of

FIGURE 2.4 Trends in real rice prices (national)

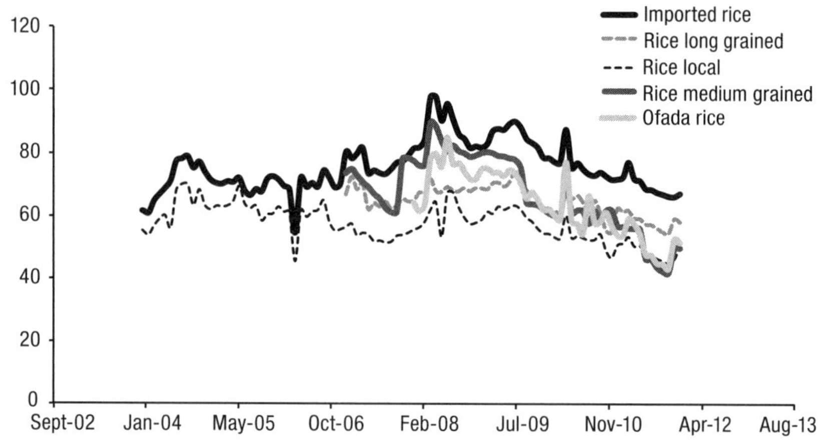

Source: National Bureau of Statistics data.

Note: The prices are real naira prices, with 2003 as the base year for a kilogram of rice. NBS data do not provide an explanation of how rice is categorized as "rice long grained," "rice local," or "rice medium grained."

imported rice and the prices of local rice moved in the same direction as international prices. The large spike in the domestic price of rice in 2008 tracks the rise in international prices resulting from the 2008 global food crisis. It appears that trends in the domestic market price for rice are related to changes in the price of rice in the international market, exchange rate, and rice trade policies in Nigeria.[8]

Generally, imported long-grain rice and other long-grain rice have consistently commanded higher prices than those of local rice, even though all prices move in the same direction. While the price of local rice is usually lower by a healthy margin than that of imported rice, the prices of some local niche rice, such as *ofada,* are closer to imported rice. Indeed, in some zones of Nigeria such as the South West, the price of *ofada* rice exceeds that of imported rice.

Prices for *ofada* rice (though only available since 2008) have also been consistently higher than prices of other local rice varieties, indicating that *ofada* is premium rice. This suggests that Nigerian consumers are willing to pay higher prices for local rice if it meets their quality and taste standards. The policy implication is that there should be increased emphasis on processing in order

8 Chapter 7 on rice trade provides a detailed discussion of the evolution and determinants of rice price in the domestic Nigerian market.

FIGURE 2.5 Enugu prices (naira/kg), 2006–2012

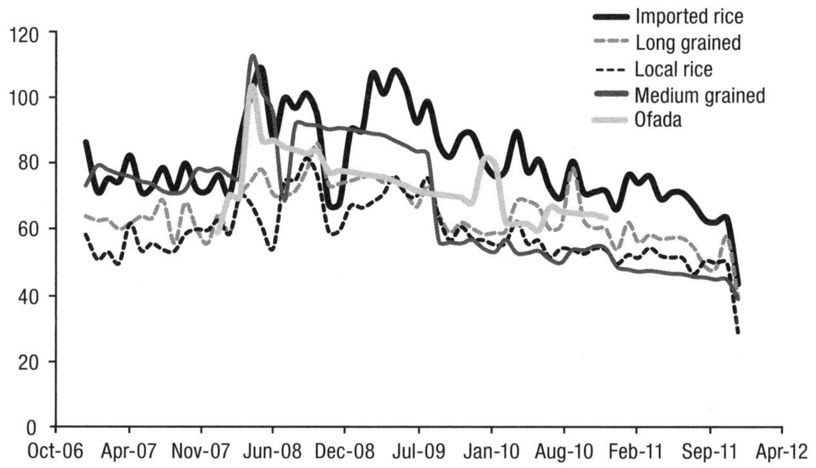

Source: National Bureau of Statistics data.

to increase the quality of local rice to attract the Nigerian consumer, especially those with higher incomes.

There have been no studies that track and explain the observed trends in real prices of rice in Nigeria over the 2002–2012 period. However, it is likely that Nigerian real rice prices are mostly driven by demand-side factors combined with (1) changes in the world price of rice and (2) exchange rate and trade policies. As discussed earlier, both per capita and aggregate demand for rice in Nigeria have been increasing at a rapid rate. Without an aggregate increase in domestic rice production to meet demand, real prices are likely to be more susceptible to global price changes and to exchange rates and trade policies as imports rise.

It is possible that the national price trends presented in Figure 2.4 mask rural/urban differences and that price changes may have different welfare implications for urban and rural households. There are few regional markets with complete time-series data for rice prices in Nigeria; no data exist for rural markets. Among regional markets, the Enugu market (located in the South East zone of the country) has detailed time-series data of monthly rice prices from 2006 to 2012. We used these series data to represent rice price trends in urban Nigeria. The price trends in the Enugu market for the average weekly prices of various types of rice in the 2006 to 2012 period are presented in Figure 2.5.

The trends in Enugu shown in Figure 2.5 display similar patterns to the national price averages that are shown in Figure 2.4. While there is a general downward trend in the prices of all types of rice, the decline was not smooth. Changes in Enugu rice price also follow changes in the international price of rice very closely. The weekly price of imported rice in Enugu, similar to the average price of imported rice at the national level, is generally higher than the average price of all types of domestic rice.

Explaining the Demand for Rice in Nigeria

The food consumption pattern of Nigerian households has been undergoing significant changes with income growth, rapid population growth, changing relative prices of items in the food basket, changes in livelihoods, and urbanization. The Nigerian diet has consistently shifted away from tubers such as yams and cassava to cereals such as rice, to the extent that today rice commands the largest share of the average household's food budget. In this section, we model the demand for rice in Nigeria in an attempt to understand what drives the growing demand for the product and to understand what policy options are available to affect the demand for rice and to ensure food security for different segments of Nigerian society.

There are several ways to model the demand for rice or any food item. One can conceive of and focus on analyzing the demand for each individual food item. This allows a researcher to control for several covariates that affect the quantities of a food item purchased by households and thus provide a rich set of environmental controls, as well as policy tools, for the policymaker. However, this approach ignores the interrelationship among goods in household budget decisionmaking, especially given that with a limited budget, increasing expenditure on one item affects the quantities of other items purchased by the household.

Alternatively, one can model the demand for food items as part of a system of demand for food or household expenditures. Given that the demand for these items is part of the household's expenditure on food (indeed part of the household's total expenditure on all goods), it may make sense to use the second approach—an approach that accounts for the interconnections among the various items in the consumer's food budget as well as the household's budget constraints. With low incomes and strictly binding food budget constraints, these interconnections may be very important for consumer well-being. While the systems approach models the interconnections among various items in the household consumption basket, its disadvantage is that it is not able to provide a rich set of environmental controls of the determinants of demand, as the single-commodity approach does.

We take the systems approach in estimating the demand for rice in Nigeria, using the Linear Expenditure System (LES) approach popularized by Stone (1954). The LES demand system models household expenditure shares on various commodity groups as a function of prices of these commodity bundles and consumer budgets. It is derived from maximizing the consumer's Stone-Geary utility function subject to the consumer's expenditure constraint. The LES model divides the consumer's expenditure shares on any commodity into two components: minimum required consumption/expenditure (which we call subsistence consumption) and additional expenditure shares over and above subsistence expenditure.

A summary of the LES model is as follows, with accompanying details of the model in Appendix B: Maximize $U = \prod_i (Q_i - \gamma_i)\alpha_i$, subject to $\Sigma_i p_i * Q_i \leq Y$, where U is the index of consumer utility, Q_i is quantity of good i, γ_i is the subsistence consumption of good i, α_i is the marginal budget share (MBS) of good i in the consumption basket, p_i is the price of good i, and Y is the consumer's total expenditure (budget). From the first-order conditions for utility maximization, subject to the expenditure constraint, and rearranging, we derive the estimation equation

$$S_i = \frac{p_i \alpha_i (Y - \Sigma p_j y_j)}{p_i Y_j} + \frac{\alpha_i}{Y},$$

where S_i is the expenditure on a good as a share of total household expenditure and all other variables are as defined in the text above.

Besides the ease of estimation, the LES model of demand has several attractive features. Because it is derived from a utility function, the parameter estimates have easy economic interpretation. It also satisfies the requirements of a demand function; among them, increasing all prices and incomes by the same proportion leaves quantity purchased (expenditure shares) unchanged, and the weighted expenditure shares of all commodities add up to unity. However, the LES model also has some disadvantages. The primary disadvantages are that it requires all commodities in the consumption basket to be noninferior and that consumer preferences be additive, hence independent in commodity bundles.

In our model, we treat the household as the unit of analysis. While economic theory models consumption decisions at the individual level, we were unable to do so because we have consumption and budget data only at the household level. However, this constraint ought not bias the results in the analysis significantly, as most consumption decisions in Nigeria are made at the household level, typically by the patriarch/matriarch of the household.

The data for estimating the share equations in the LES model were obtained from the LSMS–ISA (2011), which is a national survey on household welfare conducted by the NBS in partnership with the World Bank. The data used in the analysis are from a 5,000-household cross-section survey derived from the larger nationally representative sample of 22,000 households. Although the geographical domain of analysis is Nigeria's six geopolitical zones, the data are nationally representative as population weights are included for the sample households.

All the findings in this chapter are derived using these population weights. The sample of 5,000 households represented in the data was surveyed twice—once in 2010 and once in 2011—to gather detailed data on agricultural activities and household consumption. Post-planting (lean season) data were collected in 2010, and postharvest data were collected in 2011. However, because of many extremely large and unrealistic outliers in the 2010 data, the analysis in this chapter is based on the 2011 postharvest data.

The LSMS–ISA data contain information on cassava and cassava flour; however, we chose to aggregate both into a single category for simplicity. Where an observation was three standard deviations away from the mean, we excluded it from the analysis to avoid undue influence on parameter estimates. Sample statistics for the data are presented in Table B.1 in Appendix B. As is shown in Table B.4, Nigerian households, on average, spend about 6.6 percent of their budgets on rice. This average budget share, however, has a large variance. There are clearly regional and rural/urban differences in average expenditures on rice in the sample, but nevertheless all show the importance of rice in the Nigerian diet.

Estimates of Expenditure and Demand Elasticities

We present the results of the estimates of the budget share equations for rice (total), imported rice, and local rice as a share of a household's total expenditures. This includes both ABS and MBS, describing the average expenditure share for rice and the additional expenditure on rice following a small change in income, respectively. We begin this subsection by discussing the budget shares, followed by a discussion of price and expenditure elasticities, which describe the sensitivity of a household's demand for rice due to a small change in price and expenditures, respectively. We do this first at the national level, and for comparative purposes only, we also report the estimates for dining out. Because we are interested in the effects of policy on different groups of people, we also present and discuss the budget shares and elasticities of demand for rural and urban households, as well as by income classes.

DEMAND CHARACTERISTICS AT THE NATIONAL LEVEL

The national average budget shares and marginal budget shares, expenditure elasticities of demand, market prices of rice, average household expenditures on rice, subsistence consumption of rice, and subsistence consumption of rice as a proportion of total household expenditure are presented in Table 2.6. For the purposes of comparison, we also present data for the food expenditures with the largest ABS and MBS, which is dining out. Generally, column 5 in the table indicates that the average Nigerian household spends more on domestic rice (273 versus 203 naira, or about 34.5 percent more) than on imported rice, in absolute terms.

Table 2.6 also shows that the average budget shares of all rice, local rice, and imported rice are relatively high. While the ABS for local rice is higher than that for imported rice, the marginal budget share for imported rice is higher than its local rice counterpart. While the average budget shares of all rice, local rice, and imported rice are relatively high, their marginal budget shares are lower than their ABS, suggesting that, on average, Nigerian consumers devote less of their additional budget to rice. In summary, at the national level households will devote a decreasing share of extra expenditure to rice as a whole, with even smaller shares being devoted to local rice purchases. This suggests that the share of household expenditure devoted to rice purchases will decrease with income, even though the total amount the household spends on rice will increase. For comparison, the MBS for dining out is higher than the ABS, suggesting that Nigerian consumers in all groups will devote a larger share of additional expenditure to dining out as incomes rise.

One of the major strengths of the LES approach is that it allows the researcher to estimate the proportion of expenditures that is considered as subsistence consumption (Appendix B). This is the amount of consumption expenditures the consumer is "committed to" without regard to his or her level of income.[9] Column 5 of Table 2.6 presents the average household expenditures on the various types of rice, while column 6 presents total subsistence consumption for various commodities (referred to here as gamma). Generally, the level of subsistence expenditures on different kinds of rice in Nigeria is very high. Indeed, column 7 shows that subsistence consumption as a share of average household expenditure on various types of rice ranges from a low of 55.8 percent (imported rice) to a high of about 78 percent (local rice).

9 This concept of subsistence consumption differs from the traditional meaning of subsistence, which refers to farmers growing enough to feed themselves rather than selling their produce in the market. Also see a discussion of this concept in Appendix B.

It is also interesting to note that the share of local rice that is for subsistence consumption is far higher than the subsistence share of imported rice. On the other hand, the subsistence share of expenditure on dining out is only 19.4 percent compared to 78 percent for local rice. Policies that affect the subsistence level of purchases may have serious welfare implications for consumers who spend a large part of their household budgets on these goods, especially for those at the low end of the income distribution (i.e., the poor).

Column 3 of Table 2.6 presents the estimated expenditure elasticities of demand for the country as a whole for all rice, local rice, imported rice, and dining out. The estimated expenditure elasticities are positive and significantly different from zero, at a 99 percent confidence level, indicating that all rice, imported rice, and local rice in the demand system are normal goods whose consumption increases with income. We note that the expenditure elasticity of demand for imported rice is much higher than that of local rice; in fact, it is twice as high. This suggests that, ceteris paribus, the share of expenditure on imported rice will grow at twice the rate of that on local rice as total household expenditures (or incomes) grow.

Although the estimated expenditure elasticities of demand are positive for all types of rice, they are significantly less than unity, suggesting that expenditures on rice in Nigeria will grow at a slower pace than household expenditures, ceteris paribus. Stated another way, the results indicate that while the absolute amount of expenditure on rice will increase with total household expenditures, the percentage spent on all rice, local rice, and imported rice will decrease as total household expenditures increase, if all other factors remain unchanged. The decreased expenditure shares are consistent with the decreasing MBS of these goods noted above; they are also consistent with Engel's law: the general observation that expenditure on food proportionally decreases as income increases.

DEMAND CHARACTERISTICS OF URBAN VERSUS RURAL HOUSEHOLDS

It is possible that there are differences in the patterns of consumption between urban and rural areas and that the national averages presented in Table 2.6 mask such differences. Table 2.7 presents the budget shares, expenditure elasticities, and average household and subsistence expenditures. The table also shows, for both rural and urban areas, the ratios of subsistence expenditures to average household expenditures. Panel A presents the estimates for urban households, while Panel B presents these for rural households. Generally, urban households spend about twice as much on imported rice as they do on domestic rice, while the reverse is true for rural households, as shown in

TABLE 2.7 Household budget shares and expenditure elasticities, rural and urban locations

Commodity	(1) Average budget share (%)	(2) Marginal budget share (%)	(3) Expenditure elasticity	(4) Market price (naira)	(5) Average household expenditure (naira)	(6) Value of gamma[a] (naira)	(7) Share of gamma in expenditure (6)/(5) (%)
Panel A: urban							
All rice	5.6	2.4	0.42				
Local rice	1.8	0.4	0.20	96	176	158	89.8
Imported rice	3.8	2.0	0.53	154	364	268	73.6
Dining out[b]	12.8	20.2	1.58	921	1,217	253	20.8
Panel B: rural							
All rice	7.1	5.5	0.77				
Local rice	4.7	3.0	0.72	92	288	196	68.1
Imported rice	2.4	2.5	1.03	178	144	70	48.6
Dining out[b]	9.3	14.1	1.51	679	573	140	24.4

Source: Linear Expenditure System model results.

Note: [a]Gamma represents the part of expenditure for subsistence consumption of food. [b]The market price of dining out here represents the average cost of dining out for a family of four. The average total household expenditure on dining out is low relative to the price of dining out, suggesting that most households do not dine out.

column 5. The expenditure share on rice that is treated as subsistence consumption, in column 7, is much higher for urban households than for rural households; as much as 90 percent of the expenditure on local rice by urban households is considered to be subsistence, while only 68 percent of such expenditures are considered to be subsistence by rural households.

Also, while the MBS for local rice is relatively low among urban households (0.4 percent), it is quite high among rural households (3.0 percent), as shown in column 2. For imported rice, the MBS is slightly higher among rural households than urban households: 2.5 versus 2.0 percent, respectively. This indicates that in general, rural households devote a larger portion of additional expenditure to rice than do urban households. The implication for rice demand in Nigeria is that the expenditure on rice will grow faster among rural households than among urban households if rural and urban incomes continue to grow at the same rate. This is especially true for local rice.

The estimated pattern of consumption shown here may indicate that while urban households treat all types of rice as subsistence food, rural households may treat rice differently; they treat imported rice as a suprasubsistence food, while treating local rice as subsistence food (see how much less of rural households' subsistence consumption is spent on imported rice than local rice in

Table 2.7). As in Table 2.6, the marginal budget shares of additional expenditure on rice of all types are generally lower than the average budget shares among both urban and rural households.

The estimated expenditure elasticities for rice are presented in column 3 of Table 2.7. In general, the expenditure elasticities of demand for rice are positive for both rural and urban households and suggest that the demand for all types of rice increases with household expenditures in both rural and urban areas. For urban households, the expenditure elasticities for all types of rice (all, local, and imported) are significantly less than 1, suggesting that they are necessities. For rural households, the expenditure elasticity of demand for imported rice is slightly greater than 1, while for local rice it is significantly less than 1. These estimates suggest that while local rice is considered a necessity by both rural and urban households, imported rice is considered a necessity by urban households only; it is considered a luxury item by rural households.

For the purpose of comparison, the ABS for dining out is 12.8 percent and the MBS is 20.2 percent for urban households, while the expenditure elasticity is 1.58. For rural households, the ABS for dining out is 9.3 and the MBS is 14.1 percent, while the expenditure elasticity is 1.51. It is clear from these estimates that the expenditure share of dining out will increase for both rural and urban households with increased household expenditures. Similarly, the expenditure share of imported rice will increase for rural households as total household expenditures increase.

The responsiveness of quantity demanded to changes in the price of a particular type of rice—own-price elasticities of demand, hereafter—is presented in Table 2.8: for local rice and imported rice; and for Nigeria as a whole and for urban and rural areas. As in other tables, we present own-price elasticities of demand for dining out for comparison with the elasticity of demand for rice.

Generally, the own-price elasticities of demand for these products are negative, suggesting that budget shares decrease with increased prices, ceteris paribus. However, the absolute values of the elasticity coefficients are generally

TABLE 2.8 Household own-price elasticities of demand

Commodity	(1) National	(2) Urban	(3) Rural
Local rice	–0.39	–0.13	–0.33
Imported rice	–0.64	–0.34	–0.46
Dining out	–1.11	–1.02	–0.68

Source: Linear Expenditure System model.

less than 1 at the national level, and for both rural and urban areas, suggesting that the expenditures on these food items are not very sensitive to price changes. Another way of putting it is that consumers will simply spend more, as the quantity demanded changes little as a result of an increase in price.

Note that demand is less price elastic in both the urban and rural sectors compared to the national level. This is possibly because national-level estimates may also reflect the movement of consumers between urban and rural sectors, each of which exhibits different demand characteristics. Generally, the own-price elasticities of demand for all types of rice are much lower for urban households than for rural households, suggesting that the demand for rice is less sensitive to price changes among urban households than among rural households.

The welfare implication of these low price elasticities is that households (both urban and rural) may spend more on rice (both domestic and imported) when prices go up, as they may be unwilling to substitute away from rice products with the price increases because rice is the preferred staple. In contrast, they would be more willing to substitute away from dining out if food prices in local eating establishments rise given a higher own-price elasticity of demand for dining out in both urban and rural areas.

One policy instrument often used in Nigeria to stimulate rice production and substitute the consumption of local for imported rice is an import tariff. However, this will work if and only if there is a high degree of substitutability between imported and locally produced rice that is derived from estimates of cross-price elasticities of demand among the two types of rice and others in the market. Estimates of all own-price and cross-price elasticities of demand for domestic and imported rice and all other food commodities are presented in Table B.2 in Appendix B.

In addition to the generally low own-price elasticities of demand presented in Table 2.8, the estimated cross-price elasticities of demand between local and imported rice as well as between rice and other local food staples, as reported in Table B.2, are very low—indeed, in no case does the absolute value of the cross-price elasticity of demand approach 0.05.[10] The estimated cross-price elasticity of demand between local and imported rice is −0.003, suggesting that while these types of rice are complements in consumption, the degree of complementarity is so low that the relationship among the two is not economically significant. Clearly, however, the two types of rice are not substitutes.

10 The full range of cross-price elasticities of demand is not reported in the text but is available in Appendix B. We report only the cross-price elasticities between local and imported rice.

The lack of substitutability between imported and local rice we find here is similar to what Demont et al. (2013) find for Senegal. Similarly, the cross-price elasticities of demand between rice and other domestic food staples is very low. Perhaps the lack of substitution between the two types of rice, as well as between those two types of rice and other food staples, explains the low own-price elasticities of demand, as there appears to be no effective substitutes for these types of rice in Nigeria. Besides consumer welfare implications, the lack of substitution between local rice and imported rice has important policy implications. Unless consumers develop a taste for local rice that rivals their taste for imported rice, it will be difficult to achieve rice self-sufficiency (no rice imports, as the government defines it) through a tariff strategy because there will always be demand for imported rice. Of course, there can be policies (including marketing and information campaigns) and incentives put in place to improve the quality and taste of local rice to rival those of imported rice.

DEMAND CHARACTERISTICS ACROSS INCOME GROUPS

So far we have analyzed the income and price elasticities of demand for rice for average national, rural, and urban households, but it is possible that expenditure patterns vary systematically among different groups of consumers, especially among income classes. If policy changes have disproportionate effects on the prices of foods consumed by households that are least able to withstand price shocks or least able to substitute away from such goods, these households may be the ones that will bear the brunt of the negative effects of such policy changes. It is therefore important for policy purposes to investigate the pattern of consumption among various income groups.

We therefore estimate price and expenditure elasticities of demand for rice, including the ABS and MBS values, for households among two different income classes—poor and rich—in both urban and rural areas. We chose to use this broader definition of income classes because of the limited sample size of the LSMS–ISA (2011). Although an ideal breakdown of household incomes is to subdivide the sample of households into quintiles, doing so by urban and rural population as well produced unreliable parameter estimates for each subgroup in the LES regressions due to small sample sizes.

Instead, we found the results to be more statistically significant when the data were aggregated into two income classes. We defined the poor as belonging to the lowest three income quintiles, while the rich were defined to be in the upper two quintiles. This definition is also consistent with the actual poverty headcount ratio of 62 percent of households earning less than US$1.25

TABLE 2.9 Household own-price elasticities of demand by urban and rural location and income class

	Urban		Rural	
Commodity	(1) Poor	(2) Rich	(3) Poor	(4) Rich
Local rice	−0.52	0.00[a]	−0.24	−0.11
Imported rice	−0.32	−0.19	−0.28	−0.35
Dining out	−0.35	−0.91	−0.31	−1.05

Source: Linear Expenditure System model results.

Note: [a] The own-price elasticity of local rice for urban rich was very close to zero and statistically insignificant.

per day, adjusted by the purchasing power parity (World Bank Development Indicators 2014).[11]

Table 2.9 presents the own-price elasticities of demand estimated for the rich and the poor households, according to our definition of the poor, and in both rural and urban areas. Columns 1 and 2 in the table present the own-price elasticities of demand for urban households by income class, while columns 3 and 4 present these for rural households.

There are two important results to observe. First, generally the own-price elasticities of demand are low in both rural and urban areas and among the two income classes (poor and rich). This implies that the conclusions drawn from Table 2.8 earlier apply to both income groups as well: demand for rice is not sensitive to changes in own-prices and household expenditures, for rice will naturally move in the same direction of the price change.

Second, while inelastic overall, the demand for rice is generally more price elastic for the poor, except for imported rice among the rural poor. The opposite is true for dining out, with higher own-price elasticities among the rich in both rural and urban households. This implies that the poor tend to be more affected by rice price changes than by prices for dining out. On the other hand, the richer households are more affected by prices for dining out.

Table 2.10 presents expenditure elasticities of demand broken down by the two income groups in urban and rural areas. Columns 1 and 2 present the estimates for the poor and the rich among urban households, while columns 3 and 4 present those for rural households. In general, the expenditure elasticities of demand for imported rice tend to be higher than for local rice among

11 At the time of this writing, 2010 was the most recent year for which data on poverty rates in Nigeria were available.

TABLE 2.10 Household expenditure elasticities of demand by rural and urban location and income class

	Urban		Rural	
Commodity	(1) Poor	(2) Rich	(3) Poor	(4) Rich
Local rice	1.41	0.00[a]	0.86	0.17
Imported rice	0.92	0.26	1.06	0.57
Dining out	1.00	1.29	1.16	1.67

Source: Linear Expenditure System model results.

Note: [a]The expenditure elasticity of local rice for urban rich was very close to zero and statistically insignificant.

both rural and urban households, confirming again that the results of the expenditure elasticities in Table 2.7 apply to both income groups. The only exception is for the urban poor, whose expenditure elasticity for imported rice is less than that for local rice.

Expenditure elasticities are generally higher for the poor. Among urban households, the expenditure elasticity of demand for rice (both local and imported) tends to be higher for the poor than for the rich (1.41 and 0.92 versus 0.00 and 0.26). The pattern is similar for rural households (0.86 and 1.06 versus 0.17 and 0.57). For the poor in general, expenditure shares on imported and local rice increase with total expenditures. The reverse is true for dining out: expenditure elasticities of demand are higher for the rich in both urban and rural areas.

Table 2.11 shows the corresponding ABS and the MBS values of local rice and imported rice for all income groupings in both rural and urban areas. Generally, poor households devote a larger share of their budgets to rice than richer households do in both rural and urban areas (total ABS of local and imported rice). Urban households tend to allocate a higher budget share to imported rice than local rice, irrespective of whether they are poor or rich. On the other hand, poorer rural households allocate a much higher share to local rice (5.7 percent versus 2.0 percent). Among the rural rich, the shares are evenly split between imported and local rice.

Overall, at the national level, poorer households spend far higher shares of their budget on local than imported rice (5.3 percent versus 2.5 percent, respectively). These poor households not only have a larger ABS, they also tend to have a larger MBS for rice, especially local rice, than the richer households. Rural households tend to have higher ABS and MBS for local rice than urban households, confirming the greater preference of rural households for local rice than urban households, as discussed earlier.

TABLE 2.11 Household average and marginal budget shares by rural and urban location and income class

	Urban		Rural	
Commodity	(1) Poor	(2) Rich	(3) Poor	(4) Rich
Panel A: marginal budget shares (MBS) (%)				
Local rice	4.8	0.0[a]	4.9	0.5
Imported rice	3.9	0.8	2.1	1.6
Dining out	7.6	20.9	10.3	21.3
Panel B: average budget shares (ABS) (%)				
Local rice	3.4	1.0	5.7	2.8
Imported rice	4.2	3.2	2.0	2.8
Dining out	7.6	16.2	8.9	12.8

Source: Linear Expenditure System model.
Note: [a]The MBS for local rice for urban rich was very close to zero and statistically insignificant.

In addition, rich households do not increase their expenditures on local rice as their total expenditure increases (indicated by the very low MBS values). Therefore, rich urban households treat local rice almost like an inferior good. Depending on the income status of the household, this has important implications for household welfare, as an increase in the price of rice will hurt poorer consumers the most in both urban and rural areas by forcing them to spend an even greater proportion of their meager incomes on rice. On the other hand, decreasing the price of rice will benefit these lower-income households the most, as it will allow them to buy the same quantity of rice at a lower cost.

The combined analysis of price and expenditure elasticities, as well as budget shares, emphasizes the greater vulnerability of poorer households to rice policy changes and thus their welfare with regard to income. Because poor households spend a larger share of their additional expenditure on rice than do rich households, they are more likely to suffer much larger welfare losses from increased rice prices than rich households would. Although demand for rice by low-income households is more price elastic than rich households, demand is sufficiently price inelastic for all households that total expenditure on rice increases for all households as a result of a rice price increase.

Increasing the price of rice can have adverse effects on food consumption baskets and therefore the well-being of households. On the other hand, policies that decrease the price of rice will benefit the poor the most, as they will spend less to get the same quantity of rice. Such findings emphasize the importance of estimating these more disaggregate elasticity parameters for informing rice-pricing policies rather than those estimated without taking into account their location or income status.

Conclusion

This chapter analyzes the consumption of and demand for local and imported rice in Nigeria. We find that rice has become the most dominant food staple, at least by value, in the Nigerian diet in a relatively short period of time. Rice accounts for about 6.6 percent of the average Nigerian household budget. About 84 percent of Nigerian households eat rice regularly, consuming both locally produced and imported rice.

Although most Nigerians consume and demand both local and imported rice, there are regional and other locational differences in the pattern of rice consumption in Nigeria: urban households tend to consume more imported rice, while rural households tend to consume more local rice. More local rice is consumed in the northern zones of Nigeria than in the southern zones. While a small proportion of the rice consumed in Nigeria is grown on people's own farms (about 20 percent of local rice consumed and about 10 percent of all rice consumed in Nigeria), most of the rice consumed in Nigeria is purchased on the market.

The rapid increase in rice consumption in Nigeria has been in part determined by increased per capita income, urbanization and the lifestyle changes it engenders, decreases in the relative price of rice, and the rapid increase in total population. Rice consumption suits lifestyle changes that include both spouses working outside the home: it is easier and faster to prepare compared to other Nigerian staples, such as cassava or yam.

The regional as well as the rural urban difference in the consumption of local and imported rice is dictated, among other things, by differences in diet and customs: northern rural households prefer local rice that absorbs a lot of water so it can be made into sticky balls for local dishes, while urban dwellers prefer the parboiled long-grain rice that can be used for foreign dishes.

An estimated model of the demand for rice in Nigeria shows relatively high ABS and MBS for both local and imported rice, although there are differences in the composition of rice consumed across regions: ABS and MBS for local rice are much higher for rural households than for urban households, while average and marginal budget shares for imported rice are higher for urban households. For both rural and urban households, we find that a very large portion of rice consumption can be considered a subsistence expenditure on both local and imported rice.

Expenditure elasticities of demand for both local and imported rice are positive, but they are generally less than 1 for urban households. For rural households, the expenditure elasticity for local rice is less than 1, but the expenditure elasticity of demand for imported rice is greater than 1. These expenditure elasticities indicate that the consumption of both imported and local rice will continue to grow, especially with continued rapid population growth. The estimates, however, also suggest that household expenditure shares for rice will decrease as total expenditure rises.

Estimated own-price elasticities of demand are very low, generally far less than unity, indicating that the demand for both local and imported rice does not respond very much to price changes. The implication is that household expenditures on rice will increase when the price of rice increases and will decrease when the price of rice decreases. This has welfare implications for the average Nigerian consumer. In addition to low own-price elasticities of demand, the estimated cross-price elasticity of demand for local and imported rice is very low, indicating low substitutability between the two types of rice.

An increase in the price of imported rice may not necessarily translate into increased demand for domestic rice and vice versa. This has important implications for rice policy in Nigeria, especially as it relates to rice trade policy. We will come back to this in Chapter 7. In addition to low elasticities of substitution between local and imported rice, there is also low substitutability between both types of rice and local food staples.

Besides low expenditure and price elasticities at the national level, there are differences in the structure of demand for rice across the rural–urban divide and across income groups. We find that ABS and MBS for local rice among low-income groups are much higher than they are for higher-income groups. The average and marginal budget shares for local rice as well as the subsistence component of rice expenditures are particularly high for urban poor households. At the same time, the poor households are more sensitive to price changes. These facts suggest that the poor are likely to be hurt the most by policies that result in increased rice prices, while they are likely to benefit the most from policies that lead to reduced rice prices.

In summary, several key messages emerge from the demand analysis we conducted in this chapter. First, rice has grown to become one of the most important staples in the Nigerian diet in a relatively short period. Given income and population growth, rapid urbanization, and lifestyle changes, the demand for rice will continue to grow at a rapid pace. A large component of rice consumed in Nigeria is for subsistence consumption, and therefore changes in rice prices have important implications for food security in the country.

Second, local rice and imported rice are consumed for different purposes and are therefore not easily substitutable for each other. The estimates from the LES confirmed that the cross-price elasticity of demand between the two types of rice is very low. Third, using prices to change the type of rice Nigerian consumers buy is not likely to be successful, as it will require an unreasonably large price differential to work—a price differential that may not be politically feasible to bring about.[12] Fourth, simply focusing on increasing the supply of local rice through higher prices will only hurt low-income groups, as a large component of a poor household's rice expenditures is for subsistence.

Finally, an important policy implication is that there is good reason to improve the competitiveness of local rice brands and generate information and marketing campaigns targeted at and promoting their consumption in Nigeria. In this regard, there are some interesting lessons that can be drawn from existing niche rice markets in Nigeria, such as the illustrated case study of *ofada* rice. We now tackle the broader issue of how Nigeria can improve the competitiveness of local rice by examining in more detail the challenges and opportunities for transforming the rice sector in Nigeria in the next four chapters.

---

12 In the spring of 2014, the Nigerian government had to drastically decrease tariffs on rice imports, imposed earlier to promote the substitution of domestic for imported rice, due to increased smuggling of imported rice to avoid the high tariff. See Chapter 7 for a detailed discussion of recent rice tariff policy in Nigeria.

Chapter 3

PRODUCTION SYSTEMS: BIOPHYSICAL AND ECONOMIC ENVIRONMENT AND CONSTRAINTS

Hiroyuki Takeshima and Oladele Samuel Bakare

Rice production in Nigeria and West Africa[1] dates back 2,000 to 3,000 years ago. Such production was mostly in the flood plains of the Niger River (Figure 1.1 in Chapter 1) and involved one of only two species of cultivated rice in the world, *Oryza glaberrima*, or African rice. The introduction of the second species, *Oryza sativa*, or Asian rice, occurred after the arrival of the Portuguese in the 17th century (Linares 2002). However, it was only during the last few decades that rice production in Nigeria began to emerge as an important food security issue, as demand outstripped supply with ever-increasing gaps.

The growing gap indicates that the expanding market for rice has not been followed by a supply response among domestic producers. In particular, rice yields have grown very little in Nigeria. A crucial question, therefore, is whether Nigeria has a comparative advantage in rice production. In order to address this question, we need to examine whether biophysical and socioeconomic conditions are favorable among Nigeria's most promising rice production systems, and if they are, how the productivity gap can be reduced.

The objective of this chapter is to provide a descriptive overview of the current rice production systems in Nigeria, their underlying biophysical and socioeconomic constraints, and their potential for expanding rice production. The chapter is organized as follows.

First, we describe current production technologies and practices in use in Nigeria, such as the uses of modern inputs (improved varieties and fertilizer) and technologies (irrigation and mechanization). From this information, a typology of rice-producing households in Nigeria is determined according to the type of technologies and production environment and their capacity to expand rice production in the short to medium term.

In the second and third sections, we review some of the key biophysical and socioeconomic constraints, respectively, that affect the potential for

1 For the definition of West Africa, see footnote 2 in Chapter 1.

expanding rice production in different parts of the country. We show, for example, that the typical rice-production environments in Nigeria, such as climate, water access, and soils, may not be as favorable as they are in major rice-producing regions of Asia. Nevertheless, compared to other tropical West African countries, the prevalence of a lowland rice ecology in Nigeria offers it the greatest potential in the region for production growth, except for certain socioeconomic constraints that prevent households from adopting modern inputs and improved technologies. Based on this review, we conclude the chapter by summarizing the key characteristics of current rice-production environments in Nigeria.

The information and data used in this chapter are based on an extensive literature survey, international data sources, and data gathered from the most recent Nigeria Living Standards Measurement Study–Integrated Surveys on Agriculture (LSMS–ISA) described earlier in Chapter 2. In particular, in Chapters 3 and 4, we use not only the data from the LSMS–ISA's agricultural modules in 2011 (LSMS–ISA 2011), but the 2013 version (LSMS–ISA 2013), to which we were given advanced access at the time of writing this book.

Data from the agricultural module are based on the same sample as those used in the demand analyses in Chapter 2. They contain detailed information on production practices for various crops and the use, at the plot level, of inputs such as fertilizer, irrigation, and agricultural machinery for approximately 3,000 farm households out of a total of 5,000 households interviewed. These farm households were selected in the same way as the remaining nonfarm households, through stratified random sampling. Among the farm households interviewed, approximately 300 were rice-producing households. The module provides the most up-to-date and nationally representative picture of rice-production systems in Nigeria.

Production Systems and Farming Practices

Rice production in Nigeria has traditionally been characterized by low yields and slow growth. Yield reached a peak in the mid-1980s and has since been stagnant or even declining (row 19, Table 3.1). This is in contrast to Asian and Latin American countries, as well as many other West African countries. Though rice production has increased in the last four decades, this has been driven mostly by area expansion.

The current rice area in Nigeria is approximately 2.4 million hectares (ha), the largest within Africa south of the Sahara (SSA) and almost twice as large as that of the second largest producer, Madagascar (Seck et al. 2012). The

TABLE 3.1 Rice paddy yield trends (five-year average, tons per hectare)

(1) Periods	1961–1965	1966–1970	1971–1975	1976–1980	1981–1985	1986–1990	1991–1995	1996–2000	2001–2005	2006–2011
(2) Bangladesh	1.7	1.7	1.7	1.9	2.1	2.4	2.7	3.0	3.6	4.1
(3) Brazil	1.6	1.5	1.5	1.4	1.6	1.9	2.3	2.8	3.3	4.2
(4) Cambodia	1.1	1.3	1.3	1.0	1.2	1.4	1.5	1.9	2.1	2.8
(5) China	2.6	3.2	3.4	3.9	5.0	5.5	5.8	6.3	6.2	6.5
(6) Colombia	2.0	2.6	4.0	4.2	4.4	4.5	4.1	4.8	5.8	5.2
(7) India	1.5	1.6	1.7	1.8	2.1	2.4	2.7	2.9	3.0	3.3
(8) Indonesia	1.8	2.1	2.5	2.9	3.8	4.1	4.4	4.3	4.5	4.9
(9) Iran	2.7	2.7	3.1	3.5	3.4	3.5	4.0	4.2	4.4	4.5
(10) Philippines	1.3	1.5	1.5	2.0	2.5	2.7	2.9	2.9	3.4	3.7
(11) Sri Lanka	1.9	2.2	2.0	2.3	2.9	3.1	3.1	3.3	3.5	3.8
(12) Thailand	1.8	1.8	1.9	1.8	2.0	2.1	2.3	2.5	2.9	2.9
(13) Venezuela	1.7	1.9	2.5	3.1	2.7	3.4	4.1	4.7	4.9	5.0
(14) Côte d'Ivoire	0.9	1.1	1.2	1.2	1.2	1.2	1.1	1.6	1.9	1.8
(15) Ghana	1.0	1.2	1.0	0.9	0.9	1.3	1.9	1.9	2.2	2.2
(16) Guinea	1.7	1.7	1.7	1.7	1.7	1.7	1.7	1.7	1.7	1.9
(17) Liberia	0.5	1.1	1.2	1.2	1.3	1.2	1.0	1.3	1.0	1.3
(18) Mali	1.1	0.9	0.9	1.2	1.0	1.4	1.7	2.0	2.0	2.7
(19) Nigeria	1.1	1.4	1.7	1.7	2.1	2.1	1.8	1.6	1.4	1.7
(20) Senegal	1.3	1.2	1.1	1.2	1.9	2.1	2.3	2.4	2.5	3.2

Source: Authors' calculations based on FAO (2014).

dominant rice-production systems in Nigeria can be described as comprising lowland, upland, and irrigated rice. Among these, the rainfed lowland system covers about 64 percent of the area under rice and therefore contributes the most to national production (column 2, Table 3.2). The system is found in river valleys and swampy alluvial lowlands along the Niger River and especially in the *fadama* farming system areas in Niger and Kaduna states (for the location of states, see Figure 1.1 in Chapter 1).[2]

In contrast to the rainfed lowland system, irrigated lowlands account for only 4 percent of the national area and contribute only 7 percent to the total volume of rice produced in the country. The remaining rice production comes mostly from upland systems. Less common are the deepwater, floating rice

2 *Fadama* is a Hausa name in Nigeria for irrigable land in low-lying plains of shallow aquifers found along major river systems.

TABLE 3.2 Dominant rice-production systems in Nigeria

(1) Production system	(2) Average share of national area (%)	(3) Average share of national production (%)	(4) Average yield range per year (tons/ha)
Irrigated	4	7	> 2.0
Rainfed lowland	64	65	1.0–3.0
Rainfed upland	30	27	1.0–2.0
Mangrove/deepwater	2	1	NA

Source: Authors' estimates modified from Table 1 in Ezedinma (2005), Table 1 in Erenstein et al. (2004), and National Bureau of Statistics and World Bank (2011, 2013).

Notes: NA = figures not available. We adjusted the irrigated area downward to be consistent with the estimates from National Bureau of Statistics and World Bank (2011, 2013) and included the difference in the rainfed lowland category. We then adjusted the share of national production proportionately.

systems found in the Sokoto-Rima Valley (Drabo 1976) and mangrove rice systems in the Niger Delta's tidal mangrove swamps.

Not surprisingly, yields are highest for the irrigated systems (typically more than 2 tons/ha), followed by rainfed lowland systems (between 1 and 3 tons/ha) and finally upland systems. The North Central zone produces the most rice, accounting for 30 to 50 percent of national production (Table 3.3).[3] The North West and North East zones each produce about 15 to 30 percent, respectively. Production from the South East, South South, and South West zones combined accounts for only 10 to 20 percent of total production in Nigeria. The majority of the rice is produced by small-scale farmers on less than 1 hectare plots of land (Table 3.4). Only about 20 percent and 5 percent of rice producers cultivate plots of more than 1 ha and 3 ha, respectively.

Farming practices, modern input use, and water control practices can vary widely within each production system. In the lowland rice system, for example, land preparation techniques range from traditional practices of cultivation to the use of herbicides and mechanized tillers. Water control practices, such as bunding, are not common. Rice is either directly seeded or transplanted. In almost all cases, one crop per year is harvested, except in irrigated areas. In upland rice systems, intercropping sometimes occurs, especially with maize, cassava, yams, and vegetables (Longtau 2003a). Rice is seeded directly or broadcast on level and sloped hillsides. Much of the upland rice is grown in southern Nigeria, where rainfall is higher, which also explains the use of intercropping common in the humid forest zones.

3 See the footnote in Figure 1.1 for the location of the six geopolitical zones in Nigeria. Ranges of estimates are somewhat large due to the small sample sizes.

TABLE 3.3 Rice production shares by geopolitical zones (percent)

Geopolitical zones	NBS (average 2007–2010)	NAERLS (average 2008–2009)	LSMS–ISA (average 2010/2012, harvested between September and March) [95% CI]
North West	29	23	22 [16, 28]
North East	21	15	21 [17, 27]
North Central	31	42	41 [32, 51]
South East	8	11	15 [8, 23]
South South	2	2	1 [0, 1]
South West	2	7	0 [0, 0]

Source: National Bureau of Statistics (NBS), National Agricultural Extension and Research Liaison Services (NAERLS), and authors' calculations based on Nigeria, NBS and World Bank (2011, 2013).

Note: LSMS–ISA = Living Standards Measurement Study–Integrated Surveys on Agriculture. Numbers in brackets are 95% confidence intervals (CI). A list of the states in each geopolitical zone is provided in Figure 1.1 in Chapter 1.

TABLE 3.4 Percentiles of household-level rice area distributions in Nigeria (average of 2010 and 2012)

	Percentiles						
Categories	1%	5%	25%	50%	75%	95%	99%
Upper bound (ha)	0.02	0.05	0.2	0.4	0.8	3.7	8.3
Adjusted for multiple crops (ha)	0.01	0.03	0.1	0.3	0.7	2.7	7.6

Source: Authors' calculations based on Nigeria, NBS and World Bank (2011, 2013).

Note: ha = hectare. "Upper bound" assumes rice takes up the entire plot even though multiple crops are reported on the same plot. "Adjusted for multiple crops" assumes rice share is the area of the plot divided by the number of crops reported.

Modern input use generally involves purchased inputs such as improved seeds, fertilizer, and chemicals. In the past, the use of improved modern seed varieties was most common in irrigated lowland areas (80 percent), moderate in upland areas (67 percent), and lowest in rainfed lowland ecologies (37 percent) (Dalton and Guei 2003). The majority of the most popular improved varieties listed in Table 3.5 are not originally from Nigeria (column 5). The popular Nigerian-bred variety, FARO 15, is considered old, indicating the absence of newer superior varieties that can replace it.

Fertilizer use in rice production is also generally higher in lowland and irrigated areas, but application rates vary widely across regions and rice ecologies. Table 3.6 summarizes rice yield, fertilizer use, and nitrogen levels obtained from LSMS–ISA (2011) data in both northern and southern Nigeria. Northern and southern Nigeria are defined as in Takeshima and Edeh (2013), where the northern region consists of pastoral systems, agropastoral/sorghum and millet systems, irrigated systems, and cereal/root crop mixed systems,

TABLE 3.5 Popular rice varieties grown in Nigeria

(1) Variety	(2) Year released	(3) % share	(4) Pedigrees	(5) Origin
FARO 44	1993	11	SIPI 661044, SIPI 651020	Taiwan
FARO 15	1974	7	FARO 1 (BG 79) × IR-8	Nigeria
FARO 46	1992	6	ITA 150 (63-83/Multiline)	IITA
Ex China	1988	5	Unknown	China
FARO 52	2001	4	WITA 4 (TOX 3100–44-1-2-3-3)	IITA (Benin)

Source: Authors' modifications from ASTI (2013) and Takeshima (2014).
Note: IITA = International Institute of Tropical Agriculture.

while the southern region consists of root crop systems, tree crop systems, and coastal artisanal systems.

While fertilizer use is still low in Nigeria (Liverpool-Tasie and Takeshima 2013), its use is higher when limited to the rice plots. At the national level, around 75 kilograms per hectare (kg/ha) and 215 kg/ha of inorganic fertilizer are used on rice plots and irrigated rice plots, respectively (columns 4 and 7 in Table 3.6), although these estimates are subject to a large margin of error due to the small sample size. Assuming NPK (nitrogen, phosphate, and potash) and urea contain 15 percent and 46 percent of nitrogen, this is equivalent to approximately 67 kg/ha and 18 kg/ha of nitrogen used on irrigated and rainfed rice plots, respectively (columns 7 and 10). Average yields of irrigated rice are estimated to be around 3.7 tons/ha in the northern region and 2.8 tons/ha in the southern region, excluding outliers in the data (columns 5 and 6). The average figures for southern Nigeria for yield and nitrogen are in a similar range as authors' field observations in Nasarawa State presented in Table 4.1 of Chapter 4.

Use of irrigation is limited to areas in northern Nigeria. Out of approximately 81,000 ha of irrigated rice area, about 90 percent are in the North West (48,000 ha) and the North Central (26,000 ha) zones (columns 4 and 2 in Table 3.7). Shares of irrigated area to rice area are slightly higher in the North West zone (around 11 percent). This, however, is still well below shares in Asia witnessed during the Green Revolution. In fact, the limited use of irrigation in Nigeria has not changed much since the 1990s (Oramah 1996).

For the few areas under irrigation, the most common water control system is diverted stream or river. Other types of irrigation systems are also observed, however, particularly in the North West zone, where more than half of the rice area irrigated between January and August uses pump-based irrigation (column 4). As is discussed later, while pump-based irrigation is associated

TABLE 3.6 Average quantity (per hectare) of fertilizer used and yield of rice production in Nigeria (January–August 2010)

(1) Fertilizer use and yield	All types of production			Irrigated production			Rainfed production		
	(2) Northern	(3) Southern	(4) Country	(5) Northern	(6) Southern	(7) Country	(8) Northern	(9) Southern	(10) Country
All inorganic fertilizer (kg/ha)	104 [76, 149]	67 [39, 111]	75 [47, 114]	228 [116, 388]	195 [100, 374]	215 [125, 319]	81 [58, 121]	64 [35, 103]	67 [41, 97]
NPK (kg/ha)	60 [33, 89]	25 [11, 48]	32 [18, 50]	113 [10, 282]	56 [0, 158]	91 [25, 206]	51 [23, 75]	24 [12, 48]	29 [15, 48]
Urea (kg/ha)	44 [23, 82]	32 [14, 63]	35 [18, 65]	106 [53, 250]	138 [54, 263]	115 [62, 161]	31 [15, 62]	29 [10, 60]	29 [13, 52]
Nitrogen (kg/ha)[a]	29 [20, 47]	19 [9, 35]	21 [12, 34]	66 [37, 126]	72 [37, 136]	67 [44, 109]	22 [15, 35]	17 [8, 32]	18 [10, 29]
Yield (ton/ha)	2.2 [1.6, 3.1]	1.7 [1.1, 2.0]	1.7 [1.2, 2.0]	3.7 [1.9, 4.9]	2.8 [1.3, 3.5]	3.4 [2.4, 4.2]	1.4 [1.0, 1.6]	1.6 [1.0, 2.0]	1.6 [1.0, 2.0]

Source: Authors' calculations based on Nigeria, NBS and World Bank (2011).

Note: Figures are calculated using sample weights. Numbers in brackets are 95% confidence intervals for estimated averages. For rainfed, we dropped observations with more than 1 ton/ha of fertilizer, which are unlikely. NPK and urea are fertilizer products, while nitrogen is one of the chemical components of these fertilizer products. Yield figures exclude outliers, which are above 3 tons/ha in rainfed, 5 tons/ha in the irrigated southern region, and 10 tons/ha in the irrigated northern region. Here we include only plots where rice areas within the plot are reported. Including other plots with unknown rice area, however, does not significantly change the patterns presented. [a]Nitrogen is calculated assuming NPK 15-15-15 (with 15% nitrogen component), and urea 46%. Since manure is minimal, we ignored nitrogen from manure. Use of manure and other inorganic fertilizer is negligible based on the data.

TABLE 3.7 Rice irrigation area by water regime (average of 2010 and 2012, January–August, 1,000 ha)

(1) Rice areas by water regimes	Geopolitical zones						
	(2) NC	(3) NE	(4) NW	(5) SE	(6) SS	(7) SW	(8) Total
Rainfed rice area	533 [308, 758]	534 [331, 736]	376 [270, 482]	601 [0, 1,744]	147 [0, 414]	2 [0, 5]	2,194 [975, 3,413]
Irrigated rice area total	26 [15, 37]	5 [0, 11]	48 [20, 76]	2 [0, 4]	1 [0, 2]		81 [51, 111]
(a) Irrigated area (planted Jan.–Aug.)	24 [14, 35]	5 [0, 11]	37 [15, 59]	2 [0, 4]	1 [0, 2]		69 [44, 94]
(i) *Irrigation system (planted Jan.–Aug.)*							
Diverted stream	24 [14, 35]	3 [0, 8]	12 [4, 20]	2 [0, 4]	1 [0, 2]		42 [28, 56]
Bucket			5 [1, 9]				5 [1, 9]
Hand pump		1 [0, 3]	6 [0, 13]				7 [0, 14]
Treadle pump			0 [0, 1]				0 [0, 1]
Motor pump			15 [2, 27]				15 [2, 27]
Gravity		1 [0, 2]					1 [0, 2]
(ii) *Water sources*							
Well			2 [0, 6]				2 [0, 6]
Borehole		1 [0, 2]	15 [3, 27]				16 [3, 28]
Lake/ natural pond			3 [0, 6]	2 [0, 4]			4 [1, 8]
Created pond	0 [0, 1]	1 [0, 3]					1 [0, 4]
River/stream	24 [14, 35]	2 [0, 6]	7 [2, 13]	0 [0, 0]	1 [0, 2]		35 [22, 47]
Other		1	10		1		11
(b) Irrigated area (planted Sept.–Dec.)	1 [0, 4]		10 [0, 20]				11 [1, 22]

Source: Authors' calculations based on Nigeria, NBS and World Bank (2011, 2013).

Note: In order to improve estimates, we present averages based on 2010 and 2012 data combined. Figures are based on sample weights. Numbers in brackets are 95% confidence intervals. While some plots had multiple crops reported, we assumed the entire area of such plots was allocated for rice. This calculation method is found to give aggregated rice area estimates closer to figures in FAO (2014). The figures in 2 to 7 may not necessarily add up to the total figure in column 8 due to the rounding.

with higher production costs and yields, river-diversion irrigation is relatively low cost but results in lower yields.

Most rice farmers also grow other crops, either for their own consumption or as cash crops to supplement rice. Table 3.8 summarizes the share of rice growers in each group of geopolitical zones growing other crops. Generally,

TABLE 3.8 Share of rice growers growing other key crops (percent), by irrigation status

	All zones		North West, North East		North Central, South East, South South, South West	
(1) Crops	(2) All rice growers	(3) Rice irrigators	(4) All rice growers	(5) Rice irrigators	(6) All rice growers	(7) Rice irrigators
Cowpea	26 [22, 30]	38 [23, 52]	36 [30, 41]	59 [38, 80]	9 [5, 13]	10 [0, 21]
Cassava	19 [16, 23]	22 [13, 32]	9 [6, 11]	5 [0, 12]	38 [31, 46]	43 [26, 61]
Sorghum	56 [51, 61]	57 [42, 71]	63 [58, 69]	55 [33, 77]	43 [35, 52]	58 [40, 76]
Maize	48 [43, 52]	34 [21, 47]	64 [58, 69]	36 [16, 56]	20 [14, 25]	31 [14, 47]
Millet	17 [14, 20]	40 [25, 54]	24 [19, 29]	59 [38, 80]	5 [3, 8]	14 [2, 27]
Groundnut	17 [14, 20]	10 [3, 17]	19 [14, 23]	1 [0, 4]	14 [9, 19]	22 [7, 36]
Yam	15 [12, 18]	9 [3, 15]	4 [2, 6]	0	33 [26, 41]	21 [7, 34]
Vegetables	13 [10, 16]	14 [3, 24]	14 [10, 18]	23 [5, 41]	11 [7, 16]	1 [0, 4]

Source: Authors' calculations based on National Bureau of Statistics and World Bank (2011, 2013).
Note: Figures are based on sample weights. Numbers in brackets are 95% confidence intervals.

sorghum, maize, and cowpea are the crops that are most commonly grown by rice producers at the national level (column 2). Cassava and yam are also grown by a third of rice growers in the North Central and southern zones (column 6). Interestingly, rice irrigators (columns 3, 5, 7) are more likely to grow some of the rainfed subsistence crops (cowpea, millet) than rainfed rice producers. Rice seems to compete with maize or vegetables for irrigation in some zones. Though more rigorous assessments are needed, current cropping patterns indicate that an expansion of rice irrigation would be likely to displace crops like maize or vegetables.

In addition to low levels of irrigation, mechanization in rice production is low. According to the most recent but limited available data, only about 20 percent of the rice area is cultivated by tractors, although this share is higher than that for other crops.[4] As Table 3.9 shows, mechanization patterns also vary across geopolitical zones. This is partly because of local agroecological and topographical conditions for land clearing. In the North Central zone, which is mostly semihumid and savannah vegetation, 50 percent of the rice areas are cultivated by tractors (column 4). Labor-intensive irrigated and lowland rice production systems in this zone may be less viable (i.e., less profitable) without tractor use.

In contrast, the more humid and forested zones in southern Nigeria rarely use tractors for land clearing (column 3). Use of animal traction is

4 These results should be interpreted with caution, as the sample size used in the calculations was small and thus has a potentially large margin of error.

TABLE 3.9 Share (percent) of total rice areas cultivated by tractors and animal traction by geopolitical zones

(1) Category	(2) Total	(3) SE, SS, SW	(4) NC	(5) NE	(6) NW
% of rice area cultivated by tractor	20 [6, 33]	1 [0, 2]	50 [37, 63]	42 [29, 55]	2 [0, 3]
% of rice area cultivated by animal traction	20 [7, 33]	0 [0, 0]	0 [0, 0]	19 [12, 26]	80 [70, 89]

Source: Authors' calculations based on Nigeria, NBS and World Bank (2011, 2013).

Note: SE = South East, SS = South South, SW = South West, NC = North Central, NE = North East, NW = North West. Numbers in brackets are 90% confidence intervals.

also very limited on rice farms in these zones; according to various local experts, this is due to tsetse flies and heavier soils. In the North West, where livestock production is highest, approximately 80 percent of the rice area is cultivated with animal traction and only about 2 percent with tractors (column 6). Here, animal traction seems to provide an intermediate means for mitigating labor constraints on rice production. The North East zone is somewhat between the North Central and the North West zones with regard to tractor use (column 5).

Altogether, the varied use of technologies—particularly irrigation and mechanization, modern inputs such as improved seeds and fertilizer, and other farming practices and socioeconomic characteristics—results in distinctive types of rice producers in Nigeria. To gain an overview, a cluster analysis, as in Takeshima, Nin Pratt, and Diao (2013), reveals four typologies: (1) subsistence producers; (2) small-scale intensive irrigators; (3) wealthier mechanized producers; and (4) other market-oriented producers (Table 3.10).

Among these, 78 percent of rice producers are market-oriented farmers (columns 3 through 5), while 22 percent are subsistence farmers (column 2). The market-oriented producers include those who give out harvests as gifts.[5] Among market-oriented producers, small shares of two types of competitive producers are also identified—small-scale intensive irrigators and wealthier mechanized producers—each accounting for 7 and 5 percent of total rice producers in Nigeria, respectively (columns 3 and 4). The remaining

5 Gift giving can reflect informal income transfer (for income smoothing) as in Madagascar, where stock and gifts are considered part of marketable surpluses (Barrett and Dorosh 1996). Similar gift exchange is part of informal insurance and credit in northern Thailand (Townsend 1995). Kin landlords and tenants in the Philippines are found to exchange rice grains as gifts in case of emergency (Sadoulet, de Janvry, and Fukui 1997). Similar arrangements are observed in Nigeria, as indicated in the data.

TABLE 3.10 Characteristics of major rice producers in Nigeria, 2010

		Market-oriented producers		
(1) Characteristic	(2) Subsistence producers	(3) Small-scale intensive irrigators	(4) Wealthier mechanized producers	(5) Other market-oriented producers
Share of all rice producers (%)	22	7	5	66
Real wage (daily wage/rice price ratio)	13	9	20	23
Farm gate rice price (US$/kg)	45	59	43	51
Fertilizer (US$/ha)	13	27	67	27
Chemicals (US$/ha)	11	24	67	11
% using tractor	13	10	100	10
% using irrigation in diverted stream	3	100	42	0
% with nonfarm income source	51	78	80	54
HH non-food expenditure (annual US$ per capita)	77	75	111	45
HH assets (US$)	347	247	738	337
Distance to the nearest dam (km)	60	24	140	62
Farm size for all crops (ha)	0.7	0.5	3.0	0.7
% in northern region	70	89	0	73
Rice harvest per farm household (tons)	0.4	1.3	2.1	1.0

Source: Authors' calculations based on cluster analysis.
Note: HH = household.

market-oriented producers are less input intensive, with smaller rice production per farm households.

Small-scale intensive irrigators are those who typically operate on 0.5 hectare plots and practice labor- and input-intensive rice irrigation. They are often located within canal irrigation systems relatively close to dams in the North West, North East, or North Central zones. These producers enjoy access to irrigation or water, higher farmgate rice price, and cheaper labor, and they sometimes practice mechanized land preparation. There were approximately 68,000 producers of this type in Nigeria, producing 89,000 tons of rice in the 2010 rainy season and selling more than half of their harvest. Wealthier mechanized rice producers typically use tractors on their plots and use modern inputs (fertilizer, chemicals) intensively, together with some irrigation.

Biophysical Production Constraints

The varied pattern of rice producers in Nigeria is not surprising considering the diverse nature of the underlying biophysical and socioeconomic endowments of the country, as well as biotic and abiotic constraints, which introduce distinctive challenges for improving rice yields and production output in different parts of the country. In this section, we first review the biophysical production constraints such as climate, water access, and soils, as well as other biotic and abiotic factors that affect yields. This is followed in the next section by a review of the most common socioeconomic production constraints facing Nigerian rice farmers.

Climate

If there is one thing that most influences the growth of plants in rainfed environments like those in Nigeria, it is climate. More specifically, having the right amount of sunlight, rainfall, and favorable temperatures during the main growing period is critical. These affect physiological processes involved in grain production—vegetative growth, development of spikelets, and grain fillings (Yoshida 1978). Rainfed rice production is generally possible in areas with annual rainfall of more than 1,300 millimeters (mm) (Akpokodje, Lançon, and Erenstein 2001), and a substantial area of Nigeria falls under this rainfall regime. Most regions of the country fall within the Guinea savanna ecoregion, with a growing period of 165 days or more (Andriesse and Fresco 1991), where rainfed rice can be grown with some flexibility in planting and harvesting time.

However, dry season in Nigeria is highly arid. Even in the North Central zone of Nigeria (Minna, Makurdi), monthly rainfall is very low during the dry season (Table 3.11). Such aridity seems more pronounced compared to other major rice-growing locations in similar latitudes outside Africa. With the exception of Bangkok (Thailand), at least 40 mm of monthly rainfall are observed even during dry season in the other rice-growing regions listed in column 3 of Table 3.11 (Sri Lanka, the Philippines, and Colombia). Unless yield response to water is high, dry-season rice production may be limited in central Nigeria because irrigation costs can be high. Though little empirical evidence exists, providing 50 mm of water for 1 ha through pump irrigation in Nigeria's North Central zone can cost roughly US$70 for pumping alone, as expensive as 100 kg of fertilizer (or about 45 kg of nutrients).[6] Aridity during the dry season, coupled with a low-yield response to water, might partly explain the slower expansion of rice irrigation in central Nigeria compared to other regions.

6 Based on authors' fieldwork in 2013. Related discussions are provided in the next chapter.

TABLE 3.11 Annual and dry-season rainfalls in selected locations with distinct dry seasons

(1) Locations/regions	(2) Annual rainfall (mm)	(3) Monthly dry-season rainfall (mm)	(4) Dry-season months (5 driest months)
Kano, northern Nigeria	696	0	November–March
Maiduguri, northern Nigeria	552	0	November–March
Minna, North Central zone, Nigeria	1,210	4	November–March
Makurdi, North Central zone, Nigeria	1,237	8	November–March
Enugu, southern Nigeria	1,695	25	November–March
Ilorin, southern Nigeria	1,185	19	November–March
Port Harcourt, southern Nigeria	2,294	65	November–March
Colombo, Sri Lanka	2,404	136	December–April
Trincomalee, Sri Lanka	1,569	50	March–July
Zamboanga, Philippines	1,067	44	December–April
Bangkok, Thailand	1,498	25	November–March
Benito Salas, Colombia	1,343	58	May–September

Source: Authors' calculations based on World Meteorological Organization (2013).

Solar radiation also affects rice production and yield (Yoshida 1971, cited in Barker, Herdt, and Rose 1985). In the dry season responses to fertilizer tend to be more consistent due to high levels of solar radiation (Barker, Herdt, and Rose 1985). Where the plant has adequate water, greater solar radiation leads to better nutrient uptake and higher yields. The wet season typically has lower levels of solar radiation, hence lower yields (Barker, Herdt, and Rose 1985). In northern Nigeria, solar radiation is higher than southern Nigeria (Balasubramanian et al. 2007). Suboptimal solar radiation combined with high humidity is associated with a greater incidence of pests and diseases (Windmeijer and Andriesse 1993).

Potential rice yield in the tropics is lower than the subtropics and temperate regions. This is because tropical daylight lasts only 12 hours during the summer production season, as opposed to 16 hours in the subtropics and temperate regions, and warm nights affect respiration (Tinsley 2009). In Nigeria, the grain-filling period for dry-season rice production is around April. At this time, solar radiation in the Sahel region (which corresponds roughly to the northern half of the North West and North East zones and includes states such as Sokoto) and the North Central zone (such as the city of Badeggi) can differ by as much as 4 watts per square meter (watt/m^2) (based on authors assessment from Ogolo 2011), with 21 watt/m^2 in Sokoto and 17 watt/m^2 in Badeggi. As is shown later, irrigated rice yields are higher in northern Nigeria than central Nigeria. Differences in fertilizer use and yield between the North West or North East zone and the North Central zone may be partly explained by the difference in solar radiation levels.

Temperature also affects rice production. Yield will be drastically reduced if the rice flowers during the cold period, as many of the flowers will abort and will not be able to set seed. The dry-season cropping of rice should be planned so that the flowering will not fall within the cold period of January to March. In northern Nigeria, rainfall and seasonally low night temperature can be constraining (Windmeijer and Andriesse 1993).

When these agroclimatic factors are considered, yield potential in Nigeria is comparable to the other regions in the world. Table 3.12 summarizes agroclimatically attainable yields of *indica* and *japonica* rice under irrigated conditions in Nigeria and other regions.[7] Nigeria has similar yield potential to South America, Southeast Asia, and southern Asia for both types of rice at comparable levels of input use, although it has lower potential than North America, northern Africa, or Central Asia (columns 2–5). Realizing such potential yield, however, depends on other constraints like the cost of water, soil characteristics, and incidence of pests and diseases, which are not considered in Table 3.12.

Water Access

Rainfed rice production not only depends on sufficient rainfall but requires a rice-production landscape that can maximize water access for the plant during the growing period. Based on the water sources, rice landscape can be classified into pluvial, phreatic, and fluxial (Moormann and van Breemen 1978), and rice ecologies develop as upland, rainfed lowland, and irrigated (IRRI 1984). Because of its semiaquatic nature (Kranjac-Berisavljevic, Blench, and Chapman 2003; Windmeijer and Andriesse 1993) and heavy reliance on water, wetland or lowland is the major production environment for rice.

The biophysical potential of rice production varies across production systems. Different production practices are used and different constraints exist (Akpokodje, Lançon, and Erenstein 2001; Kranjac-Berisavljevic, Blench, and Chapman 2003; Windmeijer and Andriesse 1993). In an upland rice production system, the underground water table is below the root depth, so the

7 *Indica* and *japonica* are geographic races of the Asian rice species *Oryza sativa*. *Indica* rice is long grained, while *japonica* is short grained. For example, the most commonly known types of *indica* rice grown in Asia are the *basmati* and *jasmine* varieties.

TABLE 3.12 Agroclimatically attainable yield (tons per hectare) of rice under irrigated conditions with varying input intensity

	Indica wet rice (120 days)		*Japonica* wet rice (120 days)	
(1) Country/regions	(2) High input level	(3) Low input level	(4) High input level	(5) Low input level
Nigeria	7.4	2.1	6.7	1.7
Central Africa	7.2	2.1	6.8	1.7
Central America	7.9	2.3	7.2	1.9
Central Asia	9.4	2.8	8.1	2.1
Eastern Africa	8.0	2.4	7.0	1.8
Eastern Asia	7.9	2.3	7.0	1.8
Northern Africa	9.0	2.7	7.8	2.0
Northern America	9.1	2.7	8.0	2.1
South America	7.0	2.0	7.1	1.8
Brazil	7.0	1.9	6.4	1.6
Colombia	5.7	1.6	5.4	1.3
Southeast Asia	6.7	1.8	6.3	1.6
Indonesia	6.2	1.6	5.3	1.1
Southern Africa	8.5	2.6	7.5	1.9
Southern Asia	7.7	2.3	6.6	1.7
Sri Lanka	7.4	2.1	6.1	1.2
Sudano-Sahelian Africa	7.9	2.3	7.2	1.8

Source: Authors' compilation from Global Agro-ecological Zones (GAEZ) software (v3.0) by Fischer et al. (2012), and figures are averages within each region.

Note: Differences in maturity length do not seem to affect the relative agro-climatically attainable yield level within regions. Therefore we report figures only for 120-day maturity varieties. The definitions of regions in this table are based on the GAEZ source.

plant relies exclusively on rainfall. This mostly limits the upland rice system in Nigeria to the southern zones and parts of the North Central zone with annual rainfall greater than 1,300 mm (Akpokodje, Lançon, and Erenstein 2001). Only one crop per year is possible because of dependence on rain in the wet season. This type of system exists at higher elevations, and soil erosion can be a serious problem.

Thus, maintaining soil nutrients is challenging, and rice cannot be grown more than two consecutive years without fallowing to replenish the soil, further limiting the production potential. Upland can be further divided into wet upland and dry upland, depending on the depth of the water table. In order to exploit fully the potential in the upland environment, varieties with greater rooting depth are needed to extract water for survival and grain production. Drought-resistant varieties that are more efficient in extracting water are also needed (Moormann and van Breemen 1978).

The lowland (also called wetland) rice-production system is diverse. Lowland in Africa can be defined as areas subject to periods of complete saturation of soils, with possible flooding during part of the crop-growing period (Balasubramanian et al. 2007). The lowland environment is suited to intensification, possibly growing two or more crops per year, but excess water may limit crops other than rice (Erenstein et al. 2003). Lowland systems can be further divided into four types based on geographical characteristics: inland basins, inland valleys, river floodplains, and coastal lowlands (Table 3.13).

Among these, much lowland in Nigeria belongs to inland valleys (known as *fadama*) (column 3) and coastal lowlands typically found in the Niger Delta (column 5). More broadly, *fadamas* in Nigeria have features of both floodplains and inland valleys. They tend to be wider than 200 meters, the maximum width for a lowland to be considered as an inland valley yet still have more hydrological similarities with an inland valley than a floodplain (Akpokodje, Lançon, and Erenstein 2001). An inland valley bottom is relatively less prone to drought or submergence compared to other lowlands and is considered a next frontier for intensification of rice-based cropping systems (Balasubramanian et al. 2007). Inland valley bottoms (together with foot slopes) cover 11 million ha in total in West Africa, but they have been only marginally used (Windmeijer and Andriesse 1993), although there has been an increasing focus on this environment in many West African countries (Totin, Stroosnijder, and Agbossou 2013).

TABLE 3.13 Types of lowlands in Africa

	Types of lowlands			
(1) Category	(2) Inland basins	(3) Inland valleys	(4) River floodplains	(5) Coastal lowlands
Key characteristics	Drainage depressions and inland deltas of rivers, with imperfectly to poorly drained and potentially acidic soils (ultisols, oxisols, alfisols, entisols, and vertisols)	Known as *fadamas* in Nigeria	A wide, flat plain of alluvium bordering streams and rivers that flood it periodically	Lowlands along the coast
Typical locations	Upper Nile, Sokoto, Lake Chad, Congo Basins	Mostly located in intertropical region where rainfall > 700 mm	Along Niger and Benue Rivers	Niger Delta (Nigeria)

Source: Authors' compilation based on Balasubramanian et al. (2007).
Note: mm = millimeters.

In Asia, inland valleys are upper reaches of major or minor watersheds, where sediments (alluvial, colluvial) are derived from the adjacent uplands. This is typical in Sri Lanka. Physiographic conditions similar to Sri Lanka's dominate in many of the rice-growing areas of West Africa (Moormann and van Breemen 1978). Given the similar latitude that determines rainfall, temperature, and solar radiation, rice production in Sri Lanka provides an important comparison with the North Central and southern zones of Nigeria.

The third production system, irrigated rice, can be generally categorized in the African context into three types depending on the agroecology: (1) arid and semiarid Sahel; (2) humid forest and savannah; and (3) tropical highlands (Balasubramanian et al. 1995, 2007). In Nigeria, both 1 and 2 are found, while 3 is mostly found in East and Central Africa and Madagascar (Balasubramanian et al. 1995). System 1 is characterized by high solar radiation, lower incidence of pests, significant use of direct seeding, and high water consumption due to high evapotranspiration in the dry season. System 2 is characterized by lower yields relative to 1 due to lower solar radiation, higher pest incidence, and iron toxicity (Balasubramanian et al. 2007). In West Africa including Nigeria, system 2 may be found in inland valley bottoms where rice is rotated with maize, soybean, or vegetables (Balasubramanian et al. 2007).

Soils

Besides climate and access to water, the nature of the soil can also significantly affect the potential for rice production. Here, citing Moormann and van Breemen (1978), we summarize key aspects of soil endowments in Nigeria that can hypothetically affect the biophysical potential for rice. We emphasize "hypothetical," as this has not been investigated thoroughly in the literature, at least to the best knowledge of the authors.

Soil characteristics can affect rice production in several ways. Soils affect water-holding capacity under freely drained conditions. Types of soils can affect the comparative advantage of rice over other crops and the types of rice varieties that need to be developed. Moormann and van Breemen (1978) examine various types of soils classified in the Food and Agriculture Organization of the United Nations (FAO) soil taxonomy, describe their associations with the prevalence of rice production, and categorize soils by importance in rice production.

Using the descriptions in Moormann and van Breemen, soils in Nigeria can be roughly classified based on their suitability for lowland, upland, and transitional types (Table C.1 in Appendix C). Inceptisols (mostly Cambisols) form the most important single soil order among the rice-growing soils. Rice-growing alluvial plains—both river floodplains and deltaic areas—are to a large extent Inceptisols (66). This is followed by other soils of major importance, minor importance, and little importance.

Using world soil maps from FAO/IIASA/ISRIC/ISSCAS/JRC (2012), we calculate the area of lowland, upland, and transitional types of soils, with additional separation between Cambisols and other lowland-type soils per person in each country, using the population of the countries in 2010 (Table 3.14). In Nigeria, there are 0.0022 ha (or 22 m^2) of Cambisols per person (column 2). Figure 3.1 maps their geographical distributions. The breakdown for tropical West African countries is shown in Table 3.15.

Although these classifications are subject to more rigorous assessments, important messages emerge. Given its population density, Nigeria is less endowed with suitable soils for rice production per head of population than Asia, Latin America, and some other West African countries (column 4 in Tables 3.14 and 3.15). The need for rice intensification per hectare may be higher in Nigeria than these countries. It is, however, also important to note that the rice yield in Latin America is much higher than that in Nigeria. Africa's Niger River and its tributaries have narrow floodplains, and their accompanying level terraces are less extensive than in the major rice-growing areas in Asia (Moormann and van Breemen 1978). In order to increase rice production, Nigeria will need to increase rice yield from these soils because

TABLE 3.14 Rice ecology soil endowment per capita (square meters per person)

	Lowland type					
(1) Country	(2) Cambisols	(3) Excluding Cambisols	(4) Total	(5) Transitional	(6) Upland type	(7) Minor type
Nigeria	22	797	819	680	1,840	2,132
Bangladesh	136	657	793	31	0	83
India	421	627	1,048	68	691	623
Indonesia	699	1,266	1,965	919	525	4,332
Japan	471	238	710	0	531	1,710
Sri Lanka	88	272	360	0	1,754	
Thailand	0	2,493	2,493	287	93	4,491
Vietnam	0	869	869	0	71	2,734
Brazil	352	3,431	3,784	163	4,498	34,639
Colombia	1,589	3,815	5,404	1,108	1,439	15,971

Source: Authors' calculations based on FAO/IIASA/ISRIC/ISSCAS/JRC (2012) and FAO (2012).

Note: Per capita soil endowment calculated using 2010 population numbers. Figures in columns 2 and 3 may not add up to total figures in 4 due to rounding.

TABLE 3.15 Rice ecology soil endowment per capita in some West African countries (square meters per person)

	Lowland type					
(1) Country	(2) Cambisols	(3) Excluding Cambisols	(4) Total	(5) Transitional	(6) Upland type	(7) Minor soils
Nigeria	22	797	819	680	1,840	Ac, A, F, L, R (2,132)
Côte d'Ivoire	1,292	2,979	4,271	166	1,438	A, Ac, F (9,599)
Ghana	49	554	603	70	4,650	Ac, F, L (4,055)
Guinea	439	674	1,113	4	177	Ac, F, L (22,134)
Senegal	727	1,909	2,636	878	2,422	A, L, R (8,718)
Sierra Leone	0	701	701	150	0	F (8,197)

Source: Authors' calculations based on FAO/IIASA/ISRIC/ISSCAS/JRC (2012) and FAO (2014).

Note: A = Arenosols, Ac = Acrisols, F = Ferralsols, L = Lithosols, R = Regosols. Minor soils include types of soils on which rice is rarely grown in Asia.

the expansion potential outside of areas with these soils is limited. One way to improve yield is to increase irrigated areas within these lowland soils.

Comparison within West Africa, however, should be approached with caution. Though the per capita endowment of soils suitable for rice production is generally lower in Nigeria than other coastal West African countries (column 4 in Table 3.15), Nigeria has an advantage over these countries in that relatively more rice is currently produced on the suitable soil area than in the other tropical West African countries. Figure 3.2 indicates that much rice production in Nigeria is found along the Niger and Benue Rivers (see Figure 1.1) and the Sokoto River, which runs across Sokoto State, as well as in the South East zone, including Ebonyi, where lowland-type soils are common (Figure 3.1).

In Côte d'Ivoire, Sierra Leone, and the Republic of Guinea, main rice-production areas are often on minor soils (Figure 3.3). In fact, the dominant rice ecology in these countries is upland, while it is lowland in Nigeria (Dalton and Guei 2003; Seck et al. 2012). Conditions in tropical West Africa are in stark contrast to those in countries such as Thailand, where the rice-production area closely matches major soils (Figure 3.4). As is discussed in the following section, this mismatch may partly explain why substantial domestic breeding efforts in countries such as the Republic of Guinea and Côte d'Ivoire have not led to substantial yield growths. Nigeria seems to be in a better position than these countries to develop new varieties by crossing Asian varieties (which may have high yield response in suitable soils) and local varieties (with suitable genes for the local production environment).

FIGURE 3.1 Soils in Nigeria suitable for different rice ecologies

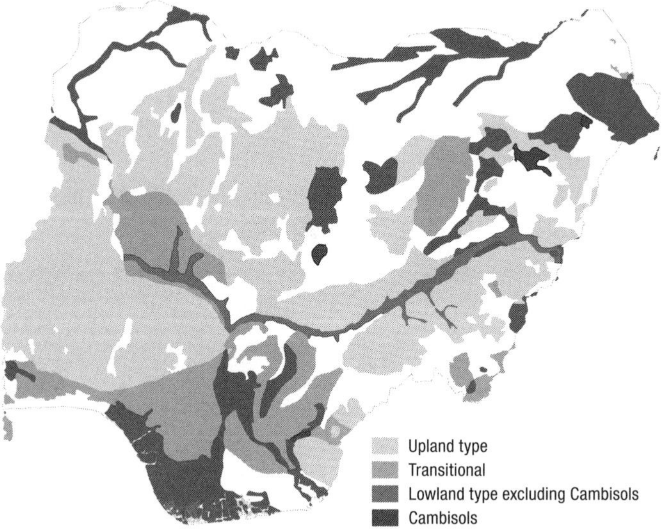

Source: Authors' analyses based on FAO/IIASA/ISRIC/ISSCAS/JRC (2012) and Moormann and van Breemen (1978).

Note: White space represents the soil less suitable for rice production.

FIGURE 3.2 Rice-production areas in Nigeria

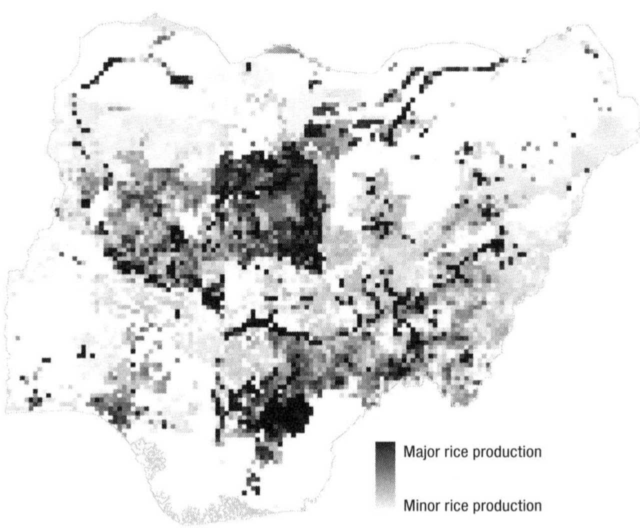

Source: Harvest Choice (2012).

Note: The relative darkness of the shading indicates the scale of rice production: very dark gray = major rice production; lighter gray = minor rice production.

FIGURE 3.3 Soils suitable for rice and actual rice production in the Republic of Guinea, Sierra Leone, and Côte d'Ivoire

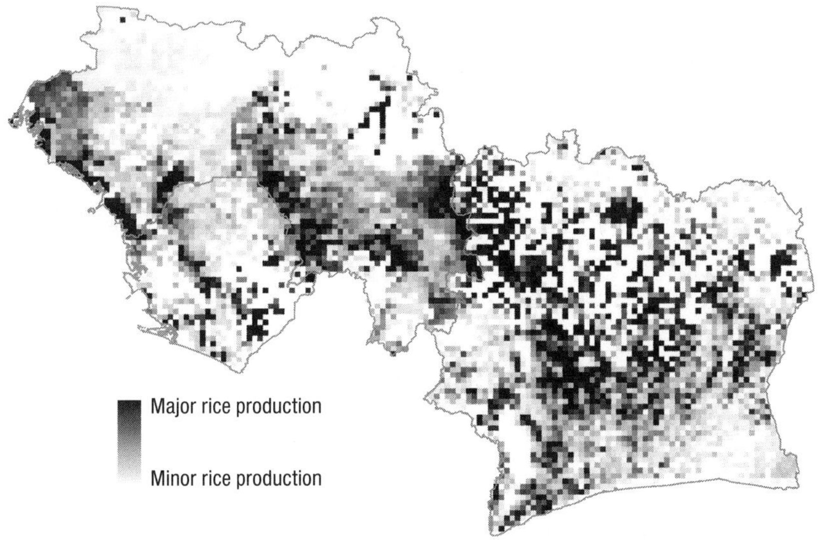

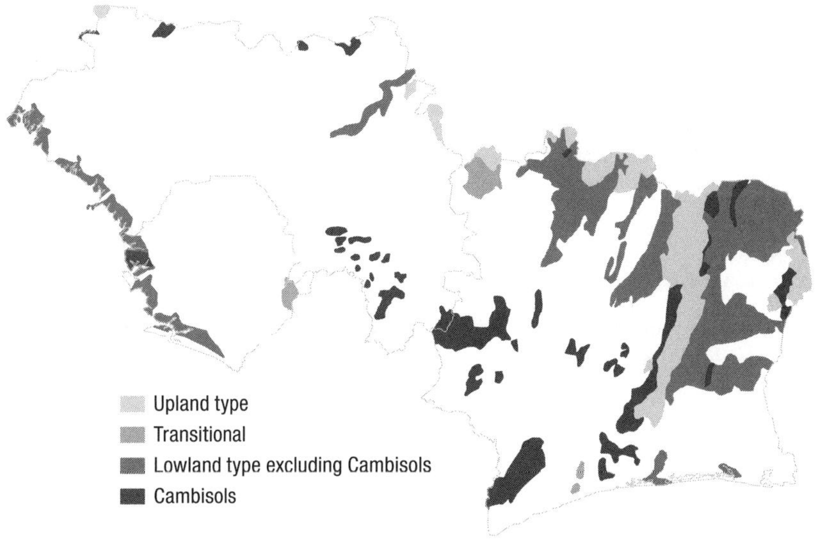

Source: Top figure is based on Harvest Choice (2012); bottom figure is authors' analyses based on FAO/IIASA/ISRIC/ISSCAS/JRC (2012) and Moormann and van Breemen (1978).

Note: In the top figure, the relative darkness of the shading indicates the scale of rice production: very dark gray = major rice production; lighter gray = minor rice production.

FIGURE 3.4 Actual rice production and rice-suitable soils in Thailand

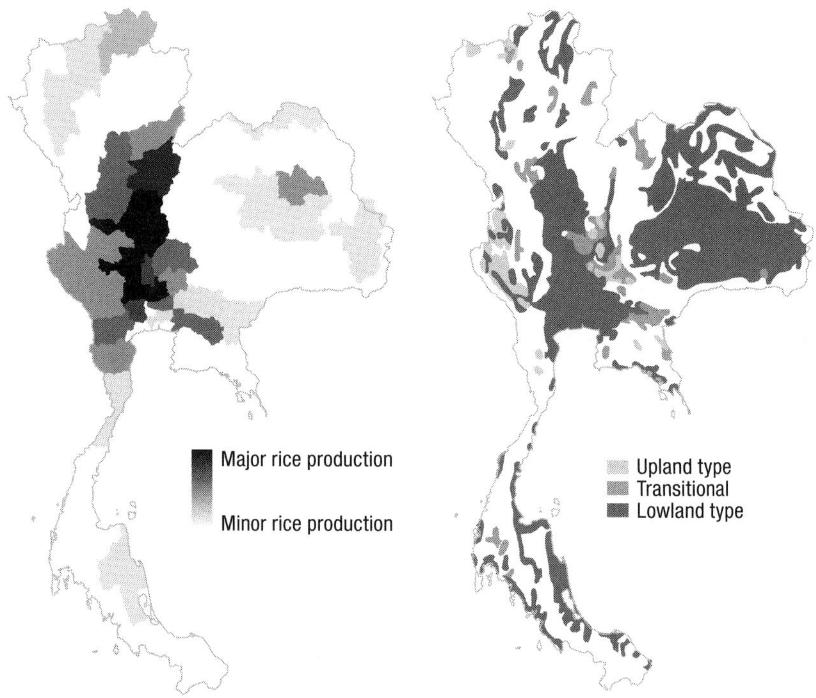

Source: Left figure is based on USDA, Foreign Agricultural Service (2013). Right figure is authors' analyses based on FAO/IIASA/ISRIC/ISSCAS/JRC (2012) and Moormann and van Breemen (1978).

Biotic and Abiotic Constraints

Rice production around the world faces various biological or biotic constraints, including pests and diseases, weeds, flooding and rainfall shortage, and iron toxicity.

PESTS AND DISEASES

The major rice pests in Nigeria include African rice gall midges (AfRGM), stem borers, stem/leaf suckers, leaf/stem feeders, vertebrate pests, birds, and nematodes (Table C.2 in Appendix C lists the types of pests, their stages of attack, and control methods). Bacteria and fungus also cause downy mildew, bacterial blight, bacterial leaf streak, and brown rot. In Nigeria, major rice diseases are blast, leaf scald, grain discoloration, sheath blight and false smut, and rice yellow mottle virus (RYMV) (Ahonsi et al. 2000; Akpokodje, Lançon, and Erenstein 2001; Alam et al. 1986) (Table C.3 in Appendix C lists the types of pathogens, symptoms of attack, and control methods).[8] Varieties resistant to pests and diseases in Asia may lose such resistance in Nigerian conditions because many of the pests and viruses are of different biotypes, given the country's diverse agroecology. Diseases such as blast can also have diverse pathogen types in tropical conditions (Rutger and MacKill 2001).

AfRGM is considered one of the most serious insect pests of lowland and irrigated rice in Africa (Ogah et al. 2012). It is distinct from the Asian rice gall midge, indigenous to Africa (Ukwungwu and Misari 1997), and was first identified by Harris and Gagné (1982). AfRGM attacks rice at the vegetative stage and destroys the growing primordial, forming a tubular gall or onion shoot. Attacked tillers are irreversibly damaged and produce no panicle (Ogah et al. 2012).

Outbreaks of AfRGM on wet-season rice crops have increased in Nigeria since the 1970s (Ukwungwu, Winslow, and John 1989; Ukwungwu and Joshi 1992a) as well as in other West African countries. In Nigeria, AfRGM had damaged over 50,000 ha, causing total yield loss in the fields worst affected (Williams et al. 1999a). Although the area damaged is relatively small compared to the total rice area in Nigeria, which currently stands at around 1.5 to 2 million ha, it has been one of the key factors affecting the performance of externally introduced varieties.

AfRGM and RYMV are not easily controlled by chemicals (Williams et al. 1999b). Screening and breeding of resistant varieties have been ways used to deal with the problems. In Asia, the similar Asian rice gall midge has been successfully treated with resistant varieties. AfRGM was serious in Ebonyi until the introduction of the Indonesian variety, *cisadane*, in the late 1990s (Maji and Fagade 2002). Though many foreign varieties break down in Nigeria due to indigenous pests and diseases, developing resistant varieties still seems the most feasible option.

Various field trials indicate that the incidence of and susceptibility to RYMV and iron toxicity may vary across varieties and locations within Nigeria. Even for the relatively newer major varieties (FARO 44, FARO 51, and FARO 52), the severity of major diseases varies significantly across locations within each of Niger, Kogi, Benue, and Nasarawa States (NCRI 2013). Even in the drier northern region (e.g., Kano), a high incidence of mottle virus has been reported (Abo, Ukwungwu, and Onasanya 2002).

8 RYMV is specific to Africa, first discovered in Kenya in 1966 (Abo, Sy, and Alegbejo 1997). It appears to have emerged after the introduction of Asian rice (*Oryza sativa*) into Africa, although this virus is absent in Asia. RYMV can also be serious in an irrigated environment, which has induced farmers in Mali to use the resistant *kogoni* variety instead of BG 90-2 (Wopereis et al. 1999).

It is difficult to combine resistance to pests with tolerance for abiotic stress. For example, among two popular lowland varieties in Nigeria, FARO 44 (*sipi*) and FARO 52 (WITA 4), FARO 44 is resistant to AfRGM but susceptible to iron toxicity, while FARO 52 tolerates iron toxicity but is susceptible to AfRGM and RYMV. African varieties (*O. glaberrima*) are generally more resistant to biotic stresses (rice diseases, insect attacks) (Diagne et al. 2011; Evenson and Gollin 1997) but less responsive to fertilizer.

WEEDS

Weeds are also known to be a key obstacle to crop production (Bakare 2007; Herdt 1991). In Nigeria especially, they are considered one of the greatest inhibitors to higher yield growth and quality improvements in rice (Ukwungwu and Abo 2004). Weeds compete with rice for nutrients, space, and light. They have caused rice yield losses of 28 to 100 percent in Nigeria (Imeokparia 1994; Akobundu 1987; Rodenburg and Johnson 2009). Recently, weeds have accounted for 3 to 75 percent of yield losses in lowland areas, depending on location, and 75 to 100 percent in upland areas (Ismaila et al. 2013). In the tropical environment of Nigeria, weeds can grow very rapidly compared to temperate regions.

Common weeds in the rice ecologies in Nigeria can be perennial, annual, or parasitic (Rodenburg and Johnson 2009). Quite a few of these may be indigenous in Africa. For example, the water lily (*Nymphaea lotus* L.) seems relatively absent outside Africa, and little information is available on its biological characteristics and appropriate controls. Tables C.4 and C.5 in Appendix C list the common weeds found in Nigeria's rice ecologies. If containing these weeds requires different technology packages (varieties, types of chemicals) than those transferrable from outside Nigeria, this may pose challenges in achieving the biophysical potential for rice in the short to medium term.

In Asia, where labor is somewhat cheaper, traditional weed control has been through manual weeding followed by herbicide application. However, rising farm wages may make such methods unprofitable. In addition, impacts of weed control measures depend on varieties grown. For example, herbicides can more effectively control weeds such as *Cyperus difformis* L. for FARO 15 and FARO 29 than FARO 27 (semidwarf variety) (Imeokparia, Lagoke, and Olunuga 1992).

African rice (*Oryza glaberrima*) traditionally has had greater tolerance against weeds. It has higher weed competitiveness through greater tillering ability, vigor, better exploitation of soil nutrients, and early seedling capabilities (Maji and Fagade 2002). It also has a high weed-suppressive ability. It can suppress weed growth and reduce weed seed production, and hence it can benefit weed management in the subsequent growing season (Saito, Azoma, and Rodenburg 2010). The key challenge is to transfer such genes into higher yielding varieties from Asia. Though development of New Rice for Africa (NERICA) varieties is promising—and some NERICA lines show greater weed competitiveness than popular Asian varieties (Rodenburg and Johnson 2009)—many of the developed lines so far have lower weed-suppressive ability than traditional African varieties (Saito, Azoma, and Rodenburg 2010).

FLOODING AND RAINFALL SHORTAGE

Besides biotic constraints, rice production systems around the world also face major abiotic stresses. Among them, flooding can significantly affect rice production, either directly by submerging rice plots or indirectly by submerging paths (roads, etc.) to the plots and rendering them inaccessible. Rice should not be submerged for more than 48 hours (Windmeijer and Andriesse 1993). Flooding also forces farmers to choose suboptimal production timing. In Nigeria, flooding is one of the major disasters that can affect the production of rice-growing households.

Table 3.16 shows the percentage of farmers in each geopolitical zone who were affected by flooding, poor rains, and pests in 2010. Due to the small sample sizes, figures for rice-growing households in southern Nigeria are not included. Flooding might have affected significant amounts of rice production in the North East zone, where about 10 percent of rice-growing households experienced flooding (column 2). Generally, rice-growing households experienced greater incidence of flooding and poor rain compared to pests in 2010.

TABLE 3.16 Share of farmers and rice farmers in Nigeria affected by flood, poor rain, and pests in 2010 (percent)

	Rice-growing households			All farm households		
(1) Geopolitical zones	(2) Flood	(3) Poor rain	(4) Pest	(5) Flood	(6) Poor rain	(7) Pest
North West	3 [0, 6]	1 [0, 3]	1 [0, 2]	3 [1, 4]	2 [1, 3]	1 [0, 2]
North East	11 [5, 18]	4 [0, 8]	1 [0, 4]	5 [3, 6]	3 [2, 4]	1 [0, 2]
North Central	4 [0, 9]	3 [0, 6]	0 [0, 1]	1 [0, 2]	1 [0, 3]	0 [0, 1]
South East				2 [0, 3]	1 [0, 2]	4 [3, 6]
South South				1 [0, 1]	0 [0, 0]	0 [0, 0]
South West				0 [0, 0]	2 [0, 2]	0 [0, 0]

Source: Authors' calculations based on Nigeria, NBS and World Bank (2011).

Note: Numbers in brackets are 95% confidence intervals. All figures are calculated using sample weights. The figures do not distinguish whether rice production itself was affected by these events. "Poor rain" is a term used in the LSMS–ISA (2011), which may indicate either drought or overall shortage of rainfall throughout the production season.

Deepwater varieties are found among both local African varieties and modern varieties such as FARO 15 (called "Water Proof" in Nasarawa State), which can tolerate flooding to some degree thanks to their high elongation and early maturing abilities. However, these varieties may not have high yield potential. For example, FARO 15 tends to lodge heavily if too much nitrogen is applied (Maji and Fagade 2002).

Drought and rainfall shortages can also affect rice production or its cost if they lead to additional irrigation. While the incidence of poor rainfall was less widespread than flooding, it still affected up to 4 percent of rice-producing households in the North East in 2010 (Table 3.16).

IRON TOXICITY

Iron toxicity is also considered a major abiotic stress on lowland rice production in Africa (Singh et al. 2004). Rice suffers from iron toxicity, resulting in discoloration of leaves, if dissolved iron in the rooting medium exceeds 300–500 parts per million (ppm) or even 30 ppm if soil nutrients (potassium and phosphorus) are low (Moormann and van Breemen 1978). The condition is often associated with nutrient deficiencies in the soil. It often emerges on acid saline soil (IRRI 1982), with pH below 5 when aerobic (Moormann and van Breemen 1978). Salinity and acidity are problems of Gleysols and Fluvisols (which are two of the major lowland-type soils listed in Table C.1 in Appendix C) in the coastal fringes, and iron toxicity can also occur on these soils in inland areas. Iron toxicity has been commonly observed in inland valleys in Sri Lanka, Orissa and Kerala States in India, and West Africa (Moormann and van Breemen 1978).

But in West Africa, where bunding and leveling are still less common than in Sri Lanka and India, the severity of iron toxicity can vary greatly across locations (Moormann and van Breemen 1978). Sri Lanka has invested in developing varieties that are resistant to iron toxicity, while Nigeria has mostly relied on importing foreign varieties such as Sri Lankan BW348-1 (Singh et al. 2004), which may not always suit local production environments. Susceptibility to iron toxicity also varies across varieties. In an advance yield trial carried out at NCRI with respect to the response of different rice lines and varieties to iron toxicity and AfRGM, the average grain yield varied between 1,965 kg/ha (*suakoko* 8) to 1,336 kg/ha (*manbefuchi*) across varieties (NCRI 2008). While iron toxicity–tolerant varieties and nutrient management showed promising results on a long-term basis (Sahrawat 2004; Sahrawat et al. 1996), their adaptability as well as the management costs are not yet known in Nigeria.

Socioeconomic Production Constraints

Despite Nigeria's modestly promising biophysical environment for rice production, various socioeconomic factors can prevent the wider adoption and diffusion of modern inputs and technology practices in the dominant rice-production systems. Access to markets, credit, and extension, for example, can affect this. But even more importantly, it is a case of the new technologies' viability in ensuring a positive return to the farmer given input and output prices, resource requirements, and costs (e.g., labor, land, and capital) and the effect on incomes.

In other words, what matters is the marginal productivity and return from the inputs and technologies being adopted, output prices of rice and other competing crops, and the opportunity costs or shadow prices of land, labor, and other household capital. Weak market linkages (or even market failures) for key inputs and technologies, access to capital, and insurance add to information and transaction costs that further affect rice-production decisions of the farmer.

In what follows, we review in more detail many of the socioeconomic constraints affecting the adoption and diffusion of modern rice technologies in Nigeria. The degree to which these are widely prevalent in Nigeria helps explain the low performance in yields relative to other countries in the region and other major developing regions of the world. It also implies a great potential to accelerate yield and output growth by removing many of these constraints through public interventions and institutional strengthening.

The Economics of Rice Production

A farm household mobilizes and allocates its resource endowments with respect to land, labor, capital, and knowledge in order to extract the maximum benefits from these endowments. It does not have to depend on farming activities alone. For example, many Nigerian farm households are engaged in other noncrop (e.g., livestock) and nonfarm income-earning activities (Oseni and Winters 2009) that are just as important in contributing to the overall income of the household. Such activities also serve to guard against risks associated with weather shocks or seasonality in agricultural earnings.

For farmers to engage in crop production, or any other activity for that matter, the production must be economically feasible or profitable. In other words, the farmer will produce up to the point when the marginal cost of an additional input is just equal to the resulting marginal utility derived from it. Under conditions of limited access to inputs such as irrigation and fertilizer in crop production, yields must be sufficiently high to ensure a positive return. Traditional staples or subsistence crops such as cassava, maize, sorghum, and millet typically fit these criteria well, as they are able to grow under harsh conditions of modest and unstable supplies of water and limited access to fertilizer, hired labor, or up-front cash payments.

For farmers to be induced to grow rice, its price must be high enough relative to other crops. They also need to be able to afford hiring labor for land preparation, weeding, bird scaring, and harvesting. As it is difficult to monitor the efforts of these workers, farmers must bear the risks of poor performance. Additionally, they must also be able to afford inputs such as fertilizer and agrochemicals, as these are needed more in rice production than in the production of other traditional staple crops.

Due to poor rural road infrastructure, modern inputs can be substantially pricier in rural areas than at the source, whether this involves fertilizer, certified seeds, hired labor, agrochemicals, irrigation pumps, tractors, or crop husbandry knowledge from extension agents. Similarly, the farmgate price of paddy rice can be substantially lower than consumer prices. These infrastructure-related transaction costs can be more serious for cash crops like rice than for traditional subsistence crops that do not require many modern inputs and do not need to be transported to the market.

Weak-performing markets (or even market failure) also complicate the economic structure of rice production. Credit market failures arise in rural areas due to the high cost of monitoring payments and long lending cycles inherent in agriculture. Farmers often lack enough cash at the beginning of production seasons to purchase modern inputs and to invest in technologies, even if these can bring higher returns. Although traders or moneylenders sometimes offer credit, yield returns must be substantially high so that taking out loans at high interest rates is justified.

Farmers therefore often continue cultivating traditional staple crops that do not require much up-front cash expense, rather than intensifying rice production. Credit for off-farm income-earning activities can be less constrained because of shorter lending cycles; a longer cycle can make lending for rice production relatively costly. Insurance market failures can also arise due to adverse selection, moral hazard, and the covariate nature of risks. While rice is subject to less price volatility than perishable crops thanks to its storability, poor infrastructure and subsequent lack of market integration can generally lead to high local price variation, and farmers may have more incentives to grow subsistence crops to avoid market risks.

Factors Constraining the Intensification of Rice Production

Once a farmer decides to grow rice, various factors can affect the adoption and use of modern inputs and technologies and therefore the rate at which production can be intensified and output expanded. Many economic factors, aside from prices alone, can influence such adoption behavior. For example, farm size, access to extension and on-farm trials, access to inputs and credit, productivity potential, and price risks are all relevant. However, other noneconomic factors such as those based on social and cultural considerations may also be important. Finally, because multiple technologies are often involved, interrelatedness or complementarity can affect adoption behavior.[9] Here we focus attention on the key factors affecting rice intensification with regard to adoption of improved rice varieties, irrigation, fertilizer, mechanization, and crop husbandry practices.

IMPROVED RICE VARIETIES

The adoption of improved rice varieties over traditional varieties depends on the superiority of traits (high-yielding genes, responses to fertilizer and water), resistance to various stresses, price premiums, required quantities and costs of additional inputs, additional crop husbandry knowledge, farmers' wealth, and the degree of risk aversion. Traditional varieties are easily diffused and adopted. Through farmers' natural selections over centuries, the traits of many existing traditional varieties have been adapted to suit the local production environment. In addition, farmers have adapted traditional varieties to match local taste preferences. They are resistant to pests and diseases, tolerate flooding or drought, and grow relatively well without external inputs like fertilizer or irrigation. Yields are modest but certain and preferred by risk-averse farmers. Modest yields can be obtained without elaborate production practices. Varieties are readily available in the informal seed sector and much cheaper than improved varieties.

The Nigerian seed system has been weak and underdeveloped for supplying improved varieties that are substantially superior to traditional varieties. The generally weak capacity of the public seed system further raises the cost of certified seeds for farmers. The pace of varietal release may be slow (Takeshima et al. 2010). Weak infrastructure also raises the costs of disseminating information on varieties. The existence of an informal seed sector partly fills these gaps, but seed qualities (variety identities and purities) are only imperfectly controlled by sellers concerned about reputations. Certified seeds are typically distributed through the Agricultural Development Project (ADP), which is a state-level agricultural extension agency to which many farmers do not have easy access. In addition, demand for rice seeds in Nigeria

9 For interested readers, a survey of the literature on the determinants of technology adoption in developing countries can be found in Feder, Just, and Zilberman (1985).

can be seasonal (Takeshima and Nagarajan 2015), but the formal seed sector may not be flexible in adjusting the timing of seed supply.

Improved varietal development, on the other hand, has typically relied either on the international agricultural research centers—such as the International Institute of Tropical Agriculture and the Africa Rice Center—or on foreign varieties. Local national agricultural research institutes (NARIs) have mostly focused on field trials and evaluations. The NCRI has the mandate for rice research in Nigeria, with seven research stations located throughout the country.[10] Although some varieties are evaluated and released by other institutions, including universities, this has been rare (Takeshima 2014). The weak capacity of NARIs has constrained the development of locally adoptable varieties. Yet Nigeria was still one of only two countries conducting varietal development in West Africa at the turn of the century, along with the Republic of Guinea (Dalton and Guei 2003).

The weak NARIs and seed systems have not prevented sufficiently good varieties from finding their way to farmers. For example, the RYMV–resistant *cisadane* variety from Indonesia had been adopted by farmers in Ebonyi State even before its official release as FARO 51 (Maji and Fagade 2002). Traits of improved varieties must be substantially superior to traditional varieties to be adopted because of various market failures and weak seed systems. In the case of maize, farmers often do not buy new varieties unless they receive a marginal return of at least 100–300 percent (US$1 in additional seed cost must return US$2–$4 in increased profits) (Pray 1997; Lopez-Pereira and Filippello 1994). Similar patterns may apply to rice seeds.

USE OF IRRIGATION, FERTILIZER, AND MECHANIZATION

Traditional varieties have often evolved resistance to water stress, and they benefit only marginally from irrigation. Improved varieties exhibit higher responses to irrigation, but this can be costly if land-leveling is insufficient. Pump-based irrigation requires substantial fuel cost for frequent water lifting compared to gravity irrigation. Affordable irrigation pumps with sufficient suction and delivery capacity are needed, unless there is access to gravity systems. Significant capital is needed to buy fuels to operate the pump frequently, as well as to buy fertilizer and herbicide. In Nigeria, substantial labor is used to smooth plot surfaces to aid water flow, in addition to labor requirements for other farm practices. Credit constraints can affect the purchase of these inputs and investments in irrigation pumps and water channel construction for river diversion.

10 Based on personal communications with NCRI staff.

Even if irrigation is feasible, rice may not be as profitable as other crops such as vegetables or sugarcane that also attract high prices in Nigeria or vegetables and maize that grow well with less irrigation water (Adeniran et al. 2010; Bello 1987). For farmers to irrigate rice, they often need to have access to highly water responsive varieties and much less costly access to water. This "tragedy of commons" is often unavoidable. If returns from irrigation are only modest, farmers may have less incentive to coordinate the water allocation or mobilize collective action for canal maintenance, further reducing irrigation returns and leading to rejection of irrigation. Low government funding for irrigation further aggravates this problem. These factors can constrain rice irrigation.

Traditional rice varieties often do not respond well to inorganic fertilizer. Historically, rice farmers in southern Nigeria maintained soil fertility through crop rotation (Welsch 1965). Improved varieties, on the other hand, respond better to fertilizer. However, fertilizer responses of existing varieties in Nigeria may be lower than newer varieties currently grown in Asia, perhaps due to the antiquity of many improved varieties. A weak extension system may also be limiting the transfer of knowledge regarding optimal fertilizer use. Fertilizer response can be optimal if the applications (dosages) are appropriate to local conditions with regard to the amount, timing, and nutrient combinations and are strictly adhered to.

Aside from fertilizer application practices, fertilizer use can be constrained by farmers' risk aversion, liquidity constraints, and a weakly functioning fertilizer market. Fertilizer is often a risk-increasing factor; even at high fertilizer use, yield can be low in some years due to both biotic and abiotic stresses. Farmers are often averse to such risks, and unless the response is exceedingly high they may not spend much on fertilizer. Because of cash constraints, fertilizer use is often suboptimal. It is generally more expensive in Nigeria—partly due to poor road infrastructure that raises domestic transportation costs—than in developed countries (Liverpool-Tasie and Takeshima 2013).

Fertilizer markets are still relatively thin in Nigeria compared to countries such as India, especially in terms of a strong retail network. While India has about 280,000 fertilizer sales points (one in every 12 km^2 of land and 6 km^2 of arable land) (Ramani 2007), Nigeria has only about 4,000 to 10,000 (one in every 90–250 km^2 of land or 36–90 km^2 of arable land) (Liverpool-Tasie and Takeshima 2013). The market is thus imperfect and inefficient in meeting demand in a timely manner. Imperfect competition and a weak regulatory enforcement capacity induce suppliers to extract rents by selling adulterated fertilizer (Liverpool-Tasie et al. 2010). These factors raise the de facto price of fertilizer for farmers and inflate the risks for its use, further discouraging the adoption and use of fertilizer for rice. However, this may be changing given recent reforms to deregulate the market and permit the formation of stronger retail networks.[11]

Finally, because rice production can be far more labor intensive than other crops, especially during land preparation and weeding, it can benefit from greater mechanization. During land preparation, the harrowing or plowing of a rice paddy field is typically done to help improve the nutrient structure of the soil and to reduce competition with weeds. If done well and combined with the use of improved varieties that are less weed competing and can take better advantage of the improved soil nutrients than traditional varieties, yield increases can be substantial.

A farmer typically will decide to mechanize these activities when yield improves from such operations, manual labor is costly, and mechanization services are available at a reasonable cost. Additionally, mechanization may also be limited if there are serious credit constraints due either to missing credit markets or high costs of credit. Availability and costs of insurance, maintenance costs, and other transaction costs may also be just as important. For example, tractors are often purchased by a few wealthy farmers with better access to low-cost credit. However, because of the difficulties in accessing credit with favorable terms in rural areas, this has often resulted in less capital investment in tractors. The exception is credit for cheaper, smaller tractors or power tillers. Using smaller tractors, however, may be less profitable if yield advantage is small, and speedy operation on larger farm plots is more practical and results in greater benefits.

Because the majority of rice plots are limited to small geographical areas that rely heavily on access to water (e.g., in the *fadama* areas), they tend to be dispersed and fragmented and to have poor road access. Providing insurance for tractors is costly in rural areas due to high monitoring cost, adverse selection, and moral hazard, further discouraging investments. All this raises the cost of supplying tractor services to farmers without some sort of government intervention to subsidize this activity, which the government is beginning to pursue. Most Nigerian farmers, therefore, do not adopt mechanization unless the yield enhancement of intensive land preparation is sufficiently high and the cost difference between manual labor and tractor services is sufficiently large.

11 A good example is the government's new Growth Enhancement Support program, which is intended to encourage the participation of private-sector dealers in distributing subsidized fertilizer to farmers. The program utilizes electronic vouchers received by farmers through mobile phones, as an "e-wallet" system. Registered farmers receive text messages alerting them that they can pick up their input package at a local redemption center operated by a private-sector agricultural input dealer.

CROP HUSBANDRY PRACTICES

Traditional rice varieties generally grow well without sophisticated production practices. For improved varieties, optimal sequencing and timing of numerous input applications and farm activities are critical for achieving a maximum yield potential on farmers' plots. Practicing such ideal crop husbandry is challenging due to its high labor requirements. For farmers to adopt improved production practices, therefore, the returns must be substantial. The cost of diffusing such technical knowledge, through extension services, for example, can also be high due to weak communication infrastructure and geographical dispersion of farmers. As a result of these constraints, farmers' access to such knowledge is very limited.

Access to extension agents is still low in Nigeria (Omotayo, Chikwendu, and Adebayo 2001). In addition, rice varieties tend to be more susceptible to local environmental conditions compared to crops like wheat (Byerlee and Traxler 2001). Extension systems must be able to inform on the right varieties and husbandry practices that are most suitable to each location—but farmers can also learn from practice. For example, rice producers in Asia have been shown to derive some of their best practices through self-learning (Munshi 2004).

Owing to such inherent challenges for raising yields in Nigeria, as discussed earlier, yields for many improved varieties are often below their full potential. Narrowing the yield gap will take time. Until new varieties can do fairly well under suboptimal input use and production practices, farmers will continue growing those rice varieties, or even other crops, with which they are more familiar and on which they have come to rely. Extension units are also often separated from research units in Nigeria, as most extension service is provided by the ADPs that are not involved in crop research. It may be difficult to incorporate farmers' feedback into varietal development research. The participatory varietal selection method implemented alongside recent variety development (such as NERICA) is promising in that regard, but its application is limited to small geographical areas (Maji and Fagade 2002).

Conclusion

Nigeria has the largest rice cultivation area in Africa south of the Sahara. Because of its size, rice production systems are highly diverse. Not only encompassing all three major ecologies, including upland, rainfed lowland, and irrigated land, they are also distributed across a range of latitudes. Each of these systems has distinctive characteristics, leading to different biophysical and socioeconomic potentials and constraints for yield increases.

Nigeria's overall biophysical potential for rice may not be as good as the potential in most of Asia. It is limited by the following factors: (1) Nigeria's tropical environment (low latitude) is not suitable for temperate varieties, although northern Nigeria has a relatively favorable environment where Asian varieties can be adopted; (2) the area with suitable soils for rice production is relatively small compared to Asia; (3) aridity in dry seasons is very high in most of Nigeria, possibly raising irrigation costs; and (4) some pests, diseases, and weeds are relatively unique to Nigeria, and transferring Asian rice varieties may often be difficult.

The biophysical potential for rice is better than in some tropical West African countries, given the dominance of lowland production, and is comparable to some tropical Asian countries such as Sri Lanka. However, Nigeria has yet to fully realize its biophysical potential for rice production, especially if improvements are made through greater investment in the sector. Among possible options for the public sector, the most essential will be investments to expand the irrigated rice area and development of superior-quality and high-yield varieties.

As in agriculture in general, rice production in Nigeria is largely represented by low-income, asset-poor, resource-constrained smallholder farmers. Larger and more commercialized rice farms are rare. At the same time, rural Nigeria remains underdeveloped with respect to markets, institutions, and physical infrastructure. Consequently, the adoption of modern technologies has been limited due to problems of underdeveloped input markets, competition with other crops and nonfarm income-earning activities, weak extension systems, poor access to credit and insurance provisions, high market transaction costs, and insufficient rural infrastructure more generally, all of which can affect the profitability of technology adoption (Feder, Just, and Zilberman 1985).

The next chapter provides a more in-depth empirical investigation into these socioeconomic constraints in order to explain current levels of profitability and, ultimately, the type of government interventions needed to promote paddy production and induce a supply response drawing on lessons and experiences from other major rice-producing regions in the developing world.

Chapter 4

POLICY OPTIONS FOR INDUCING A SUPPLY RESPONSE

Hiroyuki Takeshima

As the previous chapter notes, paddy yields in Nigeria are among the lowest in the region and well below other developing countries in Asia and Latin America. Additionally, the chapter highlights Nigeria's untapped biophysical potential for expanding rice production and output with improved water access in lowlands and other areas suitable for irrigation. The current low yields alone imply that the country has the potential to transform the sector by simply closing existing yield gaps. More specifically, a number of technological and socioeconomic constraints, together with past public investments and price policies, have inhibited the wider diffusion and use of improved technologies and farming practices, implying an untapped potential to induce a supply response and transformation of the sector.

The objective of this chapter is to examine empirically some of the key economic factors affecting the profitability of paddy rice production at the farm level and, given these current conditions, estimate whether price incentives are able to induce an aggregate supply response at the national level. The types of public-sector interventions needed to promote output growth are also explored, drawing on lessons and experiences from other major rice-producing regions in the developing world. The chapter specifically focuses on irrigated and lowland production systems because, as shown in Chapter 3, these systems have the greatest potential for rice-production growth in Nigeria. Understanding the microeconomics of these systems can provide important insights into the constraints affecting rice-production growth, despite recent government efforts to introduce price incentives through higher import tariffs.

The chapter begins by analyzing a number of representative farm budgets from selected locations in Nigeria in order to compare and contrast their profitability across different input-intensive production systems. The findings show that the constraints on profitability are related not so much to intensification but to the current set of technologies in use, which are evidently undermining the potential yield gains from such intensifications.

Next, I examine the degree to which these constraints affect the government's ability to generate a significant supply response in the sector through price incentives. This is accomplished by empirically estimating an aggregate supply response and area response of irrigated rice production using nationally representative data. Results show that the aggregate supply response to price incentives in Nigeria is low. I hypothesize that this is partially due to insufficient past investments in agricultural research and development (R&D) and irrigation infrastructure as compared with the lessons and experiences from elsewhere in the region and other major rice-producing countries in Asia and Latin America. The final section summarizes the key findings of the review and analysis before concluding the chapter.

The data used in the empirical analysis of this chapter are based on information from the Nigerian Living Standards Measurement Study–Integrated Surveys on Agriculture (LSMS–ISA) (Nigeria, NBS and World Bank 2011, 2013) data discussed in Chapters 2 and 3 and the author's own fieldwork in 2013. The supply response analysis uses the agricultural module of the LSMS data, as they contain information on the quantity of rice harvested and use of various inputs and technologies on plots where rice is grown. The fieldwork was necessary to fill in gaps, as information in the literature on the microeconomics of these production systems in Nigeria is relatively scarce.

The data from field visits were collected in the following manner. First, given distinct production environments between the two geopolitical zones that produce the bulk of paddy rice in Nigeria, North West and North Central, locations were sought in both zones. The geopolitical zones are illustrated in Figure 1.1 in Chapter 1. To select the locations, the local government areas (LGAs) where one of two possible rice irrigation methods (river diversion or lifting of groundwater) is observed were identified from the LSMS–ISA (2011, 2013). I then asked the extension agents in those LGAs to take me to the rice irrigators they know. This is because farmers are more likely to provide detailed information about their rice-production practices to the agents with whom they are familiar.

Although I am likely to have visited producers who are relatively wealthier and better connected with the extension agents, they are also more likely to have better knowledge on the most efficient production practices, and therefore they represent an upper bound of the profitability of their systems. This becomes especially important if such an upper bound is low, implying that for most other farmers, production is less likely to be profitable. In addition,

I focused on visiting a small number of rice producers in order to measure the plot size accurately, instead of surveying a larger number of producers. This is important because in African countries where plots are not regularly demarcated, farmers' measurements of plot sizes are often incorrect (Carletto, Savastano, and Zezza 2013), leading to biased estimates of yields or production costs that are measured per hectare of area.

In addition, it is necessary to ask about and validate the costs of different inputs and activities in a highly disaggregated manner, because aggregating across different cost items often leads to the underestimation of total cost figures (Deaton 1997). To address these issues, 14 samples were collected across different rice-production ecologies and geographic areas, which are small in terms of sample size but more accurately measured. However, even these conclusions would need to be interpreted with caution given the small size and nonrandom nature of the sampling procedure. Moreover, results may vary from year to year due to factors such as weather conditions or even individual specific shocks that would affect profitability in a particular location at the time of the field visit. Future studies will need to examine whether the key messages of this section, particularly those derived from fieldwork, are sufficiently representative and apply to rice production in other production environments.

Assessing the Profitability of Paddy Rice Production

As the previous chapter showed, many socioeconomic constraints affect rice production in Nigeria, including the degree of access to markets, credit, and extension. But even more important is the viability of new technologies in ensuring a positive return to the farmer, given input and output prices, resource requirements, and costs (e.g., labor, land, and capital). All these affect the profitability of rice farming. This section delves deeper into assessing the profitability of paddy rice production based on farm budgets. More specifically, it uses data from the author's field visits, as described above, to analyze some of the key characteristics of rice-production costs and their implications for profitability at the farm level within each of the selected rice ecologies.

Generally, the intensity of modern input use for rice production varies widely across different rice ecologies and regions within Nigeria. Table 4.1 summarizes performance, input use, and production costs in irrigated and

TABLE 4.1 Rice-production performance, input use, and costs in selected locations in Nigeria based on farmer interviews, 2013

Location of interviewed farmers										(11) System			
(1) State	(2) Local government area	(3) Plot size (ha)	(4) Yield (tons/ha)	(5) Nitrogen (kg/ha)	(6) Fertilizer (kg/ha)	(7) Variety	(8) Gross revenue (US$/ha)	(9) Production cost (US$/ha)	(10) Labor cost (US$/ha)	IRR (pub)	IRR (pri)	RF (d)	RF (s)
North West zone													
Zamfara	Talata-Mafara	0.28	5.5	136	536		2,480	2,237	1,134	•			
Zamfara	Talata-Mafara	0.24	6.8	158	625		2,823	2,171	859	•			
Kebbi	Argungu	1.40	5.4	300	1,071	FARO 44	2,765	2,469	1,135		•		
Kebbi	Bunza	0.25	6.1	187	406	FARO 44	3,141	2,376	1,474			•	
Kano	Karaye	0.89	4.6	185	673	TOX	NA	NA	NA				•
Kano	Karaye	0.23	3.8	160	641	yardus	NA	NA	NA				•
Jigawa	Kilikasamma	1.37	6.5	240	728	WITA 4	NA	NA	NA		•		
Jigawa	Kilikasamma	2.02	3.3	119	395	WITA 4	NA	NA	NA		•		
North Central zone													
Nasarawa	Lafia	1.51	2.7	105	331	FARO 15	1,500	1,417	501			•	
Nasarawa	Lafia	0.70	2.3	30	143	FARO 44	1,790	1,802	1,241		• (rd)		
Nasarawa	Lafia	0.75	1.6	60	268	FARO 44	NA	NA	NA		• (rd)		
Nasarawa	Lafia	1.35	1.4	34	56	FARO 44	777	393	148			•	
Nasarawa	Karu	1.00	3.0	83	350	NERICA	2,065	2,116	985		•		
Nasarawa	Awe	1.35	2.8	0	0	FARO 44	1,553	1,359	666				•

Source: Author's fieldwork, May–July 2013.

Note: IRR = irrigated; RF = rainfed; pub = publicly operated; pri = privately operated; d = deepwater; s = plots submerged much of the season from rainfall; rd = river diversion; NA = not available. NERICA = New Rice for Africa.

rainfed lowland ecologies.[1] These are ecologies that have greater potential for rice-production transformation relative to upland ecologies, which tend to be more subsistence oriented and therefore targeted more for food security and poverty reduction objectives.

An interesting outcome from analyzing the available data is the finding that in many locations, except where public irrigation and some deepwater systems exist in northern Nigeria, net profits are small relative to production costs, where net profit is calculated by subtracting production cost in column 9 from gross revenue in column 8. This pattern has important implications. First, there are rice farmers who do not seem constrained from using modern inputs. This is contrary to the general view that African farmers are typically underutilizing modern inputs due to credit constraints or other market imperfections (Awotide et al. 2013; Kelly, Adesina, and Gordon 2003; Morris et al. 2007; World Bank 2007), in which case production costs per unit of land would be much lower. In fact, fertilizer use per hectare among interviewed farmers was found to be considerably high, particularly in the North West zone.

Second, these rice farmers may actually be overusing some of the inputs, with the exception of land. This is because of the following logic: as the underlying production system begins to exhibit diminishing returns to scale given generally older technologies being used on the same plots of land, marginal costs of modern inputs rise, but because rice prices are high, farmers continue using these inputs. This is reflected in Table 4.1 by higher yields in the North West zone being observed on smaller rice plots. Consequently, the smaller net revenues are likely the results of farmers operating beyond the optimal level of input use that would maximize net revenue. This is also partially supported by the evidence presented in the next section, which demonstrates that variable input costs such as labor and irrigation pumping seem to account for a greater share of total production costs than fixed costs. An important implication of this finding is that if farmers are already overusing modern inputs, efforts to stimulate supply response through the promotion of these inputs may be limited. In other words, the price elasticity of supply would be low.

---

1 Table D.2 in Appendix D summarizes rice production costs in upland, lowland, and irrigated ecologies across Nigeria reported in other selected studies. Many studies, however, do not clarify how plot sizes were measured. The International Food Policy Research Institute's fieldwork found that farmers often over report the size of plots, leading to inaccurate figures of yield and costs. This discussion is therefore focused on the figures in Table 4.1, which were all calculated using Global Positioning System–measured plot sizes.

Partial Budget for Irrigated Rice Ecologies

Table 4.2 presents the selected farm budgets for irrigated rice ecologies. Zamfara and Kebbi States are located in the northwestern part of the country (North West zone), while Nasarawa State is located in the central part of the country (North Central zone) (see Figure 1.1 in Chapter 1 for the locations of states and definitions of the six geopolitical zones). Intensive irrigation systems comparable to Asia can be found in parts of northern Nigeria, and these production systems are rather costly (columns 3 and 4).

Typically, production practices consist of land preparation that is done either mechanically or manually, transplanting or broadcasting of seeds ("planting" in Table 4.2), applications of fertilizer and agrochemicals, and production management such as weeding and bird scaring. Harvesting is done mostly manually, while threshing is sometimes done mechanically. In certain

TABLE 4.2 Rice farm budgets in irrigated ecologies

(1) Categories Irrigation systems Varieties	(2) Unit	(3) Zamfara Public scheme FARO 44		(4) Kebbi Private-pump FARO 44		(5) Nasarawa River diversion FARO 44		(6) Nasarawa Private-pump New Rice for Africa	
Plot size (ha)		0.28		1.40		0.70		1.00	
Values		Qty	US$/ha	Qty	US$/ha	Qty	US$/ha	Qty	US$/ha
Nonlabor cost									
Land preparation									
Harrowing	Number	2	230					1	48
Plowing	Number							1	65
Fertilizer									
NPK	50 kg bag	2	253	17	431	1	51	5	161
Urea	50 kg bag	1	120	13	300	1	55	2	77
Transportation	50 kg bag	3	14						
Seeds	Kg	75	138	75	28	68	55	60	29
Chemicals									
Pesticide	Liter			1	18				
Herbicide for land preparation	Liter					5	41	10	52
Herbicide	Liter	1	35	4	16	NA	77	NA	43
Threshing (nonlabor part)	75 kg bag	20.5	47						
Empty bag	Number					18	17	40	18
Paddy transportation to the market	75 kg bag	20.5	24			18	33	1	32

(1) Categories Irrigation systems Varieties	(2) Unit	(3) Zamfara Public scheme FARO 44		(4) Kebbi Private-pump FARO 44		(5) Nasarawa River diversion FARO 44		(6) Nasarawa Private-pump New Rice for Africa	
Plot size (ha)		0.28		1.40		0.70		1.00	
Values		**Qty**	**US$/ha**	**Qty**	**US$/ha**	**Qty**	**US$/ha**	**Qty**	**US$/ha**
Labor cost									
Land preparation	Number			1	55	1	230	1	32
Plowing	Number					1	138		
Leveling	Number					1	66		
Diking/bunding	Number	1	69			1	37		
Nursery preparation	Number			1	10				
Planting	Number	1	230	1	166	1	14	1	97
Weeding	Number	2	92	1	166	1	226		
Fertilizer application	Number	3	35	1	28	1	18	1	26
Herbicide spraying	Liter	1	5	4	9	1	51	1	19
Irrigation application	Number							1	398
Bird scaring + pest control	Number	1	461	1	415	1	194	1	45
Harvesting	Number	1	115	1	111	1	166	1	129
Threshing/bagging	Number	1	115	1	176	1	332	1	129
Loading/off-loading	75 kg bag	20.5	12						
Irrigation cost	Season	1	12						
Pump	Number[b]			0.25	58			0.67	95
Hose	Number[b]							0.2	28
Fuel	Liter			1,050[c]	484			582	547
Channel construction	Number							1	45
Renting of land	Season	1	230						
Total cost (US$/ha)			**2,237**		**2,469**		**1,802**		**2,116**
Total labor cost (US$/ha)			1,134		1,135		1,241		985
Gross revenue (US$/ha)	75 kg bag	20.5	2,480	100	2,765	21	1,790[d]	40	2,065
Net revenue (US$/ha)			243		296		–12		–51
Yield (ton/ha)			5.5		5.4		2.3		3.0
Paddy prices (US$/kg)			0.46		0.51		0.78		0.69
Nitrogen (kg/ha)[a]			136		300		30		83

Source: Author's fieldwork, May 2013.

Note: The information is based on a single farmer for each location. However, I also validated each piece of information on individual activities with other farmers in each location. Blank space indicates that the interviewed farmer did not use the inputs; therefore, the related activities and payments do not apply. NPK and urea are fertilizer products, while nitrogen is one of the chemical components of these fertilizer products. [a] Calculated assuming NPK 15-15-15 (15% nitrogen component) and urea 46%. [b] The interviewee buys an irrigation pump once every few production seasons. For example, if they buy a pump every four production seasons, figures are converted into 0.25 = 1/4. Similar calculations are applied to hoses. [c] 35 liters/day, one application every 3 days for 3 months (or 30 days in total). [d] Price reported by this farmer was unusually high. I therefore used the average of the figure reported by the farmer and averages among other farmers I interacted with in the same area.

cases, payment for empty bags is made separately from threshing and bagging; in other cases, payment is made all at once. Similarly, payment for loading and off-loading of 75 kg bags of paddy is sometimes made separately from paddy transportation to the market, while in other cases it is included in the transport payment.

Irrigated rice production is labor intensive, with labor costs generally more than US$1,000/per hectare (ha), accounting for about half (or more) of total production costs. Irrigation costs are also high if the system is pump based. Fuel costs for running pumps alone are about US$500/ha per season. These costs can be affected by various factors, including irrigation efficiency and pump capacity. In the North Central zone, paddy prices are slightly higher than the North West. Despite such intensive use of labor and irrigation, however, yields are quite modest and net revenue is, at best, barely positive (after adjusting for the opportunity costs of family labor).

Columns 3 and 4 of Table 4.2 also present the exceptionally high yielding cases observed in parts of northern Nigeria: an irrigation scheme in Zamfara State and a pump-based private rice irrigation system in the *fadama* area (a Nigerian term for an inland valley bottom) of Kebbi State. Their observed yields of 5.4 to 5.5 ton/ha are quite comparable to some of the successful rice irrigation systems in Asia. As was discussed earlier, these regions share similar ecologies with the Office Du Niger in Mali or in the Senegal River Valley, where Asian varieties are found to be most attractive given their geographic latitude. These production systems are, however, costly (with per hectare production costs at around US$2,200 to US$2,500) and likely to be feasible only in highly limited geographical areas with sufficient access to water, soils, and varieties. This type of ecology accounts for only a small fraction of the northern rice-growing regions.

Partial Budgets and Fertilizer Use for Rainfed Lowland and Upland Rice Ecologies

Production practices in some rainfed lowland areas are also characterized by high production costs, with total costs exceeding US$1,000/ha (columns 3 and 4 in Table 4.3). Production is somewhat labor intensive, with labor costs exceeding US$500/ha. In the lowland rice ecology, water control has also been suggested by the farmers to be a critical constraint to intensification and maximizing yield potential. Land preparation methods in these lowland systems become very critical in this regard, including the use of herbicides and tractors. Total costs of herbicides can be around US$200/ha. In some cases, soil is fertile enough so that no fertilizer is used.

TABLE 4.3 **Crop budgets for rainfed rice (selected locations in Nasarawa State based on ecological status)**

		Ecologies					
(1) Categories	(2) Unit	(3) Deepwater		(4) Lowland		(5) Upland	
Varieties		FARO 15		FARO 44		FARO 44	
Plot size (ha)		1.51		1.35		1.35	
Values		**Qty**	**US$/ha**	**Qty**	**US$/ha**	**Qty**	**US$/ha**
Nonlabor cost							
Harrowing (tractor)	Number	1	94	1	119	1	72
Fertilizer							
NPK	50 kg bag	8	188				
Urea	50 kg bag	2	51			1.5	39
Seeds	100 kg bag	1	96	2.25	143	0.5	32
Chemicals							
Chemicals for land preparation[a]	Liter	22	150			NA	30
Herbicide	Liter	22	220	36	178	2	19
Empty bag	75 kg bag	54	23	50	24	25	14
Rope for bird scaring	Number			70	3		
Transport harvest to the market	75 kg bag	54	46	50	153	25	24
Labor cost							
Land preparation	Number	1	47	1	72	1	14
Plowing	Number					1	72
Broadcasting	Number	1	17	1	24	1	5
Fertilizer application	Number	1	17			1	10
Weeding	Number	1	81			1	19
Herbicide spraying	Liter	36	108	36	69	2	5
Bird scaring	Number			1	287		
Harvesting	Number	1	98	1	143	1	38
Threshing	Number	1	180	1	143		
Total cost (US$/ha)			**1,417**		**1,359**		**393**
Total labor cost (US$/ha)			501		666		148
Gross revenue (US$/ha)	75 kg bag	54	1,500	50	1,553	25	777
Net revenue (US$/ha)			83		194		384
Yield (ton/ha)			2.7		2.8		1.4
Paddy prices (US$/kg)			0.56		0.55		0.56
Nitrogen (kg/ha)			105		0		34

Source: Author's fieldwork in July 2013. Exchange rate = 155.

Note: Blank space indicates that the interviewed farmer did not use the inputs; therefore, the related activities and payments do not apply. [a]The quantity of chemicals for land preparation in column 5 was not available, although the respondent remembered their cost.

Production practices in upland rice ecology (column 5) are still generally primitive. Despite low yields because production is less input intensive, the system still brings in more profit than in the rainfed lowland or irrigated ecologies (US$777 in revenue minus US$393 in cost = US$384/ha, compared to lower figures in other ecologies). Tables 4.2 and 4.3 suggest that production costs in irrigated rice environments can be much higher than in upland environments, particularly due to irrigation and labor costs.

One option may be to fully exploit the limited but undeveloped irrigable areas in northern Nigeria through rehabilitation of many public irrigation schemes, such as removing sediments from the dams and canals or plot leveling, which can reduce the water delivery costs to and within the plots. Many of the existing dams were also constructed without a full assessment of their environmental impact. This has led to negative externalities created by many of these schemes, such as Bakolori Dam, located in Zamfara State (Yahaya 2002), and Tiga Dam, located in Kano State (Thomas and Adams 1999). If the rehabilitation of existing schemes is successful, this can affect the production in surrounding areas. Expansion of irrigation is needed for rice intensification and increased supply response, but it needs to be carefully designed through detailed hydrological, topographical, and socioeconomic analyses, so that the negative environmental externalities are minimized.

These partial budget analyses provide some useful insights into the supply response potential of rice in Nigeria. First, labor costs account for a significant share of rice-production costs, meaning that rice production (particularly in irrigated and rainfed lowland environments) is labor intensive. It is often difficult to expand the cultivated area of labor-intensive production because costs of supervising and monitoring hired workers increase significantly. Second, current net revenue in irrigated and rainfed lowland rice production is generally low. Some farmers earning greater income from the production of other crops or upland rice may not have much incentive to switch to rice production, unless net revenues increase significantly. However, since the costs of nonlabor inputs (particularly seeds) are generally small, subsidizing them may not substantially increase the net revenues.

Aside from the analysis of partial budgets, an examination of the degree of fertilizer use and its effectiveness in raising rice yields is also useful considering that it is one of the essential inputs for rice intensification. However, evidence of its ability to effectively induce an overall supply response through favorable fertilizer prices has been limited. Figures indicate that the rice fertilizer response in Nigeria is still low, particularly in the North Central zone and the southern (South East, South South, and South West) zones.

FIGURE 4.1 Nitrogen response for rice in Nigeria: Comparison with estimates based on Sri Lankan data

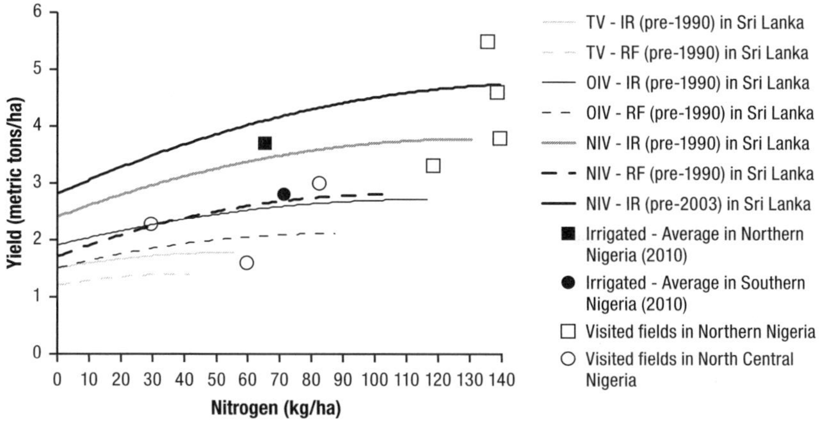

Source: Yield response lines are from Kikuchi and Aluwihare (1990) for all pre-1990 figures; Kikuchi, Maruyama, and Hayami (2003) for the pre-2003 figure. The irrigated averages are based on LSMS data from Nigeria, NBS and World Bank (2011), while data from author's own fieldwork are included to provide additional data points.

Note: IR = irrigated, RF = rainfed, TV = traditional variety, OIV = older improvement variety, NIV = newer improved variety.

For comparison, Figure 4.1 plots the nitrogen response curves of various generations of improved varieties (rainfed and irrigated) in Sri Lanka, which illustrate nicely the generational varietal improvement process starting from traditional varieties (TV), older improved varieties (OIV), newer improved varieties (NIV) introduced before 1990 (pre-1990), and newer improved varieties introduced before 2003 (pre-2003).[2] The observed yield and nitrogen levels under irrigated (IR) conditions in Nigeria are plotted against these curves. Importantly, the fertilizer response levels in the North Central zone and three southern zones in Nigeria are similar to OIV–IR (pre-1990).

Some of the dots from the North Central Nigerian zone (o) in Figure 4.1 indicate the variety FARO 44, which is relatively successful in the north. In other words, improved varieties (including FARO 44) currently used in the North Central and southern zones may be only as good as improved varieties that were already considered relatively old in Sri Lanka by 1990 for this ecology. Even in northern Nigeria, fertilizer response is close to the NIV–IR (pre-1990) and below NIV–IR (pre-2003), indicating that the improved varieties

2 While I do not have the exact sequencing of the development of these different groups of varieties, varieties developed later are likely to have better fertilizer responses.

currently used in the north are only as good as varieties that were considered new in the 1990s and are likely to be more obsolete by now. Only a handful of locations in the north achieve the similar level as NIV–IR (pre-2003). These conditions in Nigeria are consistent with the hypothesis that rice variety development in Nigeria has been trailing behind that in countries such as Sri Lanka. Given the long tradition of rice production, crop husbandry knowledge might not be the major reason for this gap.

Estimating an Aggregate Supply Response

Having looked at the farm-level profitability of rice-production systems, it is appropriate to examine the sector's overall responsiveness to price incentives. This is done by estimating the price elasticity of farmgate rice supply and the irrigated rice area, aggregated across rice producers in Nigeria. A price elasticity of rice production measures the degree to which a unit proportional change in output prices results in a proportional change in production output or area cultivated, which is an important indicator of supply responsiveness. The data used are from the LSMS–ISA (2011, 2013) discussed in earlier chapters, as well as other data indicating the spatial and temporal variations of agroecological and socioeconomic conditions. The LSMS–ISA data are appropriate because the two rounds of the survey (from 2011 and 2013) allow construction of a pseudo-panel dataset. Constructing such a dataset makes it possible to control for various district-specific unobserved factors (such as the history of rice production in the area and average farmers' production skills) that affect rice-production practices separately from observable factors.

Appendix D provides the detailed empirical specifications, econometric methods used, regression results, and procedures for calculating elasticities. This section provides a brief summary of these aspects. The underlying model is based on agricultural household models by Singh, Squire, and Strauss (1986), where a household facing credit constraints and transaction costs determines rice-production levels in a process of maximizing household utility, given the prices of rice and other competing crops, as well as prices of key inputs and access to technologies such as irrigation and mechanization.

Using a reduced-form supply-response estimation technique, the degree to which rice-production decisions have responded to recent rice-price changes in Nigeria is assessed. Farmers' decisions typically involve two steps: a decision to grow rice followed by a decision on how much to grow. Because of high transaction costs involved with market participation in rural Africa, the first decision may be made in a different mechanism from the second decision. Many

TABLE 4.4 Price elasticity of rice production and irrigated rice area in Nigeria

(1) Variables	(2) Own-price elasticity	(3) Standard error of estimated elasticity
Rice production	0.23	0.12
Irrigated rice area	0.08	0.19

Source: Author's estimation based on Nigeria, NBS and World Bank (2011, 2013).

studies have addressed these types of price-incentive responses in Africa (see, for example, Heltberg and Tarp 2002; Takeshima and Winter-Nelson 2012). Their models are extended by applying Cragg's (1971) double-hurdle model to estimate rice-production response and the Tobit model to estimate irrigated rice area response. The Tobit model is used for the irrigated rice area due to the small sample size of rice irrigators, for which the double-hurdle model is inapplicable.

The panel nature of the LSMS datasets is explored to apply a modified correlated random effects model, which can partly control for location-specific unobserved characteristics in specifications with limited dependent variables such as the double-hurdle or Tobit models (Takeshima and Nkonya 2014).[3] Because the model specification is a reduced form, it does not distinguish the effects of endogenous factors. For example, it does not model the intermediate steps such as renting additional land, transaction costs for hiring labor for land clearing, and so on that are required to start and expand rice production. Similarly, the analysis remains at the household level; although a higher price in the market would actually lower demand, I am not modeling market-level supply and demand functions.

Estimated price elasticities of rice production and irrigated rice areas are presented in Table 4.4 (detailed regression results are in Table D.1). Results in column 1 in Table D.1 indicate that a 100 percent increase in rice price may raise the probability that the farmer grows rice by 7 percent, which is calculated by summing the coefficients on "local rice/yam" (0.022), "local rice/cassava" (0.042), and "local rice/maize" (0.006) (ignoring the statistical significance for now, which is discussed below). Using the formula for the partial effects (*PEs*) in Appendix D, the same 100 percent increase in rice price is calculated to increase rice output by 0.622 tons (a number reached by summing the "local rice/yam," "local rice/cassava," and "local rice/maize" coefficients in column 3 in Table D.1) and the irrigated rice area by 0.018 ha (column 7 in Table D.1).

3 Technical details of the model and analysis are provided in Appendix D.

All three of these effects—the 7 percent probability increase, 0.622 metric tons increase in output, and 0.018 ha increase in irrigated rice area—are statistically significant. The price increase, however, has no effect on the decision to grow irrigated rice. Using the formulas shown in Appendix D, the estimated price elasticity of rice production is around 0.23. The estimated price elasticity of the irrigated rice area is 0.08, which statistically is not significantly different from zero. The estimate for the supply elasticity is close to those of others such as Rahji, Ilemobayo, and Fakayode (2008), who find a long-run estimate of around 0.24 and 0.33. Their short-run estimates were 0.06 to 0.08. Overall, the results are consistent with the view that rice production is only marginally responsive to rice prices in Nigeria. In particular, rice prices have fairly small effects on the expansion of the irrigated rice area, which is critical for yield growths.

The estimated supply elasticity is not any higher than other more advanced rice economies, and it seems to have declined recently. Price elasticities of rice supply in many Asian countries with larger rice-production sectors and more mature production technologies than Nigeria have been around 0.2 in the short run, with the exception of Bangladesh (Table 4.5, column 2). In Vietnam, the price elasticity of the irrigated rice area was around 0.17 in the early 1990s, when the share of irrigated area was already close to 50 percent (column 3).

In Nigeria, the supply elasticity of upland rice was estimated in the 1970s to be around 0.5 to 0.6 (Ngambeki and Idachaba 1985), while the average at the national level, as noted earlier, has been estimated between 0.24 and 0.33

TABLE 4.5 Price elasticity of rice supply in Nigeria and other countries

(1) Country	(2) Price elasticity of rice production	(3) Price elasticity of irrigated rice area
Nigeria	0.5–0.6 (Ngambeki and Idachaba 1985; upland rice in the 1970s)	
	0.24–0.33 (Rahji, Ilemobayo, and Fakayode 2008)	
Bangladesh	1.1 (Imai, Gaiha, and Thapa 2011)	
Thailand	0.25 (Choeun, Godo, and Hayami 2006; short run)	
	0.59 (Choeun, Godo, and Hayami 2006; long run)	
	0.2–0.3 (Behrman 1968; in the 1960s)	
Vietnam	0.22 (Khiem and Pingali 1995)	0.17 (Khiem and Pingali 1995)
Elasticity figures assumed		
West Africa	1.0 (Dalton and Guei 2003)	
Japan	0.2 (Hayami and Ruttan 1970)	
Senegal	0.3 (Fisher, Masters, and Sidibé 2001)	

Source: Author's compilations.

(Rahji, Ilemobayo, and Fakayode 2008). Current price elasticities of rice supply in Nigeria seem low considering that the rice sector is still relatively small compared to that in Asia and the yield gap is greater.

The low price elasticity of rice production and irrigated rice area in Nigeria is consistent with the hypothesis that technologies are still underdeveloped for rice production, so that price policies alone have limited effects on rice production. Moreover, a higher rice price appears to provide little incentive for farmers to expand the irrigated rice area, not only because soils suitable for irrigated rice are relatively limited in Nigeria but also because varieties are old, with quickly diminishing returns to water, and public irrigation schemes have been relatively underinvested given the size of arable land, so that costs of delivering irrigation water to the plots remain high.

Policy Implications and Role of the Public Sector

Rice production and technology adoption decisions by farmers are affected by economic returns, and these returns must be substantially high in order to overcome various market failures and weak infrastructure that constrain rice production in rural Nigeria and thus an aggregate supply response. Historically, key public interventions that raised economic returns in Asia and Latin America were investments in improved varieties, expanded irrigation, and incentives through price policies. On the other hand, the development and use of mechanization technologies and other modern inputs such as fertilizer seem to have grown endogenously (Pingali 2007), partially in response to the seed and irrigation technologies initially invested in by the public sector, but also in response to rising real wages, among other factors. In fact, public expenditures on agrochemicals during the Green Revolution era were generally small, only a fraction of fertilizer expenditure, while much of the focus was on developing pest-resistant rice varieties (Barker, Herdt, and Rose 1985, 90) and irrigation.

The design and development of machinery suitable for certain uses and local environments in rice production systems were often led by the private sector and through adaptive innovations from direct interactions with farmers. This was important, as both agroclimatic factors (soil, terrain, rainfall) and economic factors (land, labor, capital, farm size, materials available) can greatly affect the efficiency and suitability of different mechanized technologies in different locations (Binswanger 1986). Examples of such adaptive innovations in the past include those in the United States (Evenson 1982), India's Punjab region (where a diversified machinery industry emerged from

TABLE 4.6 Drivers of rice-production growth in selected Asian countries, 1965–1980

	Output increases, thousand tons							
(1) Country	(2) Modern variety effect	(3) Fertilizer effect	(4) Irrigation effect	(5) Other factors (residual)	(6) Total growth in output	(7) (2 + 3 + 4)/6 (%)	(7) Yield (tons/ha) in each year 1965	1980
Burma	647	353	685	167	1,852	91	1.7	2.8
Bangladesh	420	1,284	1,091	2,759	5,554	50	1.7	2.0
China	13,231	11,507	16,153	9,609	50,500	81	3.0	4.1
India	7,998	10,867	11,209	5,078	35,152	86	1.3	2.0
Indonesia	3,162	2,680	2,773	4,998	13,613	63	1.8	3.3
Philippines	849	1,009	801	615	3,274	81	1.3	2.2
Sri Lanka	241	215	262	316	1,034	69	1.8	2.6
Thailand	822	682	865	4,031	6,400	37	1.8	1.9
Total (above)	**27,370**	**28,597**	**33,839**	**27,573**	**117,379**	**77**		

Source: Herdt and Capule (1983). Yields are from FAO (2014).

small shops), Thailand (power tillers; see Wattanutchariya 1983), and the Philippines (Mikkelsen and Langam 1983).

Nigeria has particularly lagged behind in witnessing any such adaptations, perhaps because of the general lack historically of comparable investments in the development and diffusion of improved varieties and irrigation technologies. This section describes in more detail these investment gaps, drawing on the lessons and experiences in Asia and Latin America.

Historically, returns to rice production globally have been primarily raised through advancements in technologies that increase yields. As shown in the previous chapter, rice yield has grown steadily in Asian and Latin American countries in the past several decades. In many Asian countries, the effects of modern varieties, fertilizer, and irrigation together have accounted for a significant share of total rice-production growth, often greater than 80 percent of the total output growth between 1965 and 1980 (column 7 in Table 4.6).

Improved Varieties and Agriculture Research and Development

Successful varieties have often triggered productivity growth (Traxler and Byerlee 1993). The introduction of high-yielding rice varieties in Japan was a key driver of agricultural development in the 19th century (Jirström 2005). The use of hybrid rice in China has contributed significantly to the transformation of Chinese agriculture since the 1980s (Lin 1994).

TABLE 4.7 Number of varieties released in Nigeria and Asia by National Agricultural Research Institutes and other sources

(1) Category	(2) Pre-1970s	(3) 1971–1980	(4) 1981–1990	(5) 1991–1999	(6) 2000–2012	(7) Total	(8) Number per million ha in 1999
Nigeria							
NARI	1	5	11	0	0	17	10
IRRI, IITA, WARDA	1	5	4	6	13	29	18
Foreign	10	3	3	2	0	18	10
Total	*12*	*13*	*18*	*8*	*13*	*64*	*38*
South and Southeast Asia							
NARI	471	338	588	421		1,818[a]	18[a]
IRRI	62	71	75	14		222[a]	2[a]
Total	*533*	*409*	*663*	*435*		*2,040*[a]	*20*[a]

Source: South and Southeast Asia: Author's estimates based on Tables 5.3 and 5.5 of Hossain et al. (2003). Nigeria: Author's modifications based on Takeshima (2014).

Note: IITA = International Institute of Tropical Agriculture. IRRI = International Rice Research Institute. NARI = National Agricultural Research Institute. WARDA = West Africa Rice Development Association (now Africa Rice Center). Blank spaces indicate information was not available. [a]Based on figures up to 1999.

In many Asian countries, substantial public-sector R&D for new varieties was in existence before 1980 (Evenson 1977; Barker, Herdt, and Rose 1985). This is in contrast to Nigeria, where domestic rice research still primarily focuses on selections and transfers of imported varieties. Up to 1999, South and Southeast Asia released 18 improved varieties developed by national agricultural research institutes (NARIs) per 1 million ha of rice area. The corresponding figure for Nigeria is only about half that: 10 per 1 million ha (column 8 in Table 4.7).

Many NARI-bred varieties have been widely adopted in Asia (Hossain et al. 2003). Rice-production ecologies and preferences vary across countries, and NARIs have played an important role in successfully developing and releasing varieties suitable for local production environments. Countries such as Indonesia invested in R&D to develop varieties with superior grain qualities suitable for local milling sector and taste preferences (Unnevehr 1986). In contrast, most major rice varieties in Nigeria are of foreign origin, except the old FARO 15, which was developed in 1974.

The success of NARI R&D in Asian countries can be attributed to the size of funding, quality of human capital, and decentralization of breeding. Countries in South Asia, Southeast Asia, and Latin America had already been investing substantially in rice R&D since the 1970s, and these countries increased rice

R&D investments in the 1990s (columns 2 and 3 in Table D.3). While Nigeria spent only US$0.3–0.5 per hectare of rice area in 1998 (constant 2010 US$, purchasing power parity [PPP] adjusted)(column 4), by 1974 many Asian countries had already been spending a similar amount or more on research per hectare of rice grown.

In many Asian countries, the rice R&D system was decentralized, which is important for crops like rice whose production environment is diverse. By the 1930s, prefectural governments in Japan had become the major funders of rice R&D (Hayami and Yamada 1998). In Thailand, breeding was conducted at a provincial level as early as the 1980s (Sarkarung, Somrith, and Chitrakorn 2000). In Sri Lanka, rice is bred in four domestic stations, although the total rice area is less than half that of Nigeria (Wang et al. 2012, 96). India already had breeding stations in several states in the 1970s (De Datta 1981, 191).

Given the experiences of Latin America and Asia, Nigeria will need to increase its own R&D spending on rice if it is to boost domestic rice production in the future. Agricultural R&D was actually higher for Nigeria in the 1970s and early 1980s than at any time since (Roseboom et al. 1994; Beintema and Ayoola 2004; Flaherty et al. 2010). This situation was a common one across Africa in general: R&D for rice as a share of rice gross domestic product (GDP) peaked during the same period (Judd et al. 1986; Lipton 1988). This coincided with steady rice yield increases in Nigeria up until the mid-1980s, when yields were similar to those in Bangladesh, India, and Thailand—and even higher than those in Brazil (Table 3.1 in Chapter 3).

Domestic rice R&D is particularly important for Nigeria given its unique production environments. Compared to Sahelian West Africa, which has production environments closer to those in Asia (Dalton and Guei 2003), adaptability of imported Asian varieties is lower in Nigeria because tropical agroclimatic conditions are more diverse, and the more numerous and more diverse insects and diseases cause more severe crop damage (De Datta 1981, 182). In tropical West Africa, returns from R&D may be higher in Nigeria, where lowland ecology accounts for a significant share of rice area, than in Guinea or Côte d'Ivoire, where much of the rice area is upland ecology (Dalton and Guei 2003, Table 6.1).[4]

4 Raising yield in upland rice ecology has been challenging. Even in Brazil and Indonesia, where national average yield has risen to 5 tons/ha, upland rice yield had remained below 2 tons until recently (Bierlen, Wailes, and Crammer 1997; Jatileksono 1998), despite substantial upland rice research by the Brazilian Corporation of Agricultural Research (Empresa Brasileira de Pesquisa Agropecuária, EMBRAPA) in Brazil (Pardey et al. 2006). This may partly explain why significant NARI variety development has not increased rice yield in these countries.

In Nigeria, R&D for upland rice can bring benefits for marginal environments, for which many of the varieties developed by New Rice for Africa have been intended. However, in terms of increasing domestic rice production, the potential may not be as great as in rainfed lowland and irrigated ecologies. To achieve significant increases in output, the substantial focus of domestic R&D should be on irrigated and lowland ecologies.

Irrigation

The extent of rice irrigation in Nigeria is far below the historical levels in Asia or Latin America (Figure 4.2). Investment in public irrigation infrastructure, prior to the Green Revolution in Asia, for example, was considered to be more cost effective in the long run than some other measures such as price support policies (Hayami, Brennan, and Barker 1977; Rosegrant, Kasryno, and Perez 1998). Asia and Latin America (Colombia) invested relatively more in public irrigation per unit of arable land than Nigeria (column 3 in Table 4.8). While Nigeria invested US$8–23 per hectare (2010 US$, PPP) between the 1970s and 2000s, other countries were already investing several times more between the 1960s and 1980s. In particular, investments have declined sharply in Nigeria since 1990. Given that the rice area expanded at the same time, irrigation investment per rice area dropped to only a fraction of the levels prevailing in the 1970s and 1980s (column 4).

It is important to note that, while the investments per rice area in Nigeria have been comparable to or even higher than other countries, the stock of irrigation infrastructure is likely to be much lower, as suggested in Figure 4.2. The cost per hectare of developing irrigation schemes in Africa south of the Sahara is similar to that in Asia. If successful irrigation schemes in SSA are considered (Inocencio et al. 2007), Nigeria will not only have to increase investments in public irrigation schemes but also focus on locations with a high probability of success. Operations and maintenance (O&M) may also need to be financed largely through public resources by means of irrigation subsidies, as was done in Asia as well as in developed countries, including the United States (Gupta, Miranda, and Parry 1995).

In Asia and Latin America, the private sector was also heavily involved in investments in rice irrigation, including communal systems in Japan, the Philippines (Barker et al. 1985, 101), Malaysia (Short and Jackson 1971), Nepal (Small et al. 1986), and Sri Lanka (Chambers 1980). These systems were usually cheaper than government systems (Kikuchi, Dozina, and Hayami 1978) and were developed by mobilizing community resources or by landlords (Hayami and Kikuchi 2000, 176). In many cases, farmers were able to develop effective

FIGURE 4.2 Percent of area irrigated and irrigated area per 1,000 people

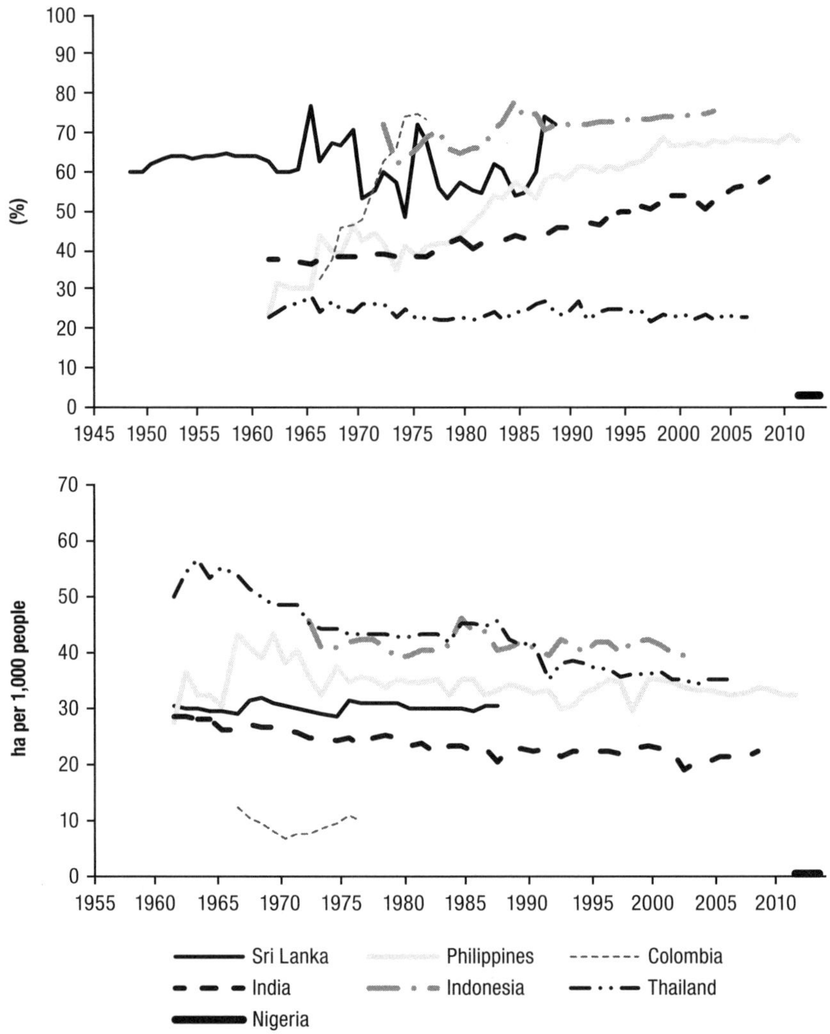

Source: Various. Sri Lanka = Inferred from Figure 11.3 of Wijayaratna and Hemakeerthi (1992) and Aluwihare and Kikuchi (1991); India = Thakkar (1999) and World Bank (1998). Indonesia = World Rice Statistics (WRS) by IRRI. WRS are also used to supplement aforementioned studies whenever possible. Philippines = Kikuchi, Dozina, and Hayami (1978). Colombia = Chandler (1979).

Note: IRRI = International Rice Research Institute.

TABLE 4.8 Annual investment in irrigation in selected countries and Nigeria

(1) Country	(2) Average annual public investment in irrigation (2010 US$million/year, PPP adjusted)						(3) Average annual public investment in irrigation per arable land (2010 US$/ha, PPP adjusted)					(4) Average annual public investment in irrigation per rice area (2010 US$/ha, PPP)				
Periods	1950–1959	1960–1969	1970–1979	1980–1989	1990–1999	2000s	1960–1969	1970–1979	1980–1989	1990–1999	2000s	1960–1969	1970–1979	1980–1989	1990–1999	2000s
Sri Lanka	229	186	332	818			249	381	924			356	464	1,013		
Philippines		36	275	711			7	58	134			11	80	207		
India	2,060	3,338	5,591	8,493	11,879		21	35	52	73		93	145	208	278	
Indonesia			1,014	1,972				56	104				122	207		
Colombia[a]		99	57	127			28	16	34			323	168	304		
Nigeria[b]			647	647	246	447		23	23	8	13		1,146	1,146	120	193

Source: Author's estimates based on various sources: Aluwihare and Kikuchi (1991) for Sri Lanka; for Nigeria, Foster and Pushak (2011) in 2001–2006, Pagiola et al. (2002) for 1990s, Pradhan (1993, 21) for 1970s and 1980s; Kikuchi, Dozina, and Hayami (1978) for the Philippines; Thakkar (1999) for India; Rosegrant and Pasandaran (1993) for Indonesia; Dinar and Keck (1997) for Colombia.

Note: PPP = purchasing power parity. Annual investment in irrigation includes new construction, rehabilitation, operation and maintenance, and other small uncategorized expenditure. Not all studies, however, provide such disaggregation, and categorization may vary across countries. Blank cells indicate missing data, not lack of investment. [a]For Colombia, approximate numbers are assessed from Figure 2 of Dinar and Keck (1997) due to the lack of access to the original data. [b]For Nigeria, Pradhan (1993, 21) mentioned only that Nigeria spent $3 billion in the two decades up to the early 1990s. Assuming $150 million was spent each year for two decades, converting each year's figure into 2010 US$, as well as making PPP adjustment, I arrived at the figure of $647 million/year (2000 US$, PPP) for the 1970s and 1980s.

community management of common resources that avoided the tragedy of commons (Hayami and Kikuchi 2000, 134; Feeny et al. 1990; Ostrom 1990, 1992; Baland and Platteau 1996).[5]

In early 20th-century Colombia, irrigation was introduced and facilities constructed by private entrepreneurs (Leurquin 1967, 227). In parts of Nigeria, similar river diversion structures have been constructed and used for rice irrigation on *fadama* areas, where farmers divert river flow using sand embankments.[6] Such areas are small, however, and yields are relatively low. The low private irrigation investments in Nigeria indicate that technologies and economic conditions are unconducive for irrigated rice production, and investment in public irrigation schemes will need to be combined with the development of sufficiently superior varieties and profitable production environments.

In particular, expansion of rice irrigation in the past has largely been to provide supplementary irrigation in the rainy season rather than in the dry season (columns 3 and 4 in Table 4.9), as rice requires substantial amounts of water. This was led by the development of good varieties with resistance to pests and diseases that are more prevalent in the wet season than dry season. This pattern of rice varietal development and irrigation expansion in Asia may have important implications for Nigeria as well.

Price Policies: Output Price Control, Input Subsidies

In addition to investments in technology, price policies were widely adopted instruments in Asia during the Green Revolution (Djurfeldt and Jirström 2005). Many Asian countries have enjoyed a higher rice price relative to fertilizer price than West African countries (Otsuka and Kalirajan 2006). If Asian governments have succeeded in achieving such favorable prices for producers, such price policies might have been important in realizing higher yield growth.

Historical patterns in Nigeria, however, contradict such a theory. Table 4.10 shows nominal rates of assistance (NRAs) for rice in selected West African countries. Table 4.11 shows the ratio of fertilizer prices to paddy prices in selected countries, another important indicator of the level of agricultural protection (Judd, Boyce, and Evenson 1986). As in Table 4.10, while many other West African countries have experienced negative NRA in much of the last several decades (except rice in Ghana between

5 The "tragedy of the commons" (Hardin 1968) refers to the situation where common-pool resources with finite reserves become exhausted by rational individual users because of the difficulty in controlling their uses.

6 Based on author's fieldwork in July 2013.

TABLE 4.9 Share of irrigated area among all rice areas in selected countries

(1) Country	(2) Periods	Irrigated area (1,000 ha) under each regime (3) DS (dry season)	(4) WS (wet season—supplementary)	(5) Rainfed	(6) % of irrigated rice area among all rice area	(7) Hectares of irrigated area per 1,000 people
Bangladesh	1970–1974	980	590	8,200	16	23
Burma	1970–1974	50	800	4,140	17	31
	1979	115	780	4,422	17	
Philippines	1970–1974	480	1,430	1,580	55	51
India	1970–1974	1,890	13,120	22,650	40	26
Indonesia	1970–1974	1,610	2,370	4,500	47	32
	1979	1,920	3,274	3,010	63	37
Thailand	1970–1974	140	630	6,260	11	20
	1979	320	866	7,491	14	26
Nepal	1970–1974	0	190	1,020	16	15
	1979	0	261	1,001	21	19
Malaysia	1970–1974	270	270	170	76	47
Vietnam	1970–1974	140	270	2,310	15	9
Sri Lanka	1970–1974	150	220	230	62	28
	1979	182	294	284	63	32
Nigeria	2010–2011	23	89	2,400	4	0.7

Source: For Asian countries, author's compilations from Chandler (1979) and Barker, Herdt, and Rose (1985). Nigerian data are from Takeshima and Edeh (2013). Population figures are taken from FAO (2014) for corresponding periods.

TABLE 4.10 Growth in nominal rate of assistance to agriculture, selected West African countries, 1960–2004

	1960–1964	1975–1979	1980–1984	1990–1994	2000–2004
Cameroon	−2.9	−14.4		−1.1	−0.1
Côte d'Ivoire	−23.5	−30.8		−19.5	−24.5
Rice		−25.0	−10.0		
Ghana	−9.0	−25.6		−1.7	−1.4
Rice		+13.0	+29.0		
Nigeria	+20.7	+6.3		+3.9	−5.4
Senegal	−9.3	−22.7		+5.6	−7.5

Source: Krueger, Schiff, and Valdès (1988); Anderson (2009); Dupraz and Postolle (2013).

Note: Blank spaces indicate that information is not available. While rates can vary if the black market exchange rate is different from the official rate, relative levels of NRA for rice during 1980–1986 were similar among countries in Asia and Africa regardless of the types of rates used for calculation (assessed from Taylor and Phillips 1991, Appendix).

TABLE 4.11 Urea-to-paddy (kg) price ratio in selected countries and Nigeria

Years	Bangladesh	Colombia	India	Indonesia	Philippines	Sri Lanka	Thailand	Côte d'Ivoire	Benin	Nigeria
1966–1970		2.1	2.9		3.2	1.4		2.7		
1971–1975	1.6	4.9	2.7	3.4	1.6	1.7		3.4	2.6	
1976–1980	2.0	3.3	2.6	1.8	3.8	1.4	3.0	2.2	1.9	1.5
1981–1985	2.4	3.3	2.6	1.3	4.1	1.7	5.1	3.3	1.5	1.6
1986–1990	1.9	1.6	2.4	1.5	2.4	1.9	2.8	2.9	1.7	1.0
1991–1995	1.9	2.7	1.6	1.5	2.4	2.6	3.8	4.2	2.2	1.3
1996–2000	2.0	1.7	1.6	1.8	1.9	1.5	2.7	3.7		1.7
2001–2002	2.0	2.2	1.7	1.7	2.2	1.6	2.9			

Source: Author's compilation based on Estudillo and Otsuka (2012), World Rice Statistics (farm harvest price of paddy) (IRRI 2014), and FAOSTAT (urea price) (FAO 2014).

Note: FAOSTAT = Statistical database of the Food and Agriculture Organization of the United Nations.

1975 and 1984), Nigeria has actually enjoyed a positive NRA except during the 2000s. These conditions are likely to hold for rice because Nigeria has been unique in the West African region, as well as the rest of Africa south of the Sahara, in terms of the level of rice import restrictions (tariffs, banning, etc.) (Akande 2001) and fertilizer subsidies.

From the late 1970s to the early 1980s, there was either a tariff of 10–20 percent or various import restrictions in Nigeria, and between 1985 through 1995, rice imports were banned altogether. After the ban was lifted in 1995, tariff rates remained between 50 to 100 percent in much of the 1990s, higher than other West African countries and the rest of SSA (Akande 2001). For example, the tariff or import tax on rice in the 1990s was 10–30 percent in Madagascar (Fafchamps and Minten 1999), 5–13 percent in Côte d'Ivoire (Abbott 2007, Appendix Table 19), and 38 percent in Cameroon, between 1996 and 2008 (Molua 2010).

While the implementation of these policies has not been effective, as shown in Chapter 7, these restrictive policies still succeeded in keeping rice prices higher in Nigeria than in other countries in Asia or West Africa. For example, the farm-harvest paddy rice price in Nigeria was in the range US$290–450 per ton (current US$/ton using parallel market exchange rates) between 1970 and 1990 (column 13 in Table D.4 in Appendix D). These prices were substantially higher than those in Côte d'Ivoire and Benin, for example (columns 9 and 10 in Table D.4). As a result, the fertilizer-to-paddy price ratio in Nigeria between 1976 and 2000 was more comparable with that in some Asian countries than that in Côte d'Ivoire

(Table 4.11). This is also consistent with the recent findings in Nigeria that fertilizer subsidies do not lead to a significant reduction in rice prices (Takeshima and Liverpool-Tasie 2015). The economic environment for rice production in terms of the fertilizer-to-paddy price ratio was therefore fairly favorable in Nigeria during this period.

For many low-income countries like Nigeria, price policies will affect production growth only if they are complemented by technical progress (Mellor 1978). In other words, technological advancement and investments in public irrigation infrastructure are more cost effective in the long run than price support policies, as has been shown in Asia (Hayami, Brennan, and Barker 1977; Rosegrant, Kasryno, and Perez 1998). On the other hand, from Asia's experience, the combination of input subsidies with investments in R&D and irrigation have been shown to result in higher yield growth (Rashid et al. 2013). These sets of evidence suggest that, in contrast to other countries in Asia, price policies have been inefficient in raising rice production and productivity in Nigeria primarily because of insufficient complementary investments in agricultural R&D and irrigation.

Conclusion

Rice production in Nigeria is characterized by low levels of improved technology adoption, especially for the superior-quality and high-yield seed varieties and irrigation. On the other hand, the diffusion in the use of other modern inputs such as fertilizer has been high at times whenever the paddy-to-fertilizer price ratio has increased, either due to fertilizer subsidies or higher rice import tariffs. Despite such incentives, the profitability of rice production has generally remained low, particularly in the irrigated and rainfed lowland ecologies, which are the major rice-production systems in Nigeria. Yield is often low despite the higher intensity of modern input use.

Evidently, higher rice prices are not sufficient by themselves to stimulate output growth for domestic rice production in Nigeria. Based on nationwide panel survey data, the price elasticity of both rice production and irrigated rice areas is low. A 1 percent increase in the price of paddy at the farmgate leads to only a small increase in rice production and has almost no effect on the irrigated rice area. This is consistent with the technological backwardness of rice production in Nigeria and results from historically low investments in rice R&D and irrigation compared to other developing countries in Asia and Latin America.

Key technologies, such as superior-quality seed varieties and irrigation and their continuous improvement, are critically important for transforming paddy production, as evident in the contrasts between Nigeria and other developing countries. Price policies are likely to be more effective in stimulating an aggregate supply response if there is greater use of improved technologies and better functioning of markets and institutions (Kherallah et al. 2002). In Asia and Latin America, governments have played a big role in investing in such technologies. Nigeria will also need to substantially increase these types of investments if it is going to successfully transform its paddy production sector.

Chapter 5

POSTHARVEST PROCESSING, MARKETING, AND COMPETITIVENESS OF DOMESTIC RICE

Michael Johnson and Akeem Ajibola

The production of domestic rice does not end with the harvesting of rice on paddy fields. The paddy rice has to undergo various processing and cleaning tasks before reaching final consumers in retail markets. At harvest, the seed grains are first loosened and separated from the unwanted parts of the harvested cereal grain (or the chaff), soaked in water, and boiled before being milled to remove the outer layer (or the husk) of the grain seed. Typically, further milling occurs to convert the grain at this stage into white rice by removing the outer bran, which defines it as brown rice. All these critical steps constitute the process of adding value and thus the competitiveness of domestic rice production with rice imports.

The purpose of this chapter is to review the postharvest processing and marketing activities of domestic rice production in Nigeria in order to evaluate domestic production's ability to compete with imports. This is accomplished by assessing the structure, conduct, and performance of the domestic rice sector. In doing so, key challenges and opportunities facing the sector are highlighted and research questions raised to address them, such as these: Is there potential to improve the quality and competitiveness of the final local rice product in domestic markets? Are there differential abilities and efficiencies among existing mill types (small to large) that affect their capacity to supply the domestic market? Are there lessons that can be drawn from elsewhere in West Africa and Asia? The chapter has been organized with this logical sequence of evaluation in mind.

The chapter begins by examining the structure and conduct of postharvest processing and marketing channels of the rice sector in Nigeria. This is followed by an evaluation of the relative performance of the different channels identified, including a comparison of the economic efficiencies of small- versus large-scale mills. The evaluation illustrates why the large-scale milling channel has the greatest likelihood of competing with imports due to its economies of scale and the quality of rice it can produce. However, it is usually unable to do so because of the inadequate and infrequent supply of high-quality local paddy. The small-to-medium-scale milling channels, on the other hand, still have room for improvement and therefore offer the potential to supply the growing demand for better-quality local rice in Nigeria.

A discussion of the challenges and constraints that emerge from this analysis follows, including the type of improvements that would be needed. Lessons are drawn from elsewhere in the region and Asia for overcoming some of these constraints. The final section concludes by summarizing the key findings and the extent to which they help answer the principal questions raised in this chapter.

Data used in the analysis are sourced from a survey of the literature, field observations, and semistructured author interviews with farmers, millers, and traders in four states: Niger, Kano, Benue, and Kwara (see map in Figure 1.1 in Chapter 1 for the locations of these states). The visits occurred in August 2012 and April 2013 and involved interviews with a group of eight to ten farmers in two states (Niger and Kwara), five large mills in Kano and Benue, four medium-scale mills (Kano, Benue, and Niger), three smaller mills (Kano, Niger, and Benue), and four retail markets (Kano, Niger, and Benue). The small mills and two of the medium-scale mills were located in milling clusters in Niger, Kano, and Benue. The information collected at each location included prices, production, input and output data, technology use and capacities, production costs (material, labor, and other inputs), and overall constraints and issues facing the dominant types of postharvest processing and marketing channels.

Based on the review and findings here, a spatial equilibrium model is adopted in the next chapter to further explore the potential for modernizing the rice milling sector, especially with regard to the appropriate mix and scale of modern technologies in order to compete effectively with imports.

Structure and Conduct of Rice Value Chain

To analyze the value chain of the Nigerian rice economy, a subsector analysis or structure-conduct-performance approach is adopted from the industrial organization literature (Holtzman 2002). Postharvest processing and marketing contributes to a large share of the rice value chain. This is because once the paddy rice has been harvested, it has to undergo various processing tasks to be converted into milled white rice before reaching consumer markets. At harvest, paddy or rough rice contains an outer layer (or husk) and a bran covering the grain seed. The removal of these outer layers constitutes the milling process. When only the husk is removed the result is brown rice, and when further milling occurs to remove the bran, this converts it into white rice. A process of parboiling the paddy rice may also occur before the milling process, as is common in Nigeria. The entire parboiling process involves washing, boiling, soaking, and steaming the paddy before drying it, which can take up to three days (Lançon et al. 2003a).

In Nigeria, the parboiling task is necessary because most consumers prefer parboiled rice for its taste and texture in preparing local rice dishes (Bamidele, Abayomi, and Esther 2010). It is also for this reason that most Nigerian consumers prefer the Thai parboiled rice among rice imports (Ogunbiyi 2011). Parboiled rice also offers other advantages: a higher recovery rate during milling and higher preservation of its nutritious properties after milling (Tinsley 2011). Milling is typically done using a mechanized de-huller (or de-husker). While traditional hand-pounding methods are still practiced, these are becoming less common and are mostly for home consumption (Akpokodje, Lançon, and Erenstein 2001).

After milling, a cleaning stage involves separating small stones from the rice, either by hand or with the use of a mechanized de-stoner. Further processing may also be carried out, especially by larger modern mills, such as polishing (buffing the white rice with glucose or talcum powder) and sifting and grading to separate any remaining small impurities and broken grains from the head rice. For the rice destined for market, a final process involves weighing and bagging the milled rice for the wholesale or retail market.

In Nigeria, the various processes described above involve many actors with varying degrees of skill and access to technologies, especially as the bulk of postharvest handling and processing is still very much a cottage industry made up of small-scale operators. Figure 5.1 describes the various channels along the entire value chain by which rice is produced, milled, and marketed in Nigeria.

The small- and medium-scale mills handle up to 80 percent of the total local rice processed.[1] Most of the operators along this channel mill paddy rice from smallholder farmers and traders for a fee, either for their own consumption or for rural markets. Based on the most recent data available, the share of rice milled for own-consumption is about 14 percent, while the rest enters the market, 63 percent going into rural and 22 percent into urban markets (LSMS–ISA 2011).

The economic activities along this chain involve many smallholder paddy farmers, small-to-medium-scale millers, middle traders, and wholesalers and

1 Borrowing from the study of Lançon et al. (2003b), we distinguish small-to-medium scale as having a milling capacity of less than 500 kg per hour. Lançon et al. define small scale as 50 to 149 kg per hour and medium scale from 150 to 500 kg per hour.

FIGURE 5.1 Domestic rice value chain in Nigeria

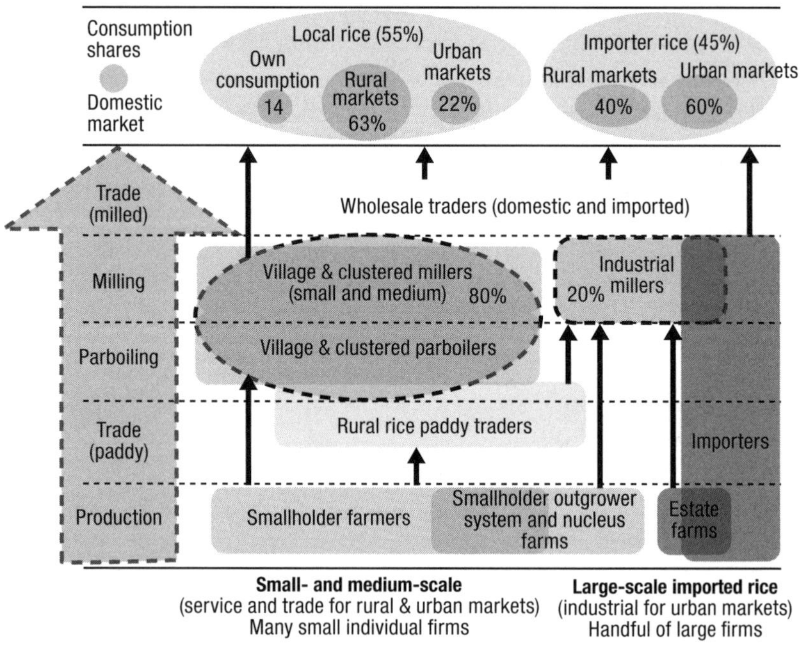

Source: Authors. Data on consumption shares are from authors' calculations based on Nigeria, NBS and World Bank (2011). The schema has been adapted from Figure 3 in DAI (2009).

retailers. It is a highly disaggregated and fragmented channel, with rice changing hands several times at points of sale from farmgate to end market. As a result, there is naturally a wide variation in the quality of the final product, which can reach consumers with unfavorable properties, such as discoloration and the presence of foreign matter (especially stones).

Most small- and medium-scale mill operators are either located in a village or small town market or within a cluster near paddy production areas. Consequently, areas with higher volumes of production usually have larger clusters of mills with higher operational capacities (medium-scale mills) and are more organized in procuring, milling, and selling of the final product, whether through the market or to middle traders (Lançon et al. 2003b). Some notable examples are the clusters found in Lafia, Otukpo, and Abakaliki, with 100 or more millers. Capacity utilization is usually highest during the months after harvest, between October and December each year. At other times of the year, paddy is procured from further afield.

Middlemen traders play a critical role in bringing paddy from farmers to rice millers and milled rice to wholesalers and retail markets. Only a few trade in both paddy and milled rice, using the services of small- and medium-scale millers for a fee to mill the paddy they purchase from farmers before resale. Others are millers themselves, usually medium-scale operators or smaller operators in the large clusters, who buy paddy from farmers or paddy traders to process and sell on the market. Though few in number, these two types of traders serve an important intermediate role in the collection and bulking of paddy. The dual role they play enables them to be selective about the type of paddy they buy, mill, and resell to the market, and in the process to help ensure homogeneity in the final product (Lançon et al. 2003a). The majority of middlemen traders serve either as paddy or milled rice traders and invest very little to improve the quality of the product itself (DAI 2009; Lançon et al. 2003a).

Despite differences in milling capacities between small- and medium-scale operators, the two compete and overlap. The main difference between the two is that medium-scale operators typically handle larger volumes, and among these are some who serve the dual role of miller and trader, wherein they buy their own paddy to mill and sell to traders downstream rather than simply milling for a fee. Although not common everywhere, such miller-traders may provide farmers with credit for inputs and even store their paddy. However, when they do store paddy, at least according to the findings of Lançon et al. (2003b), this is mostly to help smooth their milling activities during the peak and off seasons rather than to speculate against future expected prices. In their capacity as miller-traders, they do have the advantage of being more selective in the paddy variety they wish to mill, even handling the parboiling task itself before milling and in the process ensuring a better-quality product.

For the bulk of domestic rice, milling is done by smaller millers who provide the service for a fee. Many of these have varied skills and degrees of access to technologies, services, and information along the entire supply chain, and they do not engage in upgrading to better paddy varieties and processing technologies. Little vertical market integration is present in this sector, as evident from the surveys of Lançon et al. (2003b) and the authors' own field visits in Niger and Benue States, implying a weak link downstream between millers and retail markets. This is because the millers rarely deal directly with retail markets in their capacity of providing milling services and in selling to traders on-site. Consequently, there are significant inconsistencies between rice variety names and the final milled product that make it virtually impossible to link production directly with consumer preferences (Lançon et al. 2003b). It is therefore not surprising that the quality of domestic rice in market outlets varies substantially or that the product is considered inferior to imported rice, which can be relied on for consistent quality, taste, and texture.

The branding of a higher-quality grade of domestic rice varieties is limited in Nigeria. It occurs only where there is a well-established vertical link in the value chain, from a unique paddy variety to the final processed product in the marketplace, a condition that is typical among larger industrial-type milling sectors (for example, the export industries in Asia). However, this sector in Nigeria is limited to a handful of firms that mostly produce to meet demand in high-end urban markets while simultaneously importing rice for the same purpose. This is, after all, a premium product, like imported rice, that is not easily substitutable with the more standard and lower-quality rice produced in higher volumes in Nigeria. Referring to Figure 5.1, this segment of the Nigerian rice market is represented by the two channels involving industrial millers and importers.

There are only a few company brands that exist in Nigeria among the large and industrial-scale millers. These include Olam, Veetee, Stallion, Dana Foods, Isyaku Rabiu Group, and Ebony Agro (Lodestar International 2010). For these firms, the dual role of serving as an importer and a larger miller helps guard against uncertainties related to global price volatility and domestic policies that restrict rice imports (Demont and Rizzotto 2012). As a result, the sector has not always been consistent in its milling activities, producing very little or nothing at all in some years whenever it was less profitable to handle domestically produced rice relative to imports.[2] At the end of 2008, for example, one source identified only two large industrial mills in operation that year: Olam and Veetee (DAI 2009).

Although large-scale operators can potentially handle a larger share of milled rice in the country, they are usually forced to operate well below capacity due to inadequate supplies of paddy rice from year to year; this explains their limited scope to date. In 2002, for example, a survey by Lançon et al. (2003b) estimated that only 32 percent of total milling capacity of the larger millers was being utilized, compared with more than half for small- and medium-scale operators. Aside from the risks of underutilizing existing capacity, the high up-front capital investment required to set up a large industrial mill makes it prohibitive for most investors.

2 Quite a few of the large operators import parboiled brown rice directly from Asia to then mill into rice locally. This can ensure a steady supply for the millers, subject to global price trends and any import barriers.

Rice Value-Chain Performance: Cross-Country Comparisons

To measure the performance of the rice value chain after harvest, information was gathered to generate estimates of the cost and price structure associated with the different activities along the chain in order to determine margins and price competitiveness with imports. Combining data from the literature and field trips to a few of the large rice-producing states in Nigeria (Niger, Kano, and Benue), we compared these costs and prices with those reported for two Asian countries: Thailand and Bangladesh.[3]

The estimates for all three countries are mostly representative of costs and profit margins associated with smallholder paddy producers, small-to-medium-scale processors (including millers), and traders. These add up to their respective domestic market prices for paddy and milled rice. The trader and retailer's shares of the final price represent both operating costs and profit margins and therefore the differential between a buying and selling price. Paddy traders, for example, will buy from farmers and sell to millers, while milled rice traders buy from mills and sell to retail markets. Although they incur their own operating costs, this is not reported here as such information was not available.

Thailand offers a useful benchmark because the bulk of Nigeria's imports come from this country. Additionally, Thailand is considered to have one of the more efficient rice value chains in the developing world, even though average milling costs could be even lower if existing large rice mills could operate at full capacity every year (Titapiwatanakun 2012). By comparing Thailand's costs and marketing margins, we can point out parts of the value chain in Nigeria that are inefficient and need improvements. However, because we expect the rice export industry in Thailand to have a greater comparative advantage as a major supplier of rice exports in global markets, another benchmark is needed. Bangladesh serves as such a benchmark, providing an example of a net importer (like Nigeria) among major rice-producing countries in Asia.

Figures 5.2 and 5.3 present initial comparisons of the price structure of the domestic rice value chains in Nigeria and the exporter value chain in Thailand. Figure 5.2 exhibits costs and profit margins associated with the small miller rice value chain in Nigeria, which produces a standard or

3 The Nigerian data are mostly based on cost estimates in one of the major rice-producing states, Niger, Kano, and Benue. Some adjustments were made based on secondary data sources, cited under each graph or table. Data for Thailand and Bangladesh are secondary, with the source cited. These are meant only to be indicative of average performance.

FIGURE 5.2 Comparison of rice value chains between Nigeria's small milling sector and Thailand's rice export sector

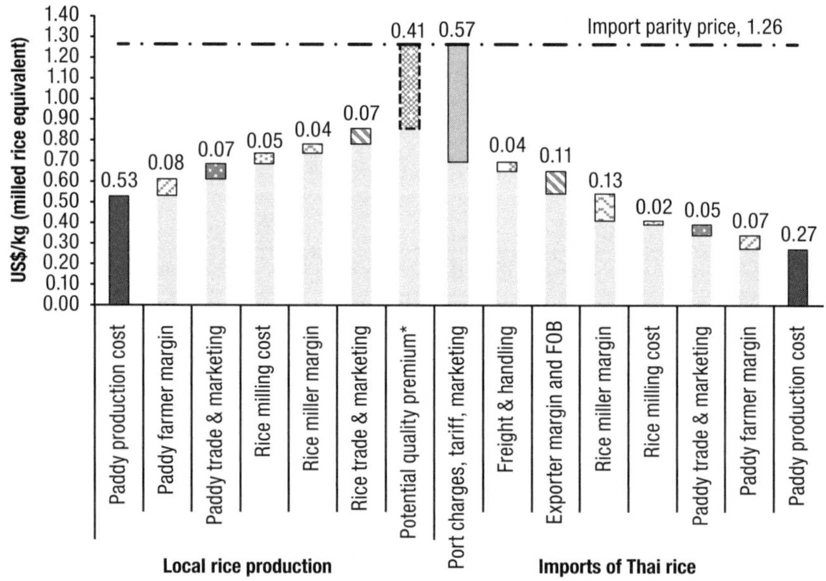

Source: For Nigeria, based on authors' own field visits in Niger State and compared with estimates by Chemonics (2009); for Thailand, based on Maneechansook (2011) and FOB data from the Thai Rice Exporters Association for the rice type known as Thai parboiled rice 100% (www.thairiceexporters.or.th).

Note: FOB = free on board. Dollar values are in constant real 2012 US dollars. Numbers may not add up exactly to 1.26 because of rounding off to two decimals. *This defines an amount consumers may be willing to pay or producers may be willing to absorb in order to bring the quality of standard local rice closer to that of premium-quality rice (from local large mills or imports).

inferior rice type. Figure 5.3 shows related values associated with the large miller value chain, which produces premium-quality rice that is more substitutable with imports. Both domestic value chains are compared with the same average rice exporter value chain in Thailand, which produces premium-quality rice. All values are measured in terms of US$ per kilogram of milled rice equivalent and in constant real 2012 US dollars to allow for a more accurate comparison.[4] The left side of each figure represents the Nigerian rice sector, while the right side represents the Thai rice sector. For

4 Rice equivalent weight accounts for the loss in paddy weight during processing (e.g., from removal of the chaff). On average, this is about 60 percent of paddy weight, although it varies with postharvest and milling technologies (see conversion factors for each miller type in Table 5.2 later in this chapter). Therefore, milled rice equivalent values for activities prior to the final milled rice output will be higher than their paddy equivalent by a factor of (1/0.6). All values in Figures 5.2, 5.3, and 5.4 were converted to constant real 2012 US dollars and in milled rice weight equivalent to allow for a more accurate comparison.

FIGURE 5.3 Comparison of rice value chains between Nigeria's large milling sector and Thailand's rice export sector

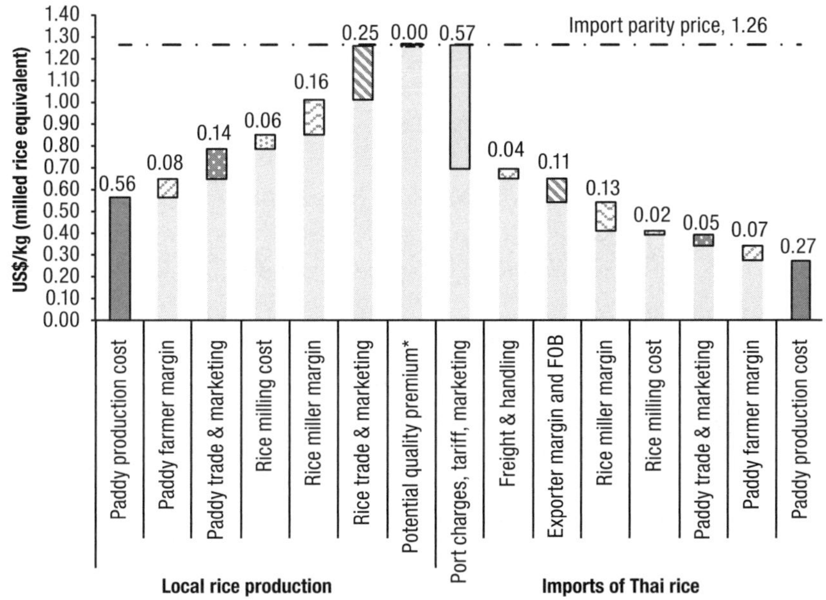

Source: For Nigeria, based on authors' own field visits in Niger, Kano, and Benue States and compared with estimates by Chemonics (2009); for Thailand, based on Maneechansook (2011) and FOB data from the Thai Rice Exporters Association for the rice type known as Thai parboiled rice 100% (www.thairiceexporters.or.th).

Note: FOB = free on board. Dollar values are in constant real 2012 US dollars. Numbers may not add up exactly to 1.26 because of rounding off to two decimals. *This defines an amount consumers may be willing to pay or producers may be willing to absorb in order to bring the quality of standard local rice closer to that of premium-quality rice (from local large mills or imports).

the latter, only the export sector is considered, as we are interested in the rice destined for Nigerian markets.

In both figures, paddy is procured from the same source—smallholder paddy farmers—but with different paddy production costs between the two figures due to the type of paddy variety sold. Improved superior (or premium) paddy is typically sold to the medium- and large-scale milling sectors; in the case of the large milling sector, as Figure 5.3 shows, this paddy fetches a price premium at farmgate of about 5 percent over more common (or standard) varieties milled by the small sector (0.64 versus 0.61 naira per kg, respectively). "Paddy production cost" refers to costs associated with growing paddy rice on the farm. The difference between the price of paddy at the farmgate and production cost is referred to as the "paddy farmer margin," while the "paddy trade & marketing" margin is the difference between a wholesale market price for paddy and the farmgate price and therefore includes traders' unknown marketing costs and profit margins.

"Rice milling costs" are associated with the costs of milling paddy procured from paddy traders. We do not include annualized fixed costs that large mills have to absorb to remain operational throughout the year, as we did not have similar estimates to benchmark against in Thailand's milling industry. However, these can be quite high in Nigeria for the large modern mills, especially given the scarcity and higher cost of capital relative to Thailand. We estimate these to be more than double the direct milling costs of the mill at maximum capacity, based on the sample of large mills visited in Kano and Benue States ($0.10 per kg in fixed costs versus $0.06 per kg for milling). We will come back to this later when comparing the relative efficiencies between small- and large-scale miller types.

For each country, the "rice miller margin" is calculated as the difference between the price received for the milled rice by the miller and the price of paddy combined with the cost of milling operations. It is therefore a miller's profit margin. The difference between the miller gate price and domestic retail price in Nigeria is referred to as the "rice trade & marketing margin." Here again, the margin represents the sum of the trader's unknown marketing costs and profit margins.

For Thailand, because we are only interested in the rice that enters Nigeria's retail markets, additional components associated with moving the rice from Thailand to final retail markets in Nigeria are introduced. First, an "exporter margin and FOB" measures the difference between the free on board (FOB) price for rice leaving the Bangkok port and the miller gate price in the country. Second, the "freight & handling" refers to the shipping costs for moving a ton of rice from Bangkok to the port of Lagos, Nigeria. Adding this to the FOB price results in the cost and freight (C&F) price recorded at the Lagos port. The freight and handling costs from Bangkok to Lagos was computed to be $44.30 per ton.[5]

The "port charges, tariff, marketing" component of Thai rice determines the final cost of the imported rice in Nigerian markets, which includes any tariff charged at the port of Lagos, plus transportation and retail marketing costs of bringing it to market. In Figures 5.2 and 5.3, given a C&F price of US$0.69 per kg for Thai parboiled rice 100 percent at the port of Lagos, plus port handling charges of $20 per ton, an exchange rate of 158 naira per US dollar, an import tariff of 50 percent, and domestic marketing costs of 20 percent of the C&F price, the final import parity price of 1 kg of Thai rice

5 From Anderson (2009) data files (also used later in Chapter 7).

in Nigerian markets is estimated at US$1.26.[6] All unit costs are in US dollars per milled rice weight equivalents.

Finally, a last term is introduced in measuring the relative price differences between local Nigerian rice and imported Thai rice, referred to in the figures as the "potential quality premium." This can be viewed as the price premium imported Thai rice enjoys over domestic rice due to its higher quality and preference among consumers. Alternatively, it can be similarly viewed as an amount that represents the additional costs that could be absorbed by domestic rice producers in order to improve the quality of local rice and compete with imported Thai rice, hence the term "potential quality premium." For the large milling sector, this is assumed to be zero, as the rice is comparable in quality to imported Thai rice.

To compare the cost of Thai rice in Nigerian markets to local rice, we need to include all costs associated with importing the rice into the country and bringing it to market, which we estimated to be $1.26 per kg (2012 US dollars), based initially on 2009 data from Titapiwatanakun (2012). Because the 2009 price in current 2012 dollars corresponded well with current prices, we compared this with field data collected by the authors in Nigeria in 2012.

Beginning with Figure 5.2 for the small milling channel, the data reveal that average paddy production cost per unit of output in Nigeria is almost double that in Thailand ($0.53 per kg versus $0.27 per kg in their milled rice weight equivalent). Nevertheless, farmers seem to be earning the same in absolute terms. Paddy trader costs are slightly higher in Nigeria. Overall, the final price for domestic rice at the mill, that is, before adding the "rice trade & marketing" component for Nigerian rice and the "exporter margin and FOB" for Thai rice, is well above the comparable price in Thailand ($0.77 per kg versus $0.54 per kg, respectively). Clearly, even after milling, the Nigerian rice industry remains less competitive. Direct milling costs in Thailand are very small—about a third of what Nigeria's small millers experience. This is not only a sign of their efficiencies and volume of operations but their large-scale nature.

Under zero tariffs, evidently, rice produced in Nigeria cannot compete with imports on price, let alone quality, because after freight and handling at the port in Lagos, the price of imported Thai rice (which is considered to be of higher quality) is still lower than the domestic rice at the small miller's gate (after the rice miller margin). Only if we include an import tariff of about

6 The C&F price is based on an FOB price at the port of Bangkok of $0.65 per kg (or $649 per ton) for Thai parboiled rice 100 percent in 2009, the year for which sufficient information on the Thai rice value chain was readily available (see Titapiwatanakun 2012), plus freight, insurance, and handling of $0.04.

50 percent and add domestic transportation and marketing costs (say about 30 percent of C&F price) does the final import parity price at the retail level rise to $1.26 per kg (199 naira per kg) and above an average retail price for domestic rice of $0.85 (135 naira per kg, respectively).[7]

For consumers, such a price differential makes sense, considering the perceived inferior quality of the bulk of domestic rice coming from the small sector in Figure 5.2 (that is, a 33 percent premium for quality alone).[8] Similar conclusions were drawn in the earlier study by Lançon et al. (2003b), who estimated the gap to be about 25 percent. The differential of $0.41 in Figure 5.2 can be viewed as an amount that the smaller domestic producers (small- and medium-scale millers) could potentially absorb as additional costs in processing higher-quality rice, therefore increasing their chances of competing with rice imports (hence the term "potential quality premium" for this element in the figure).

Turning to Figure 5.3, we compare the performance of the large milling sector in Nigeria with Thailand's milling export sector, as it provides a fairer assessment given their similarities in technology use, scale, and demand for quality seed varieties for milling. For our purposes, therefore, we can assume the milled rice by either country to be of similar quality, and thus that it can compete well when selling at the same price.

As Figure 5.3 illustrates, the large sector in Nigeria experiences much higher per-unit operating costs all along the value chain for paddy production, paddy trade and marketing, rice milling, and rice trade and marketing. The "paddy trade & marketing" component of the value chain, in particular, is significantly higher than the same component in the small milling sector in Figure 5.2. This is partially because of higher search costs and a price premium for the superior paddy variety sought after by large millers. Profit margins are still reasonable, nevertheless: about 16 percent as a proportion of the price received at the mill gate. In contrast, the smaller millers fetch only 5.6 percent. Oguntade (2011) also reports small profit margins for medium-scale millers (about 7 percent).

Beyond the mill gate, trade and marketing costs for the large millers are again larger than those of the small milling channel, probably because large millers typically target the bigger urban markets scattered throughout the country. Therefore, while the large milling sector is better able to compete

7 Retail and wholesale prices of domestic rice are based on monthly averages in 2011 collected by FEWS NET (the Famine Early Warning Systems Network) in one of the major rice marketing centers in Nigeria, in Bodija (Ibadan). The prices for imported rice in 2012 are easily around 200 naira per kg, so these figures are quite reasonable.

8 An exception is *ofada* rice, which is considered to have high-quality premium characteristics—but has only a small niche market relative to total rice demand.

with imports on quality, they cannot do so on price unless import tariffs are sufficiently high. In this case, the 50 percent levy existing at the time when the data were collected results in a comparable import parity price of $1.26 per kg, which offers positive profit margins for millers and, just as likely, wholesale and retail traders (as illustrated in Figure 5.3 by the rice trade and marketing component valued at $0.25, about 20 percent of the import parity price).

In comparing the price structure only between Nigeria, Thailand, and Bangladesh, we can gain further insight into the share of the price captured by the farmer, by paddy traders and millers, and finally, by wholesale traders and retailers. Figure 5.4 presents this price structure for all three countries and by type of rice. For Nigeria, as before, we distinguish between rice produced by the small and large miller marketing channels, which produce a standard and premium quality product, respectively. For Bangladesh, "standard" refers to the more common coarse rice brand, while "premium" refers to the fine brand targeted to higher-end urban markets and based on the work by Minten, Murshid, and Reardon (2013). For each country, the price has been broken up

FIGURE 5.4 Comparison of the rice price structures in Nigeria, Thailand, and Bangladesh

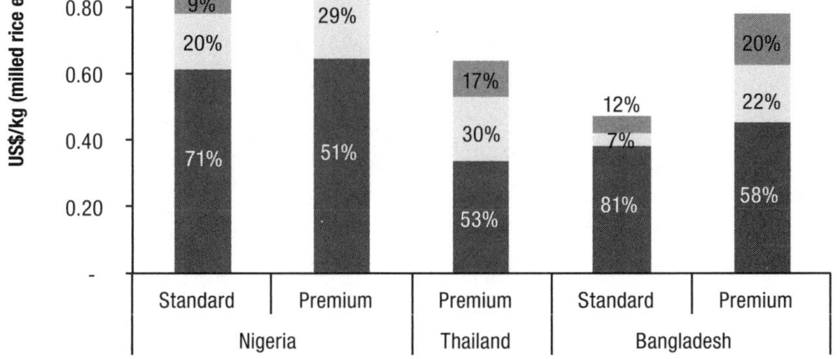

Source: For Nigeria, authors' calculations based on authors' own field visits in Niger, Kano, and Benue States and compared with estimates by others such as Oguntade (2011); for Thailand, based on Maneechansook (2011) and FOB data from the Thai Rice Exporters Association for the rice type known as Thai parboiled rice 100 percent (www.thairiceexporters.or.th); for Bangladesh, based on Minten, Murshid, and Reardon (2013) and referring to fine-quality rice grain.

Note: FOB = free on board. Prices are in 2012 US dollars and have been converted to their milled rice weight equivalents.

into three components: farmgate paddy price, paddy trade and milling, and rice trade and marketing. These capture the combined costs and profit margins among the actors involved under each component, such as farmers, millers, and traders.

As shown in Figure 5.4, the overall price structure in Asia contrasts sharply with Nigeria's. To begin with, the most striking difference is the higher farmgate paddy price and milled rice price in Nigeria compared to the two Asian countries for both standard and premium rice types. At the same time, the farmgate prices' share of the final retail prices indicates that Nigerian farmers seem to capture lower shares of the final price than farmers in Asia: 71 percent versus 81 percent for standard and 51 percent versus 53 and 58 percent for premium.

Paddy trade and milling cost is also significant in Nigeria for both types of rice: for standard rice, this cost is about five times that for standard rice in Bangladesh, and for premium rice it is more than double that for premium rice in Bangladesh and Thailand. As already pointed out earlier, one explanation is the higher trader and marketing costs in Nigeria, especially for large millers who often have to resort to procuring paddy over long distances when facing limited supplies of their preferred superior-quality paddy. For the large millers in particular, the insufficient access to paddy throughout the year implies that they are often operating well below full capacity (Oguntade 2011). If we had considered annualized fixed costs, the paddy trade and milling cost would be even greater.

As a result, the share of the Nigerian final retail price accounted for by paddy trade and milling cost is comparable with the same cost's share in Thailand's final retail price, but it is much lower than paddy trade and milling cost's share of this price in Bangladesh. This is not surprising, as large exporter mills in particular have been shown to capture large shares of Thailand's rice prices (Maneechansook 2011; Agrifood Consulting International 2005).

After milling, the share of rice trade and marketing margin is about the same across all three countries. In Bangladesh, about 20 percent of the final retail price for premium rice goes to retail traders (Minten, Murshid, and Reardon 2013). For Nigeria, this is possible given the presence of the 50 percent tariff. For the standard type of rice, these margins are much smaller in both Nigeria and Bangladesh (about 9 to 12 percent). Evidently, a key area for improving the competitiveness of domestic rice is going to be not only ensuring better product quality but lowering processing, trader, and marketing costs to increase incentives to upgrade.

Despite higher production costs, Nigerian paddy farmers seem to share the same profit margins as their Asian counterparts. Using national averages for

TABLE 5.1 Comparison of production costs, yields, and profit margins

	Nigeria		Thailand	Bangladesh
Measurement	Standard[a]	Premium	Premium	Premium
Input costs				
Production costs ($/MT)	531.0	563.0	254.0	312.0
Milling costs ($/MT)	51.0	160.0	17.0	NA
Paddy yield and returns				
Paddy yield (MT/ha)	1.9	3.0	2.9	4.2
Farmer return ($/ha)	151.0	253.0	191.0	207.0
Profit margins				
Farmer profit margin (%)	13.0	13.0	19.5	13.6
Miller profit margin (%)	5.6	16.0	24.2	NA

Source: Data sources for Nigeria are authors' calculations based on field visits in Niger, Kano, and Benue States and compared with estimates by others such as Oguntade (2011); for Thailand, based on Maneechansook (2011) and FOB data from the Thai Rice Exporters Association for the rice type known as Thai parboiled rice 100 percent (www.thairiceexporters.or.th); for Bangladesh, based on Minten, Murshid, and Reardon (2013) and referring to fine-quality rice grain. For farmer returns, the data are value-added agriculture per hectare from the World Bank (2012); for yields, FAOSTAT (FAO 2014).

Note: FAOSTAT = statistical database of the Food and Agriculture Organization of the United Nations; NA = not available; MT = metric tons. Per-unit values are in 2012 US dollars. Input costs are in milled rice weight equivalents. [a]Here, "Standard" refers both to inferior paddy varieties and lower-quality milled rice, while "Premium" refers both to superior paddy varieties and higher-quality milled rice. Trade and marketing margins were not included because of insufficient information to distinguish between operating costs and profit margins.

yields, production costs, and prices, Table 5.1 shows Nigerian farm-level profits as reasonably close to those of both Thailand and Bangladesh. Thailand's higher margins may actually be partially due to a government support program. In absolute terms, however, Nigerian farmers earn lower margins per hectare for the common varieties due to lower yields, about $151 per year compared with $253 for the local superior (and higher-yield) varieties, and with $191 and $207 for Thailand and Bangladesh, respectively (all in 2012 US dollars). With regard to production costs, both Bangladesh and Thailand enjoy much lower costs per ton for both paddy production and milling relative to Nigeria, a sign of the higher yields and efficiencies in the technology inputs being used.

The poor performance of the entire Nigerian rice value chain relative to that of Asian countries leaves a lot of room for improvement. Per-unit costs for paddy production, milling, trading, and marketing are all on average higher than those in Asia—Table 5.1 illustrates these higher costs for paddy production and milling (trading and marketing was not included because of insufficient information to distinguish between operating costs and profit). The share of paddy trader/miller activities and profit margins in the final price is especially high for the large mills (the premium rice type in Figure 5.4): three times those of smaller mills that produce the more inferior standard rice variety. As a start, simply raising the productivity of paddy production by promoting the adoption of superior varieties, as argued for in Chapter 4, can have a significant effect.

Efficiency of Small versus Large Millers

In assessing the performance of the postharvest rice value chain in Nigeria, a key question that arises is the relative economic efficiency of the small versus large milling channels that were illustrated earlier in Figure 5.1. To measure the economic efficiency of both channels, we draw on the arguments used by Timmer (1998). In his seminal study for Indonesia, he showed that small mills were more efficient than larger milling operations because of the combination of lower paddy-to-milled conversion ratios, higher capital-to-labor price ratios, higher economic discount rates, and higher price ratios of the final milled rice to paddy rice (or milled-to-paddy price ratio) that are typical in a developing-country setting.

The paddy-to-milled conversion ratio defines how much weight of milled rice is converted from a single unit weight of paddy rice (milled rice weight is always lower than paddy rice weight, because the husk and bran of the seed are removed during milling). The capital-to-labor price ratio helps determine the degree to which a mill can minimize costs—and thus compete more efficiently—by drawing on additional capital or labor, depending on which input is cheaper (for cost minimization, the marginal rate of technical substitution between capital and labor equals this price ratio).

An economic discount rate here represents the cost of a capital investment over time. An economic discount rate is the interest rate used either (1) to discount any future cash returns from a current investment or (2) to calculate the present value of any future investment. The higher the rate, the lower the present value of any future cash flows or the higher the future value of a current investment. The price ratio of final milled rice to that of paddy at the farmgate measures the price gap between the two that captures how much cost can be absorbed in processing and moving the rice to market. A small gap (hence a smaller price ratio) implies that milling costs have to be small enough for millers to remain profitable.

The paddy-to-milled conversion ratio measures the technical efficiency of the mill and can affect the ability of a mill to compete; for example, if the milled-to-paddy price ratio is small due to cheaper imports. This situation can be especially challenging for large-scale millers, who typically have added costs associated with procuring quality paddy and further processing unless they are able to capture a price premium on the quality of their brand.

Because larger mills are capital-intensive enterprises that rely more on capital inputs than labor, the capital-to-labor price ratio and discount rate are just as important in determining their economic efficiency and competitiveness. Naturally, they are more competitive when they operate under conditions of lower per-unit capital costs relative to wages. A key factor that can lower this ratio is the rate at which capacity is being fully utilized, given the scale economies of this heavily capitalized industry. As pointed out earlier, the underutilization of large mills is a common problem in Nigeria and other developing countries where the supply of sufficient quantities of paddy is not always guaranteed.

Another key factor is the economic discount rate, which simply measures the rate at which future costs and benefits are discounted relative to the present. Since large mills have higher up-front investment costs and larger benefits accruing in the future than smaller mills, a high discount rate makes the large mills less attractive than their smaller counterparts, because the present value of all future benefits is far outweighed by the total costs, translating into a negative return on investment. Under the same conditions, on the other hand, the smaller mills are more attractive as they are less capital intensive and incur only higher wage costs in the future.

Finally, higher milled-to-paddy price ratios typically imply that, relative to the price of paddy, there are higher milling and marketing costs (and/or profit margins) being captured in the final price of the milled rice entering retail markets. Conversely, a lower ratio implies higher efficiencies of these postharvest activities.

Table 5.2 provides estimated values of each of these four indicators across Nigeria, Thailand, and Bangladesh. Derived mostly from anecdotal evidence, it is meant to be indicative and form the basis for discussion. The comparison with Thailand and Bangladesh allows for consistency with the previous assessment of value-chain performance.

As seen in Table 5.1, Nigeria is evidently less efficient in milling technologies and practices, considering its lower paddy-to-milled conversion ratio and higher milled-to-paddy ratios. A large wedge between paddy and milled rice prices is an indication of higher operating costs and margins for processing and marketing, which was shown to be the case for Nigeria in Figure 5.4. It can also have the effect of introducing greater profit risks for millers and middle traders, as larger price swings are possible within this price band. This is

TABLE 5.2 Key indicators of efficiency and profitability of small- versus large-scale rice milling

Indicator	Nigeria	Thailand	Bangladesh
Paddy-to-milled conversion[a]	0.58	0.66	0.63
Milled-to-paddy price ratio[b]	1.4/2.1	1.9	1.2/1.7
Capital-to-labor price ratio[c]	2.7	1.6	2.1
Economic discount rate[d]	10.1	6.7	13.1

Source: Authors' calculations based on various sources of data.

Note: [a] For Thailand, data source is Titapiwatanakun (2012); for Nigeria, Lançon et al. (2003b); and for Bangladesh, Minten, Murshid, and Reardon (2013). [b] The two numbers represent a distinction between the quality levels of paddy and milled rice varieties: common or traditional paddy rice versus standard milled rice for the first number and superior paddy rice versus premium milled rice for the second number. Data sources for Nigeria are authors' calculations based on the author's own field visits in Niger, Kano, and Benue States and compared with estimates by others such as Oguntade (2011); for Thailand, based on Maneechansook (2011) and FOB data from the Thai Rice Exporters Association for the rice type known as Thai parboiled rice 100 percent (www.thairiceexporters.or.th); for Bangladesh, based on Minten, Murshid, and Reardon (2013) and referring to fine-quality rice grain. [c] This is the ratio of total value of agricultural capital stock per agricultural population (US dollars/person) over the agriculture gross domestic product per agricultural population (US dollars/person). Data are from World Bank (2012). [d] This is an annual average interest rate between 2010 and 2012 from the International Monetary Fund statistical database (IMF 2012). Because some banking discount rates were higher than lending rates, we chose to use the former in Nigeria. For Thailand and Bangladesh, in order to be as conservative as possible, we chose the lending rate. The four indicators here are those highlighted by Timmer (1998). All values use constant 2005 US dollars and are averaged across three years, 2005 to 2007.

possibly one reason why many small millers in Nigeria operate only on a fee basis. Even among those who buy paddy to resell as milled rice, according to Lançon et al. (2003b), the probability of making a profit is only 40 percent.

A comparison of the two price ratios and average discount rate for each country in Table 5.2 reveals that Nigeria and Bangladesh face higher investment risks from introducing capital-intensive modern rice mills (MRMs) relative to Thailand. Nigeria shows a higher milled-to-paddy price ratio for the higher-quality premium rice, indicating the higher share of milling, trade, and marketing costs captured in the final price of milled rice. This seems to be true also for the lower-quality standard rice when compared with Bangladesh (1.4 versus 1.2, respectively). The capital-to-labor price ratio is also higher in Nigeria, indicating higher per-unit capital costs relative to labor. Additionally, for Nigeria, the fact that large mills typically have to operate at well below capacity due to an insufficient paddy supply implies much higher per-unit operating costs.

The logistical costs and uncertainty associated with procuring the paddy from year to year presents an even bigger challenge in keeping operating costs down. Other operating costs stem from the heavy reliance on diesel-operated electric generators in the absence of a dependable supply of electricity throughout the year. In contrast, the dominant smaller-scale milling sector faces lower capital investment requirements and operating costs and thus has the potential to realize quick positive returns from simply upgrading post-milling activities for de-stoning, polishing, and packaging. The potential to do so exists, considering the current large price band between the standard and premium varieties of domestic rice.

Key Constraints, Opportunities, and Challenges for Rice Millers

In analyzing and comparing the two distinctive value-chain sectors for domestic rice in Nigeria—the small-to-medium- and the large-scale milling sectors—a number of constraints and opportunities associated with each emerged. To begin with, because the large industrial sector enjoys the advantage of having a higher milling capacity and modern technologies for supplying the premium-grade rice, it has greater potential to compete with imports. However, the sector is usually unable to take full advantage of its greater economies of scale, as it is often forced to operate well below maximum capacity due to insufficient access to quality paddy. Because much of the paddy is produced by smallholder farmers, this is a major challenge.

As the dominant supplier of paddy, the smallholder rice farming sector in Nigeria varies widely with regard to paddy yield and varieties of seeds grown, in addition to the type of production system and inputs used as well as distance to major processing centers and markets. All of this introduces a major logistical challenge in procuring the right quality and quantity of paddy in a timely and well-coordinated fashion.

In the past, large millers have had to resort to traveling great distances or establishing outgrower systems to procure sufficient quantities of quality paddy. However, these tactics did not always guarantee a sufficient supply and in the end proved too costly to sustain business due to both the higher per-unit operating costs of not utilizing the full capacity of the mills and the added search and administrative costs to secure paddy. To circumvent these problems, some millers have chosen to start growing their paddy on large-scale irrigated lands (Box 5.1 presents the Olam story as an example). Alternatively, the government is also proposing to set up a number of smallholder nucleus farms that will cultivate paddy on large irrigated lands in close proximity to the large modern mills (Adesina 2012).

Adding to the higher operating costs for accessing paddy, one challenge faced by large-scale operators is a periodic breakdown in utility services (such as electricity and water) and machinery parts (which have to be imported from the mill manufacturer overseas). A second challenge is the risk associated

BOX 5.1 The story of Olam in Nigeria

As a major importer in Nigeria that holds one of the largest market shares, Olam has in the past invested to process local rice using large modern rice mills. In 2006, the firm entered into a partnership with USAID (the US Agency for International Development) to establish an outgrower system with smallholder rice producers in order to promote local Nigerian rice production. The program allowed farmers to access technologies, credit, and technical assistance to produce superior paddy varieties and have assured markets and prices.

Productivity and incomes reportedly more than doubled in the first years, encouraging a Nigerian commercial bank (First Bank) to become a major stakeholder in a smallholder farmer commercial credit program. Unfortunately, the program was discontinued after 2008, when Olam failed to procure sufficient paddy after its price guarantee offering fell below other offerings of competitors in the area as world prices (and thus price of imported rice) were increasing. At the same time, credit repayment problems among many of the participating rice paddy farmers had also started to emerge.

Evidently, Olam's initial effort at procuring paddy from outgrower systems proved too risky given global price uncertainties and the logistical costs associated with maintaining varietal quality, input delivery, credit repayments, and sufficient volumes of rice each season. By 2012, Olam had started venturing into rice production (its first in Nigeria), putting 49.2 million dollars (7.675 billion Naira) into rice farming and a milling facility in Nasarawa State.

Olam intends to irrigate up to 6,000 ha of rice to supply up to 60,000 tons of paddy annually for the company's processing facility. The paddy is expected to be converted into about 36,000 tons of milled rice annually. Olam expects to focus attention on milling two superior paddy varieties in Nigeria (FARO 44 and FARO 52) and expects to get about 65 percent of paddy needed from its own farm and the rest from outgrower cooperative agreements with smallholder farmers. How it will successfully manage the latter scheme based on its past experience is not certain.

Source: Based on Olam (2011); USAID (2010); Chemonics (2009); and authors' own interviews with staff at Olam Mills, extensionists, and farmers involved in the outgrower system.

with declining domestic rice prices if the government chooses to lower the import tariff. Therefore, while the large milling sector has greater potential to compete with imports on quality, it may not be able to do so on price.

The dominant small-to-medium-scale milling sector, on the other hand, has remained quite vibrant in its ability to procure, process, and market the bulk of the paddy being produced by thousands of smallholder farmers scattered all over Nigeria. Although still very much a cottage industry, the sector employs the most within the Nigerian rice industry. Therefore, any increases in the productivity and product quality of this sector have the potential to result in larger and broader effects on overall rural employment and welfare.

The sector not only employs many processors and traders, including women, but also provides a cheaper rice alternative for millions of poorer consumers in the country, as shown by a number of research studies (see, for example, Akighir, Ngutsav, and Asom 2007; Basorun 2008; Agwu and Ibeabuchi 2011; and Ayoola et al. 2011). This is, after all, a distinctive sector of the domestic rice market relative to the higher-end markets for imported rice, as shown previously in Chapter 2. Nevertheless, for the sector to improve, many of the challenges it faces along its value chain will need to be addressed.

The principal challenge for the small milling sector in Nigeria is the many obstacles it faces for improving productivity and product quality. As still very much a cottage industry, one obvious challenge is the necessity of dealing with many producers, traders, and processors who have variable skill levels, access to technologies, and credit, and who interact only at the point of sale or for servicing. As a fragmented processing and marketing system, it has limited abilities or incentives to upgrade to better technologies, marketing, and branding, especially in the downstream part of the value chain. Relatedly, the existing marketing system has no consistent grades and standards, exhibits poor record keeping, and has poor organizational capacities. Because most of the millers mill paddy for a fee, they simply leave it up to the traders or consumers who seek their service to worry about quality. While they may wish to invest in modern equipment, many of the smaller-scale operators have poor access to credit to pay for the investment.

Simply providing credit to small-scale operators can hardly be a solution, however. Even among the medium- and larger-scale operators, access to credit has not necessarily led to investments in modern equipment. Based on their surveys in Nigeria, Lançon et al. (2003a) found that many of the larger-scale operators were not investing in upgrading their equipment because of uncertainties about paddy supply. Instead, promoting access to shared infrastructure, markets, and technologies, such as for further processing (de-stoning and polishing) within small town or village-level clusters, may be more promising.

Based on field visits by the authors to two such milling clusters in Benue and Kano States, cluster members appear to benefit from both economies of scale and location. The cluster not only enables shared access to infrastructure, technologies, and know-how but offers lower costs for accessing paddy among suppliers who prefer to deal in bulk. This is also the case for accessing buyers among rice traders who are willing to travel long distances if the cluster's reputation for quality, price, and timely delivery is perceived as good. This advantage alone seems to also attract medium-scale operators who set up shop nearby. Aside from the proximity to input and output markets, the clusters also provide members with the opportunity to adopt other technologies for further processing if such a technology already exists in the cluster. This can be of particular value to small-scale operators who are unable to afford owning the technology on their own.

Improved performance of the entire rice value chain of the Nigeria rice milling industry will require distinguishing these unique challenges faced by the small and larger milling sectors. For the larger milling sector, lowering operating and input costs, including access to adequate utility services, will be important. But even more essential is ensuring access to sufficient quantities of quality throughout the year in order to compete more effectively with imports. For the smaller sector, it is not only about improving productivity and quality but about considering the promotion of miller clusters to take advantage of economies of scale and location. Finally, for both sectors, increasing the productivity of paddy production to expand output and lower prices more generally, as well as insisting on a higher quality and standardized rice seed variety, would be beneficial.

Drawing Lessons from Elsewhere

Given the numerous challenges facing the rice economy in Nigeria, successful upgrading and improving of the milling industry will be key and will require significant investments and policy interventions, as the government was attempting to do under the Agricultural Transformation Agenda (ATA) launched in 2011. ATA was the main agricultural policy agenda under the administration of President Goodluck Jonathan and was led by the Federal Ministry of Agriculture and Rural Development. Because this has been tried elsewhere in the region and beyond, it is useful to draw on lessons from other experiences. In much of developing Asia where rice is a food staple, for example, the rice milling industry transformed itself over time in tandem with the rapid growth in paddy production during the Green Revolution. Figure 5.5 illustrates this growth for India, showing that as yields grew over time, the number of mills (whether small or large) also expanded per unit area. It is also interesting to note how the growth in rice production in India eventually led the country to switch from being a net importer to a net exporter—which

FIGURE 5.5 Evolution of modern rice mills, yields, and net exports in India

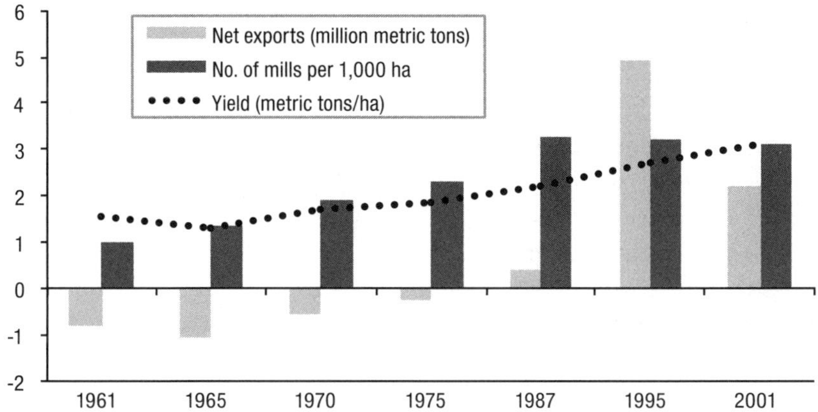

Source: For number of mills, Nayak (1996) and Harriss-White (2005); for yields and net exports, FAOSTAT (FAO 2012).

was only made possible by an expansion of the country's milling capacity for premium-quality rice destined for export markets (for example, basmati rice). Indeed, the country experienced a phenomenal growth in the utilization of large, modern rice mills (MRMs) in the 1980s and 1990s, reaching a total of about 35,000 by 2001 (Harriss-White 2005). However, MRMs never completely displaced traditional practices of parboiling, sun-drying, and using rice hullers, which continued to exist alongside the MRMs. MRMs were instead the preferred choice in both India and Thailand for milling higher-quality rice destined for export markets in both India and Thailand.

Large industrial MRMs easily handle large volumes of paddy for supplying export markets. A disadvantage of their use for domestic markets is that they often face the risk of operating below maximum capacity as they seek to procure sufficient supplies of paddy destined for domestic markets each year from thousands of smallholder farmers. Therefore, public-sector interventions are often needed to ensure that MRMs can continue to operate profitably by assisting with the procurement of paddy, guaranteeing rice purchases directly from mills (for example, for government-managed stocks), or both. In fact, much of the growth in MRMs in India could not have occurred without government support in the form of credit subsidies to promote investments in building large mills and in procuring the paddy (Harriss-White 2005). Other incentives were also introduced, such as imposing an outright ban on the use of existing small-scale mills based on the traditional huller, establishing rice procurement agencies, instituting quality-standardization agencies at the state level, and maintaining a minimum price guarantee for paddy (Shwetha 2009). Today, the Indian government continues to provide both logistical (warehousing and procurement) and price-support mechanisms for technology upgrading. Over time, these practices have introduced high fiscal and economic costs, as paddy farmers and the MRM industry have grown to rely heavily on such support. At the same time, the small and traditional millers have continued to flourish despite efforts to ban them. In fact, they still contribute a significant share of total milled rice in India (Harriss-White 2005).[9]

In contrast to India, the public sector in Thailand has not been involved in subsidizing MRMs directly, many of which are in the export sector and privately owned. Instead, the government has typically intervened by introducing price-support mechanisms for paddy to ensure a steady supply to the milling sector and, ultimately, for exports. The program is popular among politicians because it helps to garner votes among farmers and millers, who represent a large part of the population and benefit the most from the policy (Forssell 2009). Referred to either as the "rice mortgage scheme" or the "pledging program," the policy enables farmers to sell their paddy to the government via millers participating in the program at more favorable prices should market prices fall below a guaranteed price—thus benefiting both farmers and millers (Titapiwatanakun 2012). A key challenge to the policy happens whenever the guaranteed price is set higher than the market price, leading to higher domestic prices and higher holdings of government rice stocks, something that occurred in 2010, when government-owned stocks reached 5 million tons (Thongrattana 2012). Because the policy of accumulating stocks is fiscally costly and can threaten Thailand's ability to compete in international markets over time, the government has since discontinued the program and slowly begun to release the accumulated stock. In its place, a new price-guarantee program has been introduced, but this time it does not involve the direct purchase of paddy from farmers (Thongrattana 2012).

An important lesson from both India and Thailand is the failure to displace completely small- and medium-scale milling operations with large MRMs. This seems to be a general pattern elsewhere in developing Asia. Timmer's (1998) example for Indonesia provides a good explanation for this phenomenon by showing how small mills are both socially and privately profitable compared with larger mills in supplying domestic markets. Not only do they have lower per-unit operating costs, but they also do not displace rural employment, especially female labor. Although Timmer's original study was

9 The ban was finally lifted in 1996 (Harriss-White 2005).

carried out more than three decades ago, his findings continue to be relevant today. Both India and Thailand have witnessed the maintenance of a vibrant small milling sector that supplies domestic markets alongside larger MRMs that have emerged to serve high-end urban and export markets.

The experience in West Africa in general has not been very different from Asia's, except in the context of being exclusively a net importing region. Large MRMs were initially introduced with heavy state intervention as part of the import substitution strategies in the 1970s and 1980s. But these collapsed in due course as governments withdrew their support under a broad sweep of structural adjustment and market reform efforts in the 1990s. The experience in Mali is a particularly good example. After the government sold off several large mills in Office du Niger in Mali to private operators and lifted a ban on small-scale milling, the large operators suddenly found themselves unable to compete effectively with the small milling sector in procuring paddy. The latter was well positioned to operate more efficiently and flexibly with smallholder suppliers. At the time, according to Diarra et al. (1999), per-unit costs for large millers turned out to be about four times those of smaller mills (17.6 versus 4.3 West African CFA francs) because the large mills typically operated at well below capacity. This is not too different from the experience in Nigeria.

The conditions Timmer (1998) analyzed for Indonesia in the 1970s remain partially relevant in Nigeria today, as they have been in Asia in the past. However, conditions have changed as urban populations and incomes continue to grow in both Asia and Africa. Research shows demand shifting toward higher-quality and premium milled rice in Asia's emerging sophisticated urban markets (Minten, Murshid, and Reardon 2011 and 2013; Murshid 2011; and Chapter 2 in this book). In response, the postharvest rice sector has been transforming itself by adopting larger MRMs. While past adoption of higher-yield paddy varieties has certainly helped to meet the growing demand, the increased adoption of MRMs is proving more critical in meeting the demand of these new domestic markets. In some locations in Bangladesh, for example, MRMs now generate more than 90 percent of the rice marketed from the local area, according to observations by Murshid (2011). Their presence is increasing not only the milling capacity but also the quality of rice produced in the area. This shift has also led to more sophisticated institutional and marketing arrangements that involve middlemen brokers who link paddy traders with millers and the latter with rice traders and wholesalers. These actors capture a larger share of the price premium for the higher-quality rice, while farmers benefit the least (Minten, Murshid, and Reardon 2013). Evidently the added quality from improvements in processing has provided a positive return for investors as demand for premium-quality rice continues to expand in developing Asian countries. Overall, these changes are having a profound effect on local rice economies. Greater employment opportunities are appearing on the scene—such as subagents for paddy producers and other new trader entrants, medium-scale parboiling and milling firms, mini-drying yards, and new post-milling operations (for example, for puffing)—and the changes are also leading to the commodification of rice with a system of quality grades and standards (Harriss-White 2005).

For Nigeria, similar trends are poised to occur with rapid urban population growth—as is already evident in the dramatic increase of premium rice imports during the last few decades. Unlike in the Asian countries, however, the local rice sector in Nigeria has been ill prepared to respond and adapt to these changing demands. Meeting the demand for higher-quality premium rice with domestic rice, therefore, is feasible in the short run only through the promotion and use of the larger MRMs, as the government is already doing. However, the heavy government support that the larger MRMs require to achieve this goal will potentially introduce economic distortions that will undermine the health of the rice sector as a whole. As pointed out earlier, it is unlikely that the large MRMs can be economically efficient in supplying the domestic market so long as they continue to face limited access to quality paddy. Their higher up-front capital investments and overheads will lead to higher per-unit operating costs if they are forced to operate below capacity, as experiences elsewhere in West Africa and in Asia have illustrated. In fact, it is more likely that imports will not be totally displaced, while the small milling sector continues to fill the void in supplying the domestic market with a more inferior product.

Therefore, while popular attention has been on the establishment of large industrial mills, serious attention should be given to revitalizing the dominant small and medium milling sector, especially from a public policy perspective. The small sector has the potential to improve, given the current cost and price structure it faces along its entire value chain—with the potential to absorb increased costs (such as for de-stoning and colorization) to further process better-quality milled rice. The potential is even greater if there is a significant improvement in paddy production costs and yields, varieties (such as long grain), access to improved milling technologies, and lowered trader and marketing cost margins. Some of these can be addressed through the promotion of miller clusters to take advantage of agglomeration effects from economies of scale and location. Field observations showed very poor conditions of basic infrastructure such as access to electric power, water, and good feeder-road networks. Addressing these basic infrastructural problems, along with improving access to technologies, credit, and storage, will also be critical to upgrading the sector and providing options to smooth out operations outside the typical harvest season. Finally, although it is unlikely that the small milling sector can produce a premium-quality brand comparable to that of the larger and more capital-intensive milling sector, it has the potential to produce rice of sufficient quality and at a low enough cost to be able to meet the demand and price preferences of the average consumer.

Conclusion

Several key messages arise from this chapter. First, while the production of higher-quality premium rice is feasible in the short run and only through the promotion and use of the larger MRMs as the government is already doing, this strategy will likely require continued government support via an import tariff so long as Nigerian rice cannot compete with imports on price alone. It will also require an expansion in the production of superior-quality paddy if MRMs are going to access sufficient quantities to mill and the Nigerian rice sector is going to improve its productivity and competitiveness with imports over time. Otherwise, there is always the danger that in years to come the sector will still have to rely on the government to remain competitive, as experiences elsewhere in West Africa and in Asia have shown. If this dependence has to be maintained over a long period, such interventions can quickly become a fiscal burden for the government, in addition to the economic costs associated with higher consumer prices induced by import tariffs. As shown in Chapter 2, higher consumer prices will only hurt the welfare of low-income groups, as a large component of a poor household's rice expenditures is for subsistence. This could change, of course, but only if there are significant reductions in operating costs and the logistical challenge of guaranteeing a constant supply of quality paddy is removed.

Second, investments to expand the large milling sector should not come at the cost of ignoring the smaller milling sector. Serious attention should be given to revitalizing this sector as well, especially considering it has the potential to absorb 25 to 33 percent in increased costs (such as for de-stoning and polishing). Placing more emphasis on improving the productivity of small- and medium-scale milling channels, such as from the greater use of labor-saving and quality-improving technologies, will help meet a broader demand base for cheaper rice, while maintaining rural employment and income opportunities for this dominant group of millers and paddy traders in the country.

Meanwhile, trader and miller costs could also be lowered further if economies of scale and location are taken advantage of by a greater number of small-to-medium-scale millers choosing to locate within or in proximity to milling clusters. However, more research is needed to substantiate the degree to which there are potential gains to be had from such agglomeration effects.

Finally, the presence of varying qualities of domestic rice that emerge from the three different types of milling channels described in this chapter implies a more complex story about their relative competitive advantages. While the simple comparisons of production efficiencies and costs with respect to the paddy-to-milled conversion rate, milled-to-paddy and capital-to-labor price ratios, and the economic discount rate are useful for this purpose, they do not account for other policy, environmental, and economic factors that can affect their competitiveness. For example, import restrictions can help protect and thus explain the presence of large-scale operators when they would otherwise be unprofitable without the policy. High transportation and marketing costs can also affect the presence of larger-scale millers in high-production areas if they are situated too far from lucrative markets. By considering these other factors, we can test for whether they help explain the current mix of milling types in the Nigerian rice economy, as well as answer the important question posed at the beginning of this chapter: Are there differential abilities and efficiencies among existing mill types (small to large) to supply the domestic market? This is accomplished in the next chapter by using a Nigeria Rice Milling Model specifically developed for this purpose.

Chapter 6

POLICY OPTIONS FOR MODERNIZING THE MILLING SECTOR

Michael Johnson

As noted in the previous chapter, investments to expand the large milling sector should not come at the cost of ignoring the smaller milling sector. Serious attention should also be given to revitalizing this sector to help meet a broader demand base for cheaper rice while maintaining rural employment and income opportunities among this dominant group of millers and paddy traders in the country.

Another key observation in Chapter 5 was the distinctive quality differences in the final products generated by each rice milling sector. Such product differentiation implies a more complex story about comparative advantage and efficiencies between the small- and larger-scale milling sectors that produce distinctive standard- and premium-quality rice types using a different mix of technologies. Smaller de-huskers use rubber-roller technology, with either electric- or diesel-powered motors, resulting in varying degrees of quality in the final product (broken grains, discoloration, foreign debris, etc.). Medium-to-larger-scale millers use more modern and capital-intensive technologies involving parboiling, drying, milling, de-stoning, and sorting stages, resulting in a more consistent, premium-quality rice grain. This is the grain that is more apt to compete with imports. The mix of technologies, degree of capital intensiveness, scale of operations, and so forth is not only about labor and capital costs but very much a story about adequate access to paddy throughout the year, especially for the larger-capacity mills. Additionally, energy costs (fuel and electricity), transportation costs, marketing and transaction costs, and prices can all greatly affect the profit margins and viability of the modern and larger-capacity milling plants. Improving the quality of domestic rice in order to compete more effectively with imports will therefore depend on how profitable the medium-to-larger milling sector can still be if it upgrades and expands production of premium-quality rice in Nigeria.

The objective of this chapter is to assess the potential for transforming and modernizing the domestic rice milling sector in Nigeria using a mathematical programming model. More specifically, it seeks to address a number of key policy questions: Are there differential abilities and efficiencies among existing mill types (small to large) to supply the domestic market? Given these differences, what is the likely path for transforming the sector into a modernized agro-industry in order to better compete with imports and meet domestic demand? What would be the welfare implications, especially for rural employment? To answer these questions, a survey of the literature, field data, and use of a spatial equilibrium rice milling model were adopted.

The chapter begins by providing an overview of the rice milling model adopted here, including data sources and model assumptions. Details of the model—including its mathematical derivation, parameter assumptions, and validation—are provided in Appendix E. A number of policy scenarios are introduced in the model to test for each milling sector's ability to become more competitive and meet local demand for both premium and standard rice varieties. Results are measured as changes in quantity supplied, volume of imports, and employment. Findings are that the small-to-medium-scale sector is the most resilient to policy shifts and global price changes. Even with zero tariffs, these enterprises are able to stay in business and employ a larger pool of labor. The large-scale enterprises, on the other hand, require higher tariffs to stay in business and are more likely to shift to milling imported brown rice under a zero tariff policy regime. A final concluding section discusses the policy implications of these model results and how they answer the key policy questions raised above.

Introducing a Rice Milling Model for Nigeria

To assess the viability of modernizing the milling sector in Nigeria, and especially the optimal strategy for doing so, a rice milling model (RMM) for Nigeria was developed using mathematical programming techniques to reflect as much as possible the underlying economic structure of the rice milling industry in Nigeria. Adopting this approach is appropriate for a number of reasons.

First, poor data availability makes an ex ante simulation type approach more desirable. Second, the location of agro-industries is uniquely challenged by spatial constraints in accessing raw materials and product markets. Finally, the RMM is especially suitable for analyzing the recent policy focus of expanding the milling industry's large-scale subsector. The model provides empirical estimates of this policy's economic viability and implications for raising the quality of milled rice and employment in the industry relative to alternative policies that also promote improvements in either the small and medium mill subsectors or both.

The RMM builds on the literature of industrial location, industrial clusters, and industrial organization theories, studies that generally analyze the optimal scales, numbers, and locations of processing plants, given a spatial distribution of access to raw input sources, transportation and input costs, and output markets. Their application to agriculture is especially suitable because of the spatial nature of agriculture as a supplier of raw and perishable inputs to agro-industries. Fewer "traditional" location-allocation applications in agriculture have been applied in recent times (Lucas and Chhajed 2004).[1] This is despite a growing interest in the analysis of supply chains and industrial clustering. As large multinational firms become more dependent on multiple resource and input suppliers in the production process, the optimization of the complete supply chain becomes critical in maximizing the firm's profits. Optimizing the supply chain involves choosing the number, location, capacity, and types of industrial plants and/or warehouses; from whom and in what quantity to buy raw inputs; the type and volume of products to produce and which markets to target; as well as amounts to hold in inventory. Industrial organization theory introduces many of the real-world challenges typically ignored in a perfectly competitive world, such as the presence of transaction costs, imperfect information, and barriers to entry. The agglomeration effect is also important in explaining industrial clustering—for instance, lowering the cost of producing or marketing a product in a particular location due to economies of scale (Mccann and Sheppard 2003; Jones and Woods 2002). In Nigeria, the agglomeration effect can be particularly useful in explaining the appearance of milling clusters of small- and medium-scale operators in a number of states, for example.

Because of an interest in determining the optimal scale and mix of the milling industry, this chapter builds on earlier empirical work and adopts the industry profit-maximization model, as in Durham and Sexton (1996). With this model, it is possible to assess the optimal scale mix and efficiencies across the different milling sectors and to provide empirical estimates of the Nigerian rice milling industry's ability to raise the output of quality milled rice and employment in the country. Just as importantly, accessing sufficient quantities of a higher-quality (or premium) paddy variety becomes critical in this regard. While the largest cost share of milled rice is paddy, whose price is

1 One exception is econometric studies that test for the factors influencing agro-industry location in food processing in the United States, for example. The work has typically framed the problem as a profit-maximization one—relating profit to the difference between revenues (local price and quantity sold) and costs of raw inputs and plant operations, all of which vary across space. Brown, Florax, and Mcnamara (2009) provide a good review of this literature.

affected by transport costs from the farm to the mill, transport costs to product markets are expected to be just as large because destinations are typically not the same as the paddy source. Rice marketing or trader costs in Nigeria appear to be quite high, as illustrated in Chapter 5.

Referred to as the RMM for Nigeria, the mathematical programming model compares three industry scales for milling rice in Nigeria in order to determine the optimal scale mix for producing rice more efficiently and to become more competitive with imports. Both transport and marketing costs are explicitly included in the model. Basically, the model determines the optimal location and competitiveness of different types of mills across regions (small, medium, and large scale), subject to supply, capacity and storage constraints, exogenous market prices, production technology, resource costs, and transportation and marketing costs.

The RMM for Nigeria can be considered a two-commodity, spatially oriented optimization model with an objective of maximizing industry profits across three miller sectors based on current scale technologies, two differentiated paddy rice varieties (common and superior quality) and corresponding types of milled rice (standard and premium quality), and two production seasons. "Superior paddy varieties" refers to the most preferred rice seed for milling premium-quality rice among medium- and large-scale millers, such as FARO 54. "Common paddy varieties," on the other hand, refers to both traditional and other older improved rice seeds that are considered inferior for milling purposes and thus produce standard-quality rice. These are only milled by the small milling sector. The geographic unit of analysis is the state level, for all 37 states in the six geopolitical zones listed in Table E.7 in Appendix E. Figure 1.1 in Chapter 1 shows the location of all 37 states on a map. The model is calibrated to 2010 base-year quantities of total paddy production by state and milled rice output at the national level, as well as—given current levels of input and output prices—operating costs, transportation costs between state capitals, and production capacities and resources in each state. Prices are assumed to be exogenous to the model, while quantities of paddy and milled rice are treated as decision variables. The price of milled premium rice is explicitly linked to the world price (inclusive of global shipping, off-loading, and other port fees at Lagos), the rice import tariff, and any inland transportation and marketing costs. The price of premium rice faced by the medium-scale sector is assumed to lie halfway between the large- and small-scale miller prices. This assumption reflects the fact that medium-scale operators produce better-quality rice than small millers but less than the large industrial millers.

Although local paddy can be purchased from any state subject to transportation and marketing costs, this is limited to the superior paddy rice preferred by medium- and large-scale operators to produce premium-quality milled rice. The superior paddy is also assumed to capture a price premium over more common varieties, as evident from Chapter 5. The small milling sector, on the other hand, is assumed to purchase paddy only within their respective state—and then, typically, the common paddy variety to produce standard-quality milled rice. This is reasonable considering that the smaller milling sector mostly services local producers and consumers. Consequently, output is differentiated between premium and standard milled rice types. Only the medium- and large-scale millers produce the premium rice that is more comparable with imports. Therefore, to account for any changes in output as prices are made to adjust to tariff or world price changes, inverse supply functions of both paddy varieties demanded by each of the milling sectors are introduced into the model to allow prices to also adjust accordingly.

All the data and underlying assumptions are based on various secondary sources and the author's own fieldwork. Among the secondary data sources, production data were taken from the National Agricultural Extension and Research Liaison Services (NAERLS 2009) for the 2009 production season, while prices were annual averages from the National Bureau of Statistics (NBS) for 2010 and 2012 (for paddy, local rice, and imported). To be consistent with the price data, a base tariff rate of 50 percent in 2010 was used. Nigerian Agricultural Transformation Agenda (ATA) documents were relied on in defining the country's national objectives and targets for improving the milling sector.

The primary data are based on the author's own field visits to four major rice-producing states (Niger, Kano, Benue, and Kwara; see Figure 1.1 in Chapter 1 for the locations of states) in 2012 and 2013, as described in Chapter 5. As highlighted before, there are very poor data on the rice post-harvest sector, aside from rapid appraisals. The only exception is a 2002–2003 study undertaken by the Africa Rice Center and Nigerian partners (Lançon et al. 2003b). Consequently, estimation of both fixed and operational cost data relied heavily on the data collected from the field visits, including milling capacity and technologies, production costs (material, labor, and other inputs), output, and prices. These data contributed to defining the model's set of parameters in its objective (profit), production technology, and resource constraint functions. Parameter estimates for the small- and medium-scale mills were compared with those of Lançon et al. (2003a) to ensure their accuracy, as the latter involved a much larger sample. As much as possible, therefore, model parameters were estimated based on both past data and field observations by the author. As noted earlier, the model was calibrated to a baseline year of 2010 with a 50 percent tariff rate for rice imports. The baseline year of 2010 was chosen to be consistent with the price and tariff data and partially because state- and national-level production figures were not expected to change as much from the NAERLS data collected in the 2009 harvest season. Although the data used come from various sources and relate to various years, every effort was made to update to the base year where possible.

Upon validating the results of the model to the base-year levels of output, a number of policy scenarios were introduced to test for each milling sector's ability to become more competitive and meet local demand for both the premium and standard rice varieties. Further details of the model, data assumptions, and model calibration and validation are provided in Appendix E. The policy scenarios introduced in the model and corresponding results are reported by milling sector and region in the following section.

Using the Rice Milling Model: Policy Scenarios and Results

Potential effects of a number of policies and strategies, either proposed under the ATA or for future consideration, are considered in the RMM. Brown rice milling is permitted under all the policy scenarios but limited to the large-scale milling sector. Policies are either trade related (e.g., the introduction of rice import tariffs) or public investment related. For the latter, this includes (1) technology or research and development (R&D)–related investments to improve access to superior-quality paddy and the productivity of paddy production and (2) investments to improve market efficiencies through lowering of both marketing and transportation costs. More specifically, the policy scenarios introduced in the model include the following:

S1 **Trade policy (trade).** Introduction of a change in import tariff. Base is with a 50 percent tariff in place. Two additional tariffs are introduced: (i) a reduction to 0 percent tariff, and (ii) an increase to 100 percent tariff.

S2 **Technology policy (science and technology, S&T).** Investing in R&D and irrigation to promote superior-quality paddy varieties and productivity growth in overall paddy production. The scenario in the model involves increasing by 50 percent the adoption of superior-quality varieties and expansion of current irrigation area in states producing more than 50,000 metric tons (MT). Tariffs are kept at their base.

S3 **Marketing improvement (markets).** Lowering high transportation and marketing costs through improved performance (e.g., by increased investments in road infrastructure, transportation, and communications). Tariffs are kept at their base.

S4 **All combined (all).** The combined effect of all three scenarios is also considered (i.e., S1 + S2 + S3).

Aside from evaluating these policy effects under the current state of milling capacities and technology in the country, I also consider their effects under a number of investment strategies targeted to improve each of the three milling sectors, either separately or collectively, in order to expand their capacities or improve their efficiencies. Overall, four milling strategies are considered:

M1 **Large-scale strategy.** Expanded milling capacity (doubling the number of large-scale mills, increasing them from 17 to 34, as under the ATA plan).

M2 **Medium-scale strategy.** Improving further processing (de-stoning and polishing) for premium rice (by a 20 percent increase in milling costs), plus gaining a price premium of 10 percent more in milled rice prices.

M3 **Small-scale strategy.** Doubling milling costs (because small mills' operating costs are so low they would need to at least double to produce a notable improvement in processing) and capturing a price premium of 10 percent more in milled rice prices from further processing (e.g., de-stoning) for standard rice.

M4 **Comprehensive strategy.** Combining all three strategies (M1 + M2 + M3).

Evaluating the three policy effects under each of these strategies will help weigh the benefits and costs for choosing one over the other. The benefits and costs are measured in terms of changes in total milling output from the base year by miller sector and type of rice; changes in the mix of miller types from the base year with respect to their contribution to total national output; changes in the mix of premium- versus standard-quality rice produced in the domestic rice economy; and changes in employment by the rice milling industry from the base year by miller sector and type of rice.

Tables 6.1 and 6.2 summarize the results with respect to the effect of a round of policy shocks on total rice output. From Table 6.1, it is clear that the variant of policy scenario S4—which includes combined investments in rice

TABLE 6.1 Effect of policies on national output by types of miller and rice (million MT)

Policy Scenarios		By type of miller (scale)					By type of rice		
		Large[b]	Large[b]	Med	Small	Total	Premium	Standard	Total
Scenario	Tariff (%)[a]	Brown	Local only				Brown & local	Local	All
				Output (million MT)					
Base	*50*	*0.6*	*0.5*	*0.3*	*2.3*	*3.1*	*1.4*	*2.3*	*3.7*
S1 – Trade	0	1.1	0.0	0.2	2.3	2.5	1.3	2.3	3.6
	100	0.0	0.7	0.6	2.3	3.6	1.3	2.3	3.6
S2 – S&T	50	0.6	0.5	0.5	2.4	3.4	1.6	2.4	4.0
S3 – Markets	50	1.0	0.1	0.6	2.8	3.5	1.7	2.8	4.5
S4 – All	50	0.9	0.2	0.6	3.2	4.0	1.7	3.2	4.9
	0	1.1	0.0	0.6	3.2	3.8	1.7	3.2	4.9
	100	0.0	0.9	1.0	2.6	4.5	1.9	2.6	4.5
				Percent change from base (%)					
S1 – Trade	0	83.3	–100.0	–33.3	0.0	–19.4	–7.1	0.0	–2.7
	100	–100.0	40.0	100.0	0.0	16.1	–7.1	0.0	–2.7
S2 – S&T	50	0.0	0.0	66.7	4.3	9.7	14.3	4.3	8.1
S3 – Markets	50	66.7	–80.0	100.0	21.7	12.9	21.4	21.7	21.6
S4 – All	50	50.0	–60.0	100.0	39.1	29.0	21.4	39.1	32.4
	0	83.3	–100	100.0	39.1	22.6	21.4	39.1	32.4
	100	–100.0	80.0	233.3	13.0	45.2	35.7	13.0	21.6

Source: Rice Milling Model results.

Notes: [a]Base year includes a 50 percent tariff. [b]For large millers, they have an option to mill brown rice (large brown) or local paddy rice (large local only). Notice that a complete switch occurs when tariffs are removed or raised to 100 percent (favoring brown rice for the former and local paddy for the latter). MT = metric tons. S&T = science and technology.

TABLE 6.2 Effect of policies on shares in national output by types of miller and rice (percent of total rice output)

Policy Scenario		Type of miller (scale)					Type of rice	
Scenario	Tariff (%)[a]	Large brown[b]	Large local[b]	Med	Small	Total[c]	Premium	Standard
Base	*50*	*16.2*	*13.5*	*8.1*	*62.2*	*100.0*	*37.8*	*62.2*
S1 – Trade	0	30.6	0.0	5.6	63.9	100.0	36.1	63.9
	100	0.0	19.4	16.7	63.9	100.0	36.1	63.9
S2 – S&T	50	15.0	12.5	12.5	60.0	100.0	40.0	60.0
S3 – Markets	50	22.2	2.2	13.3	62.2	100.0	37.8	62.2
S4 – All	50	18.4	4.1	12.2	65.3	100.0	34.7	65.3
	0	22.4	0.0	12.2	65.3	100.0	34.7	65.3
	100	0.0	20.0	22.2	57.8	100.0	42.2	57.8

Source: Rice Milling Model results.

Note: S&T = science and technology. Numbers might not always add up to 100 because of rounding. [a]Base year includes a 50 percent tariff. [b]For large millers, they have an option to mill brown rice (large brown) or local paddy rice (large local). Notice that a complete switch occurs when tariffs are removed or raised to 100 percent (favoring brown rice for the former and local paddy for the latter). [c]Total includes large brown as well as the three local rice types.

R&D (referred to in the table as "S&T") and improved market performance (referred to here as "markets") but does not include a tariff change—has major effects on overall output. Local production of rice can potentially rise from 3.1 to 4.0 million MT at the current 50 percent base tariff rate. Simply increasing the tariff rate to 100 percent (policy scenario S1) does not produce as much, rising to 3.6 million MT, although higher than S2 for S&T alone and S3 for markets alone. There are several reasons for this. First, prices do not rise as much as the increase in the tariff—a result that is also validated from actual price data in Chapter 7. Moreover, while prices for premium rice in the large-scale sector rise by a 0.7 fraction of the tariff change in the model, prices for the bulk of standard and premium rice flowing from the medium milling sector were assumed to rise much more slowly (about 20 percent of the change in tariff). Paddy prices were also assumed to increase at about the same rate as the rise in standard rice prices. A reasonable assumption, this also implies a weak supply response in paddy production based on the previous findings in Chapter 4. Finally, the higher tariff simply replaces brown rice milling with milling of local paddy only in the large-scale sector.

A rise in premium rice prices due to the higher 100 percent tariff benefits local rice milling by raising the price margin between paddy and milled rice for both large- and medium-scale operators. As a result, as shown in S1 in Table 6.2, both sectors respond with higher output from milling local paddy—although only the smaller medium-scale sector is adding anything to total output. For the large-scale sector, its share of milled rice actually declines from 28.9 (16.2 plus 12.7) in the base to 20.2 percent (from both brown and local) as the sector shifts away from milling any imported brown rice. Alternatively, increasing investments in either R&D (paddy S&T) in S2 or market improvement (markets) in S3 also produces a higher supply response in the medium milling sector, which almost doubles its output under the current tariff regime (50 percent tariff base). For the large-scale sector, improved markets (e.g., through lowered transportation costs) in S3 actually favor the milling of imported brown rice when there has been no change in the tariff regime, as millers farther from the port of entry face lower transportation costs. The smaller milling sector seems to benefit more from the combined effect of both R&D and market improvements, as in S4, with its share of output rising above 65 percent (Table 6.2) regardless of whether the current tariff is removed or not. Increasing tariffs to 100 percent, on the other hand, reduces the sector's share from 63.3 in the base to 58.5 percent. The higher favorable prices for rice producers and millers benefit all (see increased output volumes in Table 6.1), but especially the medium- and larger-scale millers, who are able to increase their shares of national output. The combined effect of the high tariff and investments in R&D and markets naturally produces a greater supply response, especially from the medium- and large-scale sectors (or for milled premium rice). The increase in the supply of superior-quality varieties from R&D and adoption, lowered transportation and marketing costs, together with more favorable rice prices, drives such a response. I will come back to this later.

Another way to evaluate which milling sector benefits or is more resilient to a policy shock is to examine their resulting industry shares in the national total after the shock. Table 6.2 summarizes this for both miller types and the rice type on aggregate. In the base case, the large sector contributes about 29 percent to total rice milled in the country (this is including brown rice). The medium-scale sector is the smallest, contributing only about 8 percent to total rice in the country, with the rest of the rice (the more inferior standard rice) coming from the small milling sector (about 63 percent). Again, the introduction of a 100 percent tariff regime makes only a very little dent on the supply of local rice, either premium or standard. While large millers switch completely to local paddy milling, they simply replace the brown rice they would otherwise mill under a lower tariff regime. This is also a sector whose total milling capacity is constrained in the model by the actual number of mills and locations where they are situated in the country. For example, the bulk of milling brown rice occurs in Lagos (estimated to have about six large rice mills).

Altogether, the shares of local premium rice do not rise much with either higher tariffs or improvements that lower transportation and marketing costs unless there are significant investments to expand production of higher-quality premium varieties (S2 and S4 row). The lower costs not only result in lower mill gate prices for paddy but for brown rice, too. The result is that many of the large millers switch to brown rice milling, even in northern Nigeria.

The effects of the different policies on output and industry shares among types of rice millers and types of rice can vary widely at the subnational level, especially given the spatial nature of the model. Table 6.3 shows the policy effects on regional output. Because much of the production of rice in Nigeria occurs in the North Central, North East, and North West zones (for Nigeria's six geopolitical zones, see the footnote in Figure 1.1), a majority of the increase in total output comes from these three regions. All three also benefit more from policies focused on increasing, through S&T investment, productivity and growth in the supply of quality rice paddy than under the higher tariff regime: in S4, with the base tariff, total output rises by between 37 and 57 percent. The

TABLE 6.3 Effect of policies on zonal output by types of rice

Policy Scenario		Zone					
Scenario	Tariff %[a]	NC	NE	NW	SE	SS	National
		Quantity (million MT)					
Base	50	0.97	0.77	1.04	0.40	0.01	3.63
S1 – Trade	0	0.96	0.76	1.03	0.40	0.01	3.59
	100	1.02	0.89	1.19	0.40	0.04	3.63
S2 – S&T	50	1.06	0.91	1.14	0.40	0.01	3.96
S3 – Markets	50	1.23	0.95	1.38	0.46	0.04	4.51
S4 – All	50	1.36	1.01	1.63	0.49	0.04	4.98
	0	1.33	1.06	1.64	0.43	0.01	4.91
	100	1.15	1.22	1.34	0.49	0.04	4.44
		Percent change from base (%)					
S1 – Trade	0	−1.0	−1.3	−1.0	0.0	0.0	−1.1
	100	5.2	15.6	14.4	0.0	300.0	0.0
S2 – S&T	50	9.3	18.2	9.6	0.0	0.0	9.1
S3 – Markets	50	26.8	23.4	32.7	15.0	300.0	24.2
S4 – All	50	40.2	31.2	56.7	22.5	300.0	37.2
	0	37.1	37.7	57.7	7.5	0.0	35.3
	100	18.6	58.4	28.8	22.5	300.0	22.3

Source: Rice Milling Model results.

Note: NC = North Central zone; NE = North East zone; NW = North West zone; SE = South East zone; SS = South South zone; SW = South West zone. MT = metric tons. S&T = science and technology. [a]Base year includes a 50 percent tariff.

South West region is particularly sensitive to higher tariffs, as millers in this region are unable to switch to local paddy. Instead, they would have to purchase it from excess paddy producing states such as Niger and Benue.

The changing mix in the operating scale of the rice milling industry can have important implications on employment outcomes within each miller sector. To estimate employment by milling type, calculations were made based on the average of field observations with regard to the number of full-time persons employed by each milling type's capacity during peak operations and calculated as the number of employees per maximum MT capacity that can be produced in a single year. For large-scale milling operations, this is 0.002 for milling imported brown rice (i.e., 400 employees/[24 MT/hr * 24hrs * 330 days/yr]) and 0.003 for milling local paddy (i.e., 400 employees/[18 MT/hr * 24 hrs * 330 days/yr]); for medium-scale operations, this is 0.005 (i.e., 20 employees/[2 MT/hr * 8 hrs * 270 days]); and for small-scale operations, this is 0.017 (i.e., 10 persons/[0.5 MT/hr * 8 hrs * 150 days]).

National employment in the rice milling sector for large-, medium-, and small-scale mills adds up to a very rough figure of about 43,000. For our purposes, however, the accuracy of this absolute number is not as important because we are only interested here in the proportional change and direction (up or down) in employment due to different policy scenarios.

Table 6.4 reports the change in total employment due to each policy scenario introduced in the model by each miller type and by rice type (premium or standard). Table 6.5 shows the shares of labor employed by milling type. In total, it is estimated that the small milling sector employs over 90 percent of

TABLE 6.4 Effect of policies on change in employment by miller and rice type (percent change from base)

Policy		Miller type (scale)			Type of rice		
Investments	Tariff %[a]	Large[b]	Med	Small	Premium	Standard	Total
S1 – Trade	0	–9.2	–20.2	–1.4	–13.0	–1.4	–2.4
	100	–18.8	104.3	0.3	23.5	0.3	2.4
S2 – S&T	50	2.2	83.6	3.5	30.2	3.5	5.9
S3 – Markets	50	–7.6	109.9	22.1	32.8	22.1	23.1
S4 – All	50	–3.0	123.2	40.7	40.4	40.7	40.6
	0	–9.2	104.9	40.0	30.0	40.0	39.1
	100	–5.8	247.2	12.9	81.2	12.9	19.1

Source: Rice Milling Model results.

Note: S&T = science and technology. [a]Base year includes a 50 percent tariff. [b]Large here includes both brown and local paddy rice milling. The reduction in employment for large scale under the 100 percent tariff is primarily because of the closure of large mills in Lagos, as they find it unprofitable to mill local paddy as a substitute for brown rice after tariffs rise to 100 percent.

TABLE 6.5 Effect of policies on changes in employment shares by miller type (as percent of national employment in the rice milling industry)

Policy		Miller type (scale)		
Investments	Tariff %[a]	Large[b]	Medium	Small
Base[a]	50	6.0	3.1	90.9
S1 – Trade	0	5.6	2.6	91.8
	100	4.8	6.3	89.0
S2 – S&T	50	5.8	5.4	88.8
S3 – Markets	50	4.5	5.4	90.1
S4 – All	50	4.1	5.0	90.9
	0	3.9	4.6	91.5
	100	4.7	9.2	86.1

Source: Rice Milling Model results.

Notes: S&T = science and technology. Totals across miller types may not always equal 100 because of rounding. [a]Base year includes a 50 percent tariff. [b]Large here includes both brown and local paddy rice milling.

the labor force in the country's rice milling industry. The medium milling sector employs the least—although the position of the lowest-employing sector can quickly shift from the medium- to the large-scale sector as policies affect millers' operations. An important implication of this is that any small change in output within the small milling sector has more significant effects on overall employment. The other sector most affected by small changes in output is the medium sector, simply because of its higher dependence on labor per metric ton of output than the large sector. In fact, Table 6.4 shows that much of the gain in employment is experienced by the medium sector, as the production of premium rice rises following increases in R&D investments and improved markets; employment more than doubles as a result, although this is still a small rise as a share of the total employed in the industry (from 3 to 5 percent, as shown in Table 6.5).

Turning to the milling technology and expansion strategies, four different strategies are introduced to test for their effect on total output, employment, and shares of the premium versus the standard varieties of local rice. Results are summarized in Tables 6.6 through 6.8. The first strategy is the large-scale one that involves introducing the 17 new large integrated mills proposed under the ATA strategy for rice. The three largest rice-producing zones (North Central, North East, and North West) naturally gain the largest share of new mills. Among current capacities, the South West is dominated by mills that mostly process brown rice. Although capacity utilization in the base case is about 96 percent in the large-scale sector, the introduction of 17 new mills results in a much lower capacity utilization ratio of 58 percent. This may explain the small effect on total output illustrated in Table 6.6: with no tariff change, brown rice milling and paddy rice milling increase only 0.7 and 3.4 percent, respectively, in the large sector (S1 row, column under the large-scale strategy and base 50 percent tariff, and with no change in paddy productivity and improvement of markets). Despite the expanded milling activities in the large-scale sector, total rice output in the country changes little due to a limited supply of superior local paddy varieties. The superior varieties are demanded by both medium- and large-scale millers, so that an increase in the capacity of large-scale milling under the ATA strategy simply diverts the paddy away from medium-scale millers (notice the 3.5 percent decline in output for this milling sector in Table 6.6). All in all, milled output of premium-quality rice (produced by the medium- and large-scale sectors only) turns out to increase by only 0.6 percent.

At the current tariff rate, combining the large-sector strategy with improvements in paddy rice productivity and markets seems to have negligible effects on output (S4, with no tariff change and with S&T and markets). While large-scale millers actually increase output of milled brown rice by 9.4 percent, they decrease their milling of local paddy by 12.7 percent. While this may seem counterintuitive at first glance—one would expect the milling of local paddy to increase as the supply of paddy increases—the lowered transportation and marketing costs allow more large millers to partially shift to milling imported brown rice, as moving it from the port of Lagos becomes cheaper. Moreover, brown rice milling is cheaper because it does not require any parboiling and is not constrained by limited supplies during lean seasons, as is the case for local paddy.

Rice import tariffs can have significant effects on the large-scale sector. The biggest increase in large-scale output of premium rice actually occurs when tariffs are removed and a productivity change in paddy and improved market performance also occur (S4 scenario in Table 6.6 under large-scale strategy and a zero tariff rate). Production of premium-quality rice increases by 36.3 percent compared to a higher tariff of 100 percent, when it increases by only 18.0 percent. Evidently, with zero tariffs, most large millers switch to imported brown rice milling, adding a significant amount of premium rice in the marketplace and above the total supply of local paddy. The smaller milling sector contributes the most to the increase in total local rice under zero tariffs—an increase of 25.6 percent.

Examining across the four strategies in Table 6.6, a medium-scale strategy has a similar positive response in supplying premium-quality rice as does a

TABLE 6.6 Effect of different miller strategies on total output by miller and rice type (percent change from base)

		M1. Large–scale strategy			M2. Medium–scale strategy			M3. Small–scale strategy			M4. Comprehensive strategy		
	Sector	$\tau=0\%$	$\tau=50\%$*	$\tau=100\%$	$\tau=0\%$	$\tau=50\%$*	$\tau=100\%$	$\tau=0\%$	$\tau=50\%$*	$\tau=100\%$	$\tau=0\%$	$\tau=50\%$*	$\tau=100\%$
S1. No S&T and markets	*L–brown*	*40.7*	*0.7*	*−16.2*	*13.9*	*3.1*	*−16.2*	*13.9*	*0.4*	*−16.2*	*40.7*	*3.9*	*−16.2*
	L–local	−12.7	3.4	21.5	−12.7	−3.4	3.9	−12.7	−1.7	7.5	−12.7	−4.7	11.5
	Med	−1.6	−3.5	−3.7	−5.2	5.1	15.8	−1.6	−0.2	8.2	−5.2	3.7	8.2
	Small	−0.3	0.3	0.1	1.2	−1.2	0.1	1.7	3.1	0.2	3.2	2.3	0.1
	Total local	*−14.6*	*0.2*	*17.9*	*−16.7*	*0.5*	*19.8*	*−12.6*	*1.2*	*15.9*	*−14.7*	*1.3*	*19.8*
	Premium	26.5	0.6	1.7	−4.0	4.8	3.5	−0.4	−1.5	−0.5	22.8	2.8	3.5
	Standard	−0.3	0.3	0.1	1.2	−1.2	0.1	1.7	3.1	0.2	3.2	2.3	0.1
S4. With S&T and markets	*L–brown*	*40.7*	*9.4*	*−16.2*	*13.9*	*7.9*	*−16.2*	*13.9*	*9.9*	*−16.2*	*40.7*	*12.3*	*−16.2*
	L–local	−12.7	−12.7	24.7	−12.7	−6.8	6.5	−12.7	−8.7	10.2	−12.7	−9.5	11.7
	Med	8.3	−5.2	9.5	−1.1	9.6	31.3	1.7	8.6	17.5	−2.5	8.8	21.2
	Small	25.6	1.2	4.1	32.2	25.7	0.6	43.9	37.6	12.6	48.0	38.1	8.1
	Total local	*21.2*	*−16.7*	*38.3*	*18.4*	*28.5*	*38.4*	*32.9*	*37.5*	*40.3*	*32.8*	*37.4*	*41.0*
	Premium	36.3	−8.6	18.0	0.1	10.8	21.6	2.9	9.8	11.6	25.6	11.5	16.7
	Standard	25.6	1.2	4.1	32.2	25.7	0.6	43.9	37.6	12.6	48.0	38.1	8.1

Source: Rice Milling Model results.

Note: S&T = science and technology. L-brown and L-local refer to large brown and large local, respectively. *Note that $\tau=50$ percent is really the base period tariff; therefore, for the large-sector strategy, the change in output is measuring the effect of introducing 17 new mills without any change in tariffs. For the medium- and small-sector strategies, under the same $\tau=50$ percent tariff, the output changes reflect improvements in their current milling technologies, with no tariff change, but with alternative S1 and S4 scenarios (S&T here stands for improved paddy varieties). Changes from base are weighted by the share of the sector in total output.

large-scale strategy, especially under a protective tariff regime of 100 percent. The medium milling sector strategy with 100 percent tariff and no changes in paddy production of market improvements (S1) would have the effect of increasing output of the premium-quality rice by 3.5 percent compared with only 1.7 percent under the large-scale strategy of adding 17 new large mills. Under a small-scale strategy, on the other hand, the small sector would raise national output of local rice by only 1.2 percent with no change in tariffs, while adding little to increase the output of premium rice. Higher tariffs under this strategy mainly benefit the large- and medium-scale sectors. Altogether, rice output for both premium and standard rice types increases the most under a comprehensive strategy that invests in all three milling sectors, with improvements in paddy productivity and markets (S4) and with zero tariffs. National output of premium-quality rice would increase by 25.6 percent, while the supply of standard-quality rice would increase by 48.0 percent. However, much of the increase in premium rice is derived from the milling of brown rice. Ensuring sufficient growth in the supply of domestically produced premium-quality rice would occur under a more liberal tariff regime (zero tariffs) that permits the milling of brown rice. The milling of local paddy is otherwise unprofitable for a majority of large millers at zero tariffs.

Disaggregating the effects on output by zones mostly reflects the distribution of major rice-producing regions. The exception is the South West zone: this zone's output most consistently and dramatically fluctuates as tariffs change. Table 6.7 illustrates this by showing a typical 10 to 11 percent reduction in its contribution to national output at the 100 percent tariff level. The explanation for this is the going out of business of several mills, as the milling of brown rice is no longer profitable at this tariff level. This occurs as other regions, especially in the three northern zones (NC, NE, and NW), expand their large-scale milling of local paddy, with some regions even reaching maximum capacity utilization.

The effect on employment across the various strategies can also be examined. Table 6.8 reports the estimated effects of each strategy on employment, weighted by each sector's share in total employment in the country's milling industry. Naturally, employment rises as output rises in most cases. As in the case of increased output, for example, the effects on employment are greatest in S4, with increased R&D investments and market efficiencies. Within S4, a comprehensive strategy that focuses attention on all the sectors, combined with zero tariffs and permitting the milling of brown rice, produces the greatest employment effects. Because the removal of tariffs from the base of 50 percent can also be similarly viewed as a reduction in the global rice price

TABLE 6.7 Effect of different miller strategies on total output by zone (percent change from base)

		M1. Large-scale strategy			M2. Medium-scale strategy			M3. Small-scale strategy			M4. Comprehensive strategy		
	Zone	$\tau=0\%$	$\tau=50\%$*	$\tau=100\%$	$\tau=0\%$	$\tau=50\%$*	$\tau=100\%$	$\tau=0\%$	$\tau=50\%$*	$\tau=100\%$	$\tau=0\%$	$\tau=50\%$*	$\tau=100\%$
S1. No S&T and markets	NC	6.2	0.8	3.2	–0.2	1.8	4.5	–0.1	0.8	1.4	6.2	1.8	5.7
	NE	3.2	0.2	1.8	–2.0	–0.2	4.8	0.0	0.2	3.3	1.2	–0.1	4.7
	NW	6.0	–0.2	5.6	–0.5	1.5	5.3	1.4	0.3	4.1	7.7	2.8	4.8
	SE	9.5	0.0	1.3	0.0	0.3	–0.8	0.0	0.1	0.0	9.5	0.4	–1.3
	SS	0.1	0.1	0.8	0.1	0.1	0.8	0.1	0.2	0.8	0.1	0.2	0.8
	SW	1.3	0.0	–10.9	–0.3	0.0	–11.0	–0.1	0.1	–9.8	1.4	0.1	–11.0
S4. With S&T and markets	NC	16.0	10.6	6.6	12.8	10.4	9.5	11.7	13.2	6.7	18.3	13.7	9.6
	NE	11.4	6.7	10.4	4.5	6.8	12.5	12.1	12.1	12.4	12.8	12.1	10.3
	NW	22.8	16.1	11.1	14.3	16.2	9.6	18.4	16.6	8.6	27.0	16.8	12.7
	SE	10.2	4.0	3.7	0.7	2.4	0.6	4.6	4.5	2.6	13.8	6.2	2.3
	SS	0.1	0.8	0.8	0.1	0.8	0.8	0.1	0.8	0.8	0.1	0.8	0.8
	SW	1.4	0.1	–10.4	–0.1	0.1	–10.8	–0.1	0.1	–7.1	1.4	0.1	–10.9

Source: Rice Milling Model results.

Note: NC = North Central, NE = North East, NW = North West, SE = South East, SS = South South, SW = South West. S&T = science and technology. *Note that $\tau=50$ percent is really the base period tariff; therefore, for the large-sector strategy, the change in output is measuring the effect of introducing 17 new mills without any change in tariffs. For the medium- and small-sector strategies, under the same $\tau=50$ percent tariff, these are improvements in their current milling technologies, with no tariff change, but with alternative S1 and S4 scenarios (S&T here stands for improved paddy varieties). Changes from base are weighted by the share of the sector in total output.

TABLE 6.8 Effect of different miller strategies on total employment in the milling industry by miller and rice type (percent change from base total)

		M1. Large–scale strategy			M2. Medium–scale strategy			M3. Small–scale strategy			M4. Comprehensive strategy		
	Sector	$\tau=0\%$	$\tau=50\%$*	$\tau=100\%$	$\tau=0\%$	$\tau=50\%$*	$\tau=100\%$	$\tau=0\%$	$\tau=50\%$*	$\tau=100\%$	$\tau=0\%$	$\tau=50\%$*	$\tau=100\%$
S1. No S&T and markets	*L–brown*	*7.4*	*0.1*	*–2.9*	*2.5*	*0.6*	*–2.9*	*2.5*	*0.1*	*–2.9*	*7.4*	*0.7*	*–2.9*
	L–local	–3.1	0.8	5.2	–3.1	–0.8	0.9	–3.1	–0.4	1.8	–3.1	–1.1	2.8
	Med	–0.6	–1.4	–1.5	–2.1	2.0	6.3	–0.6	–0.1	3.3	–2.1	1.5	3.3
	Small	–0.4	0.4	0.1	1.8	–1.7	0.1	2.5	4.5	0.3	4.7	3.4	0.1
	Total local	*–4.1*	*–0.2*	*3.8*	*–3.4*	*–0.5*	*7.3*	*–1.2*	*4*	*5.4*	*–0.5*	*3.8*	*6.2*
	Premium	3.7	–0.4	0.8	–2.6	1.8	4.3	–1.2	–0.4	2.1	2.2	1.0	3.1
	Standard	–0.4	0.4	0.1	1.8	–1.7	0.1	2.5	4.5	0.3	4.7	3.4	0.1
S4. With S&T and markets	*L–brown*	*7.4*	*1.7*	*–2.9*	*2.5*	*1.4*	*–2.9*	*2.5*	*1.8*	*–2.9*	*7.4*	*2.2*	*–2.9*
	L–local	–3.1	–3.1	6.0	–3.1	–1.6	1.6	–3.1	–2.1	2.5	–3.1	–2.3	2.8
	Med	3.3	–2.1	3.8	–0.4	3.8	12.5	0.7	3.4	7.0	–1.0	3.5	8.5
	Small	36.9	1.8	6.0	46.3	37.0	0.9	63.1	54.0	18.2	68.9	54.8	11.6
	Total local	*37.1*	*–3.4*	*15.8*	*42.8*	*39.2*	*15*	*60.7*	*55.3*	*27.7*	*64.8*	*56*	*22.9*
	Premium	7.6	–3.5	6.8	–1.0	3.6	11.1	0.1	3.1	6.5	3.3	3.4	8.3
	Standard	36.9	1.8	6.0	46.3	37.0	0.9	63.1	54.0	18.2	68.9	54.8	11.6

Source: Rice Milling Model results.

Note: S&T = science and technology. L-brown and L-local refer to large brown and large local, respectively. *Note that $\tau=50$ percent is really the base period tariff; therefore, for the large-sector strategy, the change in output is measuring the effect of introducing 17 new mills without any change in tariffs. For the medium- and small-sector strategies, under the same $\tau=50$ percent tariff, the output changes reflect improvements in their current milling technologies, with no tariff change, but with alternative S1 and S4 scenarios (S&T here stands for improved paddy varieties). Changes from base are weighted by the share of the sector in total output.

of 50 percent, this scenario could be particularly relevant given the current downward trend in global rice prices.

Conclusion

Chapter 5 emphasized that improving the quality of domestic rice in Nigeria in order to compete more effectively with imports will depend on how much the country's milling industry can be transformed to increase its productivity and expand production of premium-quality rice. The objective of this chapter is to assess whether this is feasible and how it could be done given the current technology mix and scale of the different miller types in Nigeria. More specifically, it seeks to address a number of key policy questions: Are there differential abilities and efficiencies among existing mill types (small to large) to supply the domestic market? To what extent does increasing tariffs help the industry? This is accomplished through the development and application of a simulation model representing the rice milling industry in Nigeria—the RMM.

The empirical analysis in this chapter further validated the importance of having sufficient access to quantities of superior-quality paddy among the large- and medium-scale millers in order to increase the supply of premium-quality local rice in Nigeria. Results also show that the reduction in transportation and marketing costs can play a critical role in both the location and mix of milling types by affecting the prices of paddy and final rice product. The exception is the large-scale sector, which is likely to switch to milling brown rice if tariffs are low enough (or even absent) and the cost to ship it overland from Lagos becomes cheaper. This is particularly true for large millers farther away from paddy producers, but it is just as relevant for all millers whenever the seasonal supply of paddy is at its lowest.

Import tariffs are shown to be particularly important for protecting the large-scale sector. At zero tariffs, these enterprises would go out of business unless they are able to substitute for milling imported brown rice. The option to mill brown rice offers large-scale millers a way to manage production risks associated with inadequate and infrequent supplies of superior-quality paddy, as well as fluctuations in the price of imported rice. The price risk comes from either a reduction in the tariff rate or a fall in global prices. As the model results show, the large sector faces the risk of going out of business if the tariff is removed altogether and the miller is not allowed to mill brown rice. This is because the milling of imported brown rice does not require any parboiling and therefore offers a profitable alternative at very low tariff rates. Moreover, when facing the same tariff rates, the ratio between the price for imported brown rice and imported milled rice tends to change little across the country, as both types of rice experience the same transportation costs from the port of Lagos. Price movements for both types of rice in world markets generally move in the same direction as well. Hence, millers can be assured a positive return most of the time.

In contrast, findings show the small-to-medium-scale sector to be more resilient to policy shifts and global price changes. These enterprises are able to stay in business even with zero tariffs. It is also a sector that has more to gain from improvements in paddy productivity, markets, and processing technologies. Although the small-scale sector cannot compete directly with imports, it plays a critical role in supplying a cheaper alternative for poorer consumers. Moreover, larger employment effects for the whole milling industry are shown to occur following improvements and expansion in the small milling sector. After all, this is a sector that currently employs over 90 percent of the total number of people employed in the industry (that is, across all three milling types). The medium-scale sector, on the other hand, offers a more immediate potential for improving productivity and product quality. Like the large-scale sector, it could serve a similar purpose of helping increase the competitiveness of local rice with imports.

Finally, model results also show that a comprehensive strategy that includes targeted investments among all three milling sectors and a policy regime of zero tariffs has the potential to contribute the most to output and employment. The modeling exercise also shows, however, that for such a strategy to be successful in meeting the demand for premium-quality rice, the milling of imported brown rice should be allowed. Allowing brown rice milling is necessary if large-scale millers are to remain in business under a zero tariff regime.

Chapter 7

RICE IMPORTS, PRICES, AND CHALLENGES FOR TRADE POLICY

Paul A. Dorosh and Mehrab Malek

As discussed in Chapter 2, demand for rice in Nigeria has outpaced domestic production due to rapid population growth, increasing incomes, urbanization (and lifestyle changes that accompany urbanization), and a decline in the relative price of rice. At the same time, Chapters 4 and 6 also showed how domestic production has been only relatively modest because of problems with input supply (superior-quality seed and fertilizer), inadequate extension services, poor quality of postharvest processing and milling technologies, and inadequate (and highly variable) price incentives for farmers. This widening gap between demand and domestic production has been met by rice imports, which have increased from 590,000 tons per year in the 1970s to an average of 2.42 million tons from 2010 to 2012—a fourfold increase (USDA 2013).[1] In this chapter, we focus our attention on the trade aspect of Nigeria's rice economy.

In an effort to limit the volume of imports and to provide protection to domestic rice producers to spur production, the Nigerian government has placed substantial import tariffs on rice, although these tariffs have varied widely over time. For example, the import tariff on rice was increased from 50 percent in 2012 to 110 percent in January 2013. These tariffs are likely to be passed on to the consumer in the form of higher market prices. In addition to higher tariffs that may affect the domestic price of imported rice, there have been large fluctuations in nominal and real exchange rates that have also had substantial influence on incentives for imports as well as the domestic price of imported rice. Further complicating the price incentive picture of Nigeria's rice imports is substantial tariff evasion through underpayments for official imports through Lagos and other ports, as well as smuggling across Nigeria's borders with Benin and Niger (for the location of Lagos, see Figure 1.1 in Chapter 1).

1 In this chapter, all tons are metric tons.

The objective of this chapter is to analyze Nigeria's recent rice import policies in terms of their effectiveness and tradeoffs with regard to affecting the volume of imports, domestic prices, and tariff revenues in the country. More specifically, we present an analysis of Nigeria's international rice trade, covering international market price and quantity evidence, Nigeria's trade (and exchange-rate) policy, and the effectiveness of tariffs in influencing rice imports and domestic prices of imported and local rice. We begin by presenting various estimates of the volume of rice imports, including estimates derived from data on rice-exporting countries, which show that official figures for imports have substantially understated actual imports. Next, we discuss the price incentives for imports, as determined by exchange-rate and rice trade policy (import tariffs and periodic import bans). Price distortions created by these policies are summarized in calculations of nominal rates of protection. We also analyze the relationship between the estimated import parity price and domestic prices of imported rice and local rice. The chapter concludes by providing a brief summary of findings.

Alternate Estimates of Rice Imports

There is substantial uncertainty about the volume of Nigeria's rice imports in recent years. Table 7.1 presents summary data on recent rice imports from various sources. The large percentage of Nigeria's rice imports not recorded in the country's customs data suggests widespread tariff evasion. Nigeria's customs data indicate that an average of 342,000 tons of milled rice (384,000 tons of rice of all types) were imported each year (column 1 of Table 7.1) from 2009 to 2011. For that same time period, the United Nations Commodity Trade Statistics Database (COMTRADE) data show that Nigeria reported an average of 517,000 tons (column 2 of table 7.1) of imported rice per year.[2] However, COMTRADE data from exporting countries, shown in column 3, indicate that total exports of rice to Nigeria averaged 1.9 million tons per year. United States Department of Agriculture (USDA) estimates of Nigeria's rice imports in this period are quite similar: 2.1 million tons per year (column 4 of table 7.1). Thus, both COMTRADE and USDA data suggest that on average, only about one-quarter of Nigeria's rice imports were recorded by customs in the 2009–2011 period (Table 7.1 and Figure 7.1).

According to the COMTRADE data, in most years the bulk of Nigeria's imports originate from India and Thailand, though there have also been

2 COMTRADE data for Nigerian imports are based on statistics reported by Nigeria. Thus, there is an inconsistency between these two Nigerian data sources.

TABLE 7.1 Rice imports by Nigeria and world rice exports to Nigeria, 2008–2012

	(1) Official imports (thousand tons)	(2) COMTRADE Nigeria imports (thousand tons)	(3) COMTRADE Exports to Nigeria (thousand tons)	(4) USDA Nigeria imports (thousand tons)	(5) Official imports vs. COMTRADE exports (percent)
2008	318	160	1,021	1,800	31.1
2009	330	398	1,397	2,000	23.6
2010	551	711	2,053	2,000	26.8
2011	270	441	2,299	2,300	11.7
2012	230	—	—	2,200	—
Average 2009–2011	384	517	1,916	2,100	20.0
Jan–Aug 2012	169	—	—	—	
Sept–Dec 2012	61	—	—	—	

Source: UN Statistics Division (2012), National Bureau of Statistics (NBS), and USDA Data.
Note: COMTRADE = United Nations Commodity Trade Statistics Database. USDA = United States Department of Agriculture.

FIGURE 7.1 Rice imports by Nigeria and world rice exports to Nigeria, 2008–2012

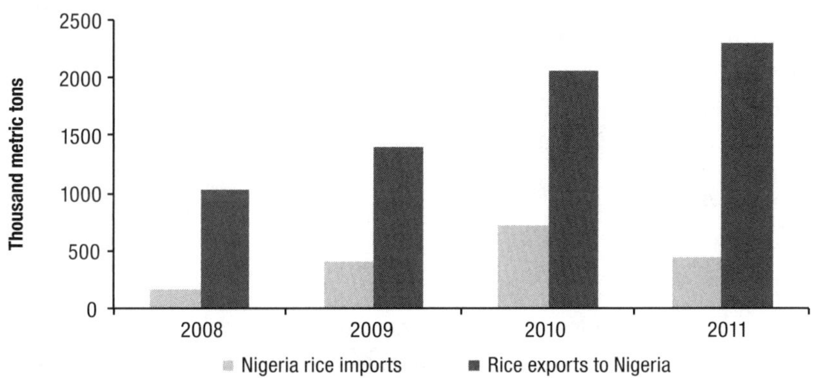

Source: United Nations Commodity Trade Statistics Database.

sizable imports from Brazil in recent years (Figure 7.2). In addition, rice smuggled from Benin or Niger into Nigeria may not be included in the COMTRADE data in all years. COMTRADE data on Benin's rice exports to Nigeria vary sharply from year to year but are relatively small in most years compared to total rice exports to Nigeria. For example, COMTRADE data for 2009 show 140,000 tons of rice exported by Benin to Nigeria (10.1 percent of COMTRADE figures for total exports to Nigeria), 552,000 tons in 2010

FIGURE 7.2 Rice imports by Nigeria and world rice exports to Nigeria by country, 2010 and 2011

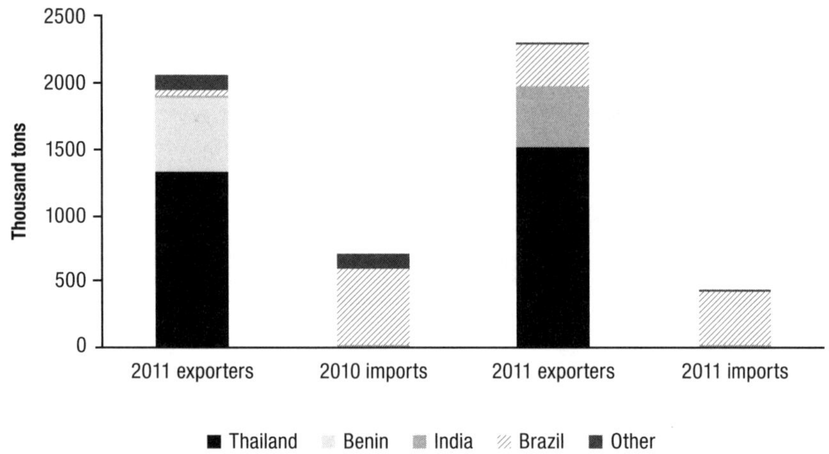

Source: United Nations Commodity Trade Statistics Database.

(26.9 percent of total exports to Nigeria), and zero in 2011 (Figures 7.2 and 7.3). Assuming about 150,000 tons of rice were smuggled from Benin to Nigeria in 2011 as well, the cross-border trade with Benin averaged about 300,000 tons per year, or about 15 percent of Nigeria's total rice imports.[3] Thus, the evidence suggests that over the period from 2009 to 2011, the bulk of rice from Benin that passed through Nigeria's ports was unrecorded.

Trends in Rice Imports

There are two major international sources for time-series data on Nigeria's rice imports: USDA and the Food and Agriculture Organization of the United Nations (FAO). In principle, these sources take into account available data on rice imports from Nigerian government sources but make major adjustments for unrecorded rice imports. Both data series show that Nigeria's rice imports have increased rapidly over the past two decades. USDA data indicate that average annual imports of milled rice have almost tripled in the past two decades, rising from 590,000 tons in the 1990s to 1.68 million tons per year in the first five years of the 2000s (Table 7.2, column 1). FAO estimates

3 Local sources report that around 8,000 bags of rice are smuggled into Nigeria every day through waterways between Nigeria and Benin. Assuming a weight of 50 kilograms per bag, the annual amount smuggled into the country would be 146,000 tons.

FIGURE 7.3 Benin's total rice trade and rice exports to Nigeria, 2006–2011

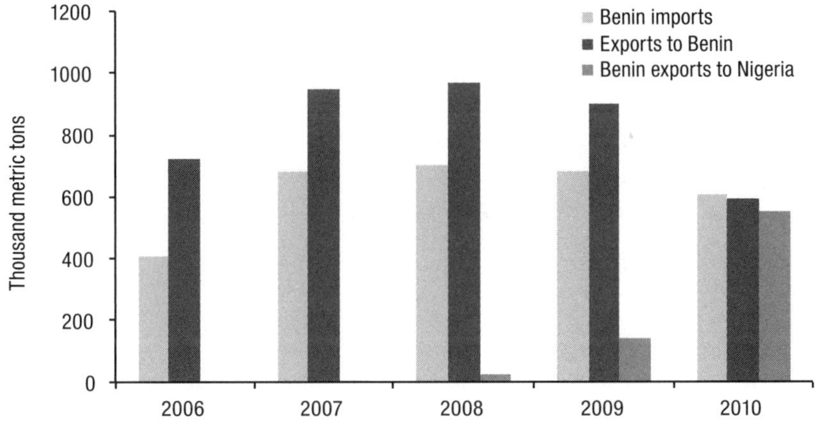

Source: United Nations Commodity Trade Statistics Database.

suggest that annual average imports doubled over the same period, increasing from 432,000 tons in the 1990s to 1.36 million tons in the first five years of the 2000s (Table 7.2, column 1). Both sources show rice imports that are substantially larger than the data from official Nigerian sources. Both sources also show a consistent increase in rice imports over time.

The growth rate in rice imports was especially high in the 1990s: an annual average rate of about 17 percent using USDA data (over 13 percent according to FAO estimates). Data from the two sources begin to diverge significantly around 2005, while maintaining a broadly consistent direction of movement (Figure 7.4). Both USDA and FAO data show that rice imports grew rapidly during the 2005–2012 period, although slower than they did during the 1990–1999 period (Table 7.2). Average annual production of milled rice rose gradually during these periods from 1.8 million tons per year in the 1990s to 2.7 million tons per year in the 2010s, growing at a much slower rate than imports. As a result, the share of imports in domestic availability of rice has risen sharply from 26.5 percent in the 1990s to almost 50 percent in the first decade of the 2000s (FAO data show the share of imports in availability rose from about 19 percent in the 1990s to about 43 percent in the 2010s).[4] Note that although imports account for a large

4 Domestic availability of rice is calculated as availability = milled production less 10 percent losses plus imports.

TABLE 7.2 Nigeria's average annual rice imports, production, and domestic availability, 1960–2012

	(1) Rice imports (thousand tons)	(2) Rice production (thousand tons)	(3) Rice domestic availability (thousand tons)	(4) Rice imports vs. rice domestic availability (percent)
USDA estimates				
1960–1969	1.00	222.00	201.00	0.5
1970–1979	199.00	357.00	520.00	38.3
1980–1989	529.00	866.00	1,308.00	40.4
1990–1999	590.00	1,818.00	2,226.00	26.5
2000–2004	1,679.00	1,851.00	3,346.00	50.2
2005–2009	1,790.00	2,131.00	3,653.00	49.0
2010–2012	2,417.00	2,725.00	4,869.00	49.6
Growth rates (1990–1999)	16.69	1.76	5.59	
Growth rates (2000–2004)	1.43	–0.06	0.49	
Growth rates (2005–2012)	5.70	4.23	4.90	
FAO estimates				
1960–1969	1.00	138.00	139.00	0.8
1970–1979	161.00	275.00	436.00	36.9
1980–1989	413.00	971.00	1,384.00	29.8
1990–1999	432.00	1,821.00	2,253.00	19.2
2000–2004	1,358.00	1,852.00	3,210.00	42.3
2005–2009	1,099.00	2,038.00	3,266.00	33.7
2010–2012	2,035.00	2,713.00	4,748.00	42.9
Growth rates (1990–1999)	12.71	1.71	3.79	
Growth rates (2000–2004)	13.67	–0.07	4.53	
Growth rates (2005–2012)	7.15	3.90	5.14	

Source: USDA and FAO data.

Note: All quantities are milled rice. Growth rates are based on logarithmic regression coefficients. USDA = United States Department of Agriculture. FAO = Food and Agriculture Organization of the United Nations. Due to rounding, figures in column 4 may slightly differ from the quotients of dividing figures in columns 1 and 3.

share of rice supply, these figures suggest they account for a relatively small share of calories (213 kcal/capita/day or 8 percent of daily per capita calorie intake in 2009 [FAO 2012]) and value of imports (2.4 percent). At a world price of $400/ton C&F (cost and freight) Lagos (rounded to the nearest hundred), 2.0 million tons of rice would cost 0.8 billion dollars, 2.4 percent of total imports in 2009 (IMF 2012; USDA 2014).

FIGURE 7.4 Rice imports by source of data

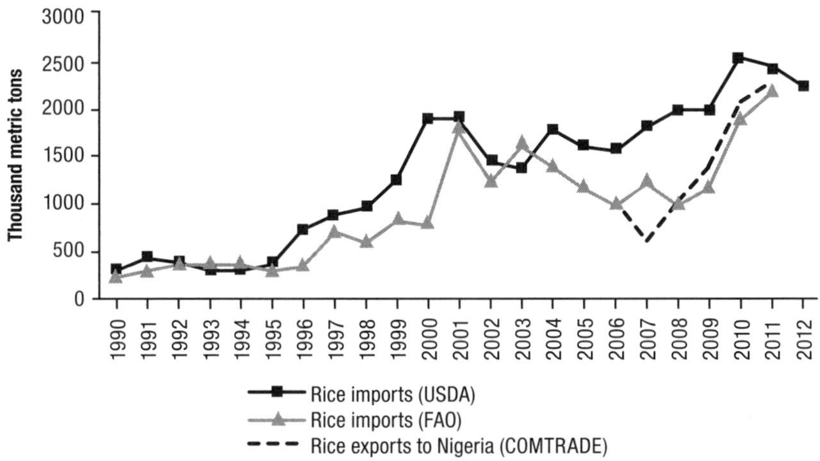

Source: USDA (2012); FAO (2012); UN Statistics Division (2012).

Note: COMTRADE = United Nations Commodity Trade Statistics Database. FAO = Food and Agriculture Organization of the United Nations. USDA = United States Department of Agriculture.

Exchange Rates and Milled Rice Import Tariffs

Although import tariffs have been the main policy instrument influencing the price of imported rice in Nigeria, changes in exchange rates, the extent of foreign-exchange rationing, and the world price of rice are also important components in the cost of imported rice (the import parity price). Fluctuations in all of these components have contributed to substantial volatility in import parity and domestic rice prices in Nigeria. In this section, we briefly review exchange-rate and rice tariff policies and discuss the impact of the import tariff and exchange-rate policies on the domestic price of imported rice.

Exchange-Rate Policy

Exchange-rate policy in Nigeria has been heavily influenced by developments in the petroleum sector. In the 1950s and 1960s, the agriculture sector was the principal foreign-exchange earner, with 3–4 percent annual average output growth rates for export and food crops. During this period, macroeconomic policies were formulated on the principle of taxation (explicit and implicit) of the agricultural sector to fund investments in other sectors of the economy (manufacturing, infrastructure development, education, etc.). With the emergence of oil as a major factor in Nigeria's economy in the 1970s, the need for explicit taxation of agriculture to fund government expenditures lessened, as did indirect tax rates. However, real exchange-rate appreciation resulting from increased oil exports led to substantial indirect taxation of agriculture.

The increase in foreign currency earnings stemming from increased oil exports led to a sharp appreciation of the domestic currency (naira) in real terms. As massive oil export revenues were spent in the domestic economy in the 1970s and 1980s, nominal wages and the domestic prices of non-tradeable goods and services rose sharply, while prices of tradeable goods such as cocoa, palm oil, rice, and many manufactured goods that were largely determined by the world price—converted to naira at the official nominal exchange rate—generally remained stable or rose only slowly, resulting in a decline in their prices relative to prices of non-tradeables. As a result, growth in the non-oil tradeable goods sector suffered, a phenomenon known as the Dutch Disease (named for the stagnation of the industrial sector in the Netherlands following a rapid increase in natural gas export revenues in the 1970s). While some governments attempted to mitigate the impact of Dutch Disease using exchange-rate protection (adopting policies to prevent the tradeable/non-tradeable price ratio from continuing to decline as the oil boom proceeded), the Nigerian Central Bank adopted a policy of gradual nominal appreciation of the naira against the US dollar and British pound in the 1970s in order to maintain naira exchange rates that reflected the country's balance-of-payments position.[5]

Table 7.3 presents the evolution of Nigeria's real exchange rate (RER), a measure of the relative price of tradeables to non-tradeables, here calculated as the nominal exchange rate multiplied by an index of the price of tradeables (the US wholesale price index) and divided by a measure of the price of non-tradeables in Nigeria.[6] During the 1970s and early 1980s, the real exchange rate appreciated by more than 64 percent (from 94.8 in 1970 to an average of 36.3 in 1985) as the nominal exchange rate (naira/US dollar) depreciated by only 25 percent from 1970 to 1985 (0.71 to 0.89 naira/US dollar), while domestic inflation (which raised prices more than 800 percent) greatly outpaced inflation in international (US dollar) prices (which rose by 180 percent).

Moreover, following the collapse of global oil prices in 1982 and the rise in real interest rates on international markets, Nigeria experienced increasingly severe foreign-exchange shortages. As its foreign-exchange reserves dwindled,

5 Throughout this period, Nigeria maintained an adjustable peg exchange-rate policy in the official market.

6 More generally, the real exchange rate is defined as RER = ER * PT / PNT, where ER is the nominal exchange rate measured in units of local currency per international currency, PT is the world price of tradeables in a common international currency (typically an index such as the weighted average of producer prices index of trading partners), and PNT is a measure of domestic prices of non-tradeables, such as the domestic consumer price index.

TABLE 7.3 Average annual nominal and real exchange rates for Nigeria, 1980–2010

	Nominal exchange rate (naira/$)	Parallel nominal exchange rate (naira/$)	Parallel nominal exchange-rate premium (percent)	US WPI (2010 = 100)	Nigeria CPI (2010 = 100)	Nigeria inflation (percent)	Real exchange rate (2010 = 100)	Parallel real exchange rate (2010 = 100)
1970–1974	0.7	1.0	52	23.1	0.1	9	85.4	130.4
1975–1979	0.6	1.0	67	36.1	0.3	21	55.0	91.0
1980–1986	0.9	2.8	217	53.8	0.7	16	43.0	131.0
1987–1992	8.5	11.3	35	60.6	2.3	27	142.1	189.8
1993–1998	21.9	74.5	240	67.1	17.5	39	71.9	219.0
1999–1906	118.5	123.7	5	76.5	45.1	13	139.2	146.5
2007–2008	122.2	122.2	0	98.0	74.7	8	106.7	106.7
2009–2011	151.0	151.0	0	100.8	99.6	12	102.0	102.0
2012	157.5	157.5	0	109.4	124.4	12	92.2	92.2
2013	157.3	157.3	0	110.1	134.9	8	85.4	85.4

Source: International Monetary Fund International Financial Statistics and authors' calculations.

Note: US WPI = US wholesale price index; CPI = consumer price index.

the Nigerian government resorted to rationing foreign exchange in the official market by implementing an import licensing system. As a result of these restrictions, a parallel foreign-exchange market emerged, with substantial premiums: they rose from an average of 67 percent from 1975 through 1979 to more than 200 percent from 1980 through 1986.

Macroeconomic reform—including introduction of a Second-Tier Foreign Exchange Market in 1986 to bring the official exchange rate close to the parallel market rate and a gradual depreciation of the nominal exchange rate thereafter—brought about a substantial depreciation of the real exchange rate and greatly reduced the spread between the official and parallel real exchange rates from 1987 through 1992. During this period, the real exchange rate was on average 142.1 (2010 = 100), 50 percent above its 1970 level (Table 7.3).

Unfortunately, this period of favorable incentives for tradeable goods production did not last. The official exchange rate was kept essentially fixed at about 22 naira/US dollar from 1993 through 1998, a period when domestic prices rose by nearly 320 percent, as domestic inflation accelerated to an average of 39 percent per year. Given only moderate increases in international dollar prices, the real exchange rate appreciated again and averaged only 71.9 for the period.

Macroeconomic stability was largely restored in 1998, and the naira was again allowed to depreciate substantially. Foreign-exchange rationing was greatly reduced, as evidenced by a sharp fall in the parallel market premium to only 5 percent. Over the period 1999 through 2006, the real exchange rate

FIGURE 7.5 Nigeria's nominal and real exchange rates, 1990–2013

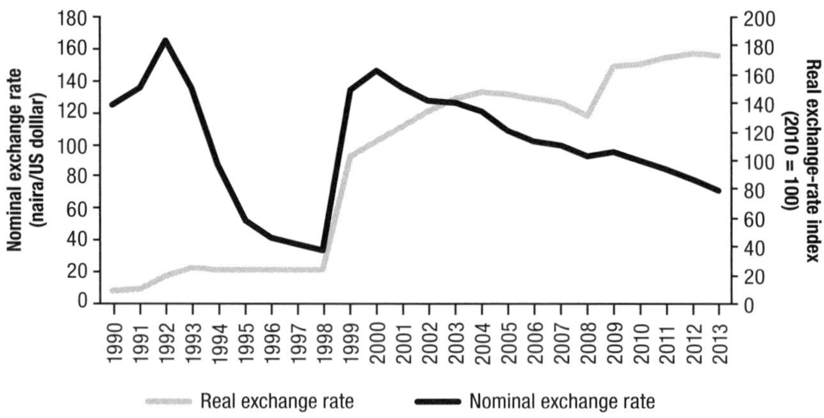

Source: Real exchange rates are authors' calculations based on IMF data; nominal exchange rates are based on monthly IMF exchange-rate data.

Note: IMF = International Monetary Fund.

averaged 139.2, essentially the same as from 1987 through 1992. After 2006, however, the real exchange rate gradually appreciated to 91.8 in 2012 (a total appreciation of 19 percent between 2006 and 2012), although in the absence of rationing, there was not a major parallel market for foreign exchange. This real exchange-rate appreciation implies a substantial reduction in real import prices (a major disincentive for domestic rice production).

Thus, inconsistent macroeconomic policy in Nigeria has led to very large swings in real exchange rates (and in the incentives for production of tradeable goods) since the surge in oil export revenues of the early 1970s. For rice and other "exportable" agricultural commodities (like palm oil and groundnuts), the appreciation of the domestic currency and the resultant shift in terms of trade meant that it was cheaper to import these products than to produce them at home. However, to some extent, rice trade policy (in the form of import tariffs) has served to partially offset the effects of real exchange-rate appreciation in recent years, as described below.[7]

Rice Import Tariffs

In the absence of foreign-exchange rationing, the cost of imports (the import parity price, P_m) is determined as the world price (cost and freight [C&F], Lagos) measured in foreign currency times the exchange rate, plus tariffs, transport, and marketing costs to the domestic market. In mathematical terms,

$$P_m = EP_w(1 + \tau) + \text{mktg}, \tag{1}$$

where P_m is the import parity price in naira, P_w is the world price (C&F, Lagos) in foreign currency units, E is the exchange rate (naira/foreign currency), τ is the ad valorem tariff rate, and mktg is total transport and marketing costs.

In many years, effective rationing of foreign exchange through restrictions on foreign-exchange availability (for imports or for foreign-exchange transfers) resulted in a significant spread between the official and parallel market exchange rates. For rice imported with foreign exchange acquired through the parallel market, the relevant import parity price is

$$P_m = E^*P_w(1 + \tau) + \text{mktg}, \tag{2}$$

where E^* is the parallel market exchange rate (naira/foreign currency) and all other variables are as defined earlier. However, for rice smuggled into Nigeria, either through under-invoicing of imports or evasion of customs procedures,

7 For further discussion of Nigeria's macroeconomic policies, see Sackey et al. (2012).

no import tariffs are paid, and if the foreign exchange for these imports is obtained at the parallel market rate, the relevant import parity price is

$$P_m^* = E^* P_w + \text{mktg}^*, \quad (3)$$

with the cost of smuggling included in the marketing cost (mktg*). On the other hand, if the smuggled rice is purchased with foreign exchange obtained through official sources, then the import parity price is given as

$$P_m = EP_w + \text{mktg}^*. \quad (4)$$

In the analysis that follows, we include measures of import parity at the official exchange rate with imports (P_m from equation 1) and import parity at the parallel exchange with no tariffs (P_m^* from equation 3).[8]

Up until the early 1990s, Nigeria kept its import tariff on milled rice relatively low: 20 percent. Incentives for rice imports nonetheless varied according to the changes in the real exchange rate discussed above, as well as the changes in the world price of rice. Rice tariffs became a major policy instrument starting in 1993, when they were increased to reduce the level of imports, which had been encouraged by the appreciation of the real exchange rate.

As shown in Table 7.4, at official exchange rates, the C&F price of rice in real terms (i.e., adjusted for domestic inflation) averaged 53.5 naira per kilogram (n/kg) from 1993 through 1998, less than half its average level from 1987 through 1992 (111.7 n/kg). Even though the average import tariff in the 1993–1998 period was higher than the average of the 1987–1992 period (60 percent and 20 percent, respectively), the average import price (including tariffs) was still lower in the 1993–1998 period (77.9 n/kg) than the previous period (134.1 n/kg), implying substantially greater incentives for imports in the official market. Since 1998, in spite of relatively high tariffs, rice prices at the wholesale level have remained stable relative to the prices in 1993–1998 and even somewhat lower than prices in the 1987 through 1992 period, in part because of widespread smuggling.

An analysis of changes in rice tariffs and the levels of premiums in the foreign-exchange market (which reflected the extent of foreign-exchange rationing in the official exchange market) shows that the incentives for and levels of smuggling have varied substantially over time. In the first half of the 1980s, the level of domestic prices relative to import prices in the official and parallel markets suggests that domestic prices were determined largely by the

8 Note that in any given market, the source of rice imports may vary over time, so that the prevailing market price (the minimum of P_m^* and P_m) may fluctuate. The *average* market price of imported rice over time thus may be between P_m^* and P_m.

TABLE 7.4 Import prices, import tariffs, exchange-rate premiums, and wholesale prices of rice, 1980–2013

	Import price official ER (2010) n/kg	Import tariff (percent)	Import price official ER w/ tariff (2010) n/kg	Parallel exchange-rate premium (percent)	Import price parallel ER (2010) n/kg	Wholesale market price (2010) n/kg
1970–1974	131.5	20	157.8	52	176.7	246.6
1975–1979	74.6	20	89.5	67	110.0	158.9
1980–1986	44.0	20	52.8	217	100.1	145.3
1987–1992	111.7	20	134.1	35	140.3	198.3
1993–1998	53.5	60	77.9	240	137.2	118.9
1999–1906	69.5	97	136.9	5	73.7	120.0
2007–2008	69.8	72	116.3	0	74.9	126.8
2009–2011	66.1	60	105.6	0	71.4	125.0
2012	73.9	72	127.2	0	79.2	110.8
2013	65.7	110	131.5	0	71.0	110.8

Source: Oyejide (1986a, 1986b) for 1970–1976; Robertson (1983) for 1977–1981; Federal Office of Statistics data for 1982–2000 (Walkenhost 2009); Nigeria, NBS (2013b).

Note: Import price is calculated as 5 percent of the Thai broken rice price (with the broken rice price taken from Anderson [2009] data files) plus freight costs—which are estimated as 17.8 percent of the Thai broken rice price—multiplied by the appropriate exchange rates. The wholesale price for 1970–1976 is a spliced series constructed by Oyejide (1986a, 1986b); the wholesale price for 1977–1981 is from Robertson (1983) and the 1982–2000 wholesale price is from the Federal Office of Statistics, reported by Walkenhorst (2009); the 2001–2013 wholesale price is the retail price of imported rice in Enugu (Nigeria, NBS 2013b) divided by 1.5 (the estimated markup from wholesale to retail). Enugu is a city in the South East zone. ER = exchange rate. n/kg = naira per kilogram.

cost of obtaining foreign currency on the parallel market to import rice. From 1980 to 1986, Nigeria kept rice import tariffs fixed at a relatively low level of 20 percent. Wholesale market prices averaged three times the cost of imported rice at Lagos port (145.3 n/kg versus 44.0 n/kg), suggesting that most imported rice was sold at prices reflecting the parallel market exchange rate that was on average 217 percent above the official exchange rate (Table 7.4).[9]

Import tariffs remained at 20 percent during the 1987 to 1992 period of relatively small premiums in the parallel exchange-rate market that followed the partial liberalization of the foreign-exchange market in late 1986. Import prices at the parallel exchange rate were generally close to the official market import price (including the tariff), suggesting relatively little incentive, on average, for smuggling.

9 Average import prices in the parallel market are determined by the world rice price, as well as the official exchange rate and the parallel market premium. Parallel market premiums were especially high in 1983, 1984, and 1985 (456, 341, and 324 percent, respectively), but since these were years of relatively low international rice prices (as compared to 1980 and 1981), the average import price at the parallel exchange rate largely reflects the lower premiums of the early years.

The period from 1993 through 1998, like the early 1980s, was again characterized by high domestic inflation, limited depreciation of the nominal exchange rate, and foreign-exchange rationing. In an effort to stem the flow of rice imports, the government gradually raised import tariffs. They reached 100 percent in 1996 and averaged 60 percent for the period as a whole. Given that the average premium in the parallel market (240 percent) far exceeded the average import tariff and that average wholesale prices were less than the cost of imports at the parallel exchange rate, the data suggest that most rice imports in this period likely came through official channels. But customs data on monthly tariff payments provide evidence of substantial evasion of import tariffs. While the custom data on the rate of tariffs paid on rice imports are broadly consistent with official tariff schedules (although data by month vary sharply), the total quantity of rice imports in the customs data is much lower than the COMTRADE data based on exports of rice to Nigeria. For example, the implicit ad valorem tariff rate for milled rice (Table 7.5, column 2), calculated as monthly data of import tariffs collected divided by declared customs values, was 60 percent from 2009 to 2011. Using COMTRADE data on quantity of rice exports to Nigeria (which indicate that on average only about 19 percent of imports were recorded from 2009 through 2011) and the average reported price in the NBS customs data for recorded imports, dividing estimated tariff revenues by the value of total imports results in a rate of only 12 percent (Table 7.5, column 4). Likewise, from January to August 2012, NBS customs data show that import tariff revenues divided by the official cost and freight import value (C&F Lagos [$/ton]) equaled 58 percent and subsequently rose to 100 percent for September to December 2012. If only about 20 percent of rice imports paid the tariff, however, the average effective tariff rate would have ranged from 12 to 20 percent.

Figure 7.6 shows trends in import parity price with and without tariffs, as well as the domestic market price of imported rice at Enugu, the regional market with the most complete price series in Nigeria. Enugu is a city in the South East zone (see Figure 1.1 in Chapter 1). Generally, the market price of imported rice moves in the same direction as import parity price including the tariff, although the former follows the latter with a lag. Second, beginning in 2008, the domestic price of imports seems to deviate from import parity price including the tariff, and this deviation has increased; this is likely to be partly because tariffs and other trade restrictions have increased since 2000. The figure suggests that some of the increased tariffs are passed to consumers in the form of higher domestic price of imported rice, but the pass-through rate is only partial and seems to be decreasing in recent years, probably due

TABLE 7.5 Nigeria's rice imports, tariffs, and market prices, 2008–2013

	(1) C&F Lagos ($/ton)	(2) Implicit ad valorem tariff rate (percent)	(3) Official/total imports (percent)	(4) Average effective tariff rate on total rice imports (%)	(5) Pm Enugu (n/kg)	(6) Imported price Pm (w/ tariff) Enugu (n/kg)	(7) P_Enugu-urban imported (n/kg)	(8) Nominal Rate of Protection (7)/(5)–1 (%)	(9) Percent difference between Pdom and Pm (w/tariff) (7)/(6)–1 (%)
2008	550.2	44	18	8	112.9	156.6	161.6	43	3
2009	373.4	64	17	11	100.5	183.8	194.7	94	6
2010	430.0	64	28	18	116.4	126.1	179.4	54	42
2011	508.2	52	12	6	138.9	182.0	179.1	29	–2
2012	583.9	72	10	8	162.1	261.6	206.7	28	–21
2013	577.5	110	—	—	162.1	298.4	—	—	—
Av. 2009–2011	437.2	60	19	12	118.6	164.0	184.4	55	12
Jan–Aug 2012	581.8	58		—	161.3	241.2	204.3	27	–15
Sept–Dec 2012	588.1	100		—	163.6	302.4	211.5	29	–30

Source: Authors' calculations based on Nigeria, NBS (2013b) and UN Statistics Division (2012).

Notes: NBS = National Bureau of Statistics; Pdom = NBS retail price of imported rice in Enugu. The figures indicate a very low rate of official tariff collection; however, the high prices in retail markets suggest substantial rents (possibly including unofficial payments). Annual averages are calculated based on monthly data, so figures in column 8 will differ slightly from those calculated using the figures in columns 5 and 7.

(1) C&F Lagos is calculated as 5 percent Thai broken rice plus $44.28/ton freight costs (from Anderson [2009] data files).

(2) The implicit ad valorem tariff is based on monthly data of import tariffs collected divided by declared customs values (2008–2011) and on official tariff rates (2012–2013).

(4) The average effective tariff is the estimated total revenues (implicit ad valorem tariff × official imports) / total (United States Department of Agriculture est.) from rice imports. USDA estimates of rice imports are not substantially different from United Nations Commodity Trade Statistics Database data for this period.

(5) Pm Enugu is calculated as [C&F Lagos + port handling charges and transport and marketing costs to Enugu] × (1 + 50% retail margin).

(6) Imported price Pm (w/tariff) Enugu is calculated as [C&F Lagos + tariff (column 2) + port handling charges and transport and marketing costs to Enugu] × (1 + 50% retail margin).

(7) P_Enugu-urban imported is the NBS retail price of imported rice in Enugu, also known as "Pdom."

(8) The Nominal Rate of Protection for rice is calculated as annual averages of monthly values of [P_Enugu-urban imported (Pdom) in column 7 / P_m Enugu in column 5] – 1. Therefore, figures may be different from those calculated using figures in columns (5) and (7), which are annual averages of respective prices.

(9) The percent difference between P_Enugu-urban imported (Pdom) in column 7 and the imported price P_m (w/tariff) Enugu in column 6.

FIGURE 7.6 Nigeria's domestic and import parity rice prices and rice imports, 2005–2013

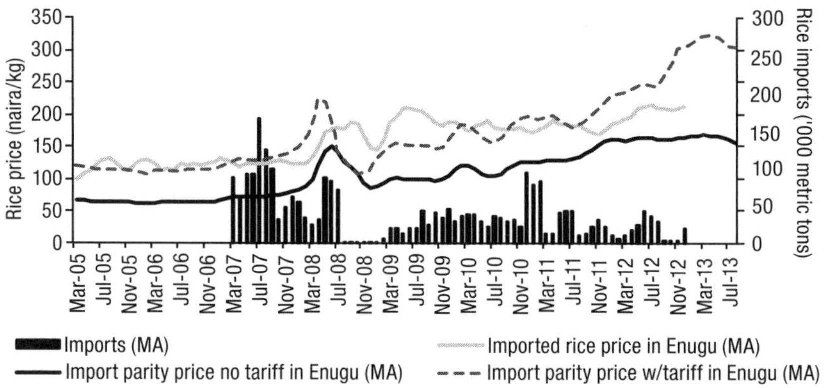

Source: Authors' calculations based on Nigeria, NBS (2013b).
Note: Data are three-month lagged moving averages (MA).

to increased tariff evasion. Since 1998, there is little evidence of rationing of foreign exchange, as parallel market rates have been close to official exchange rates. Rice import tariff rates were kept at 100 percent from 1996 to 2007 (except for a one-year reduction to 75 percent in 2002). Wholesale prices of rice in this period appear to be somewhat less than the cost of imports plus transport and marketing margins to wholesale markets, suggesting that there may have been substantial tariff evasion in the 1996 to 2012 period.[10]

As shown in Figure 7.6, import parity prices (including the 100 percent import tariff) were broadly stable between 2005 and mid-2007. International rice prices rose sharply in 2007 and 2008, as did wheat and maize prices, particularly following restrictions on rice exports by India and Vietnam. In order to prevent a large increase in domestic prices of rice, Nigeria reduced its rice import tariff to zero in April 2008.

The rice import parity price, without tariff, dropped to below N 130 (Nigerian naira) per kg by August 2008, while domestic prices remained high. Imports thus appear to have been constrained during the period from mid-2008 through late 2009, a time when there were no official trade restrictions. The large gap between import parity prices and domestic prices suggests that there were substantial excess profits (rents) in this period. By early 2010, import parity prices with tariff had risen and were again approximately equal to the domestic price (at Enugu) until almost the end of 2011. Import

10 During the global food crisis of 2008, tariffs on rice were temporarily removed in an effort to mitigate the effects of high world rice prices.

tariff rates averaged about 50 percent from mid-2009 through mid-2012, with potential tariff revenues (including any possible rents) estimated to have been about $800 million per year (or N 120 billion, at N 150/$1).

There was a substantial increase in the tariff rate toward the second half of 2012. Because of tariff evasion, however, this large jump in tariffs did not result in corresponding increases in domestic prices in major rice markets. Given that international prices were broadly stable during this period, the lack of response could be explained by an increased evasion of tariffs (in percentage terms) and no increase in total marketing margins between C&F and retail markets. But another explanation is that domestic sales came from ample imported rice stocks accumulated prior to the tariff increases. The bill introducing the new tariffs on rice was signed into law in April 2012. This law increased tariffs on polished/milled rice to 50 percent effective on July 1, 2012, and to 110 percent effective January 1, 2013. In May 2012, there was a large increase in rice imports (Figure 7.7), which may have been the result of increased purchasing in order to avoid the new tariff.

In August and September of 2012, recorded quantities of rice imports fell to near zero at the time of the tariff increase. By October 2012, the volume of rice imports had recovered in the COMTRADE data but remained very low in the NBS/customs data, suggesting even greater tariff evasion after the tariff increase (Figure 7.7).

Though cross-border land imports appear to have been relatively small in the 2009–2011 period, they may have increased substantially in 2013

FIGURE 7.7 Nigeria's monthly rice imports, COMTRADE and NBS/customs data, 2010–2013

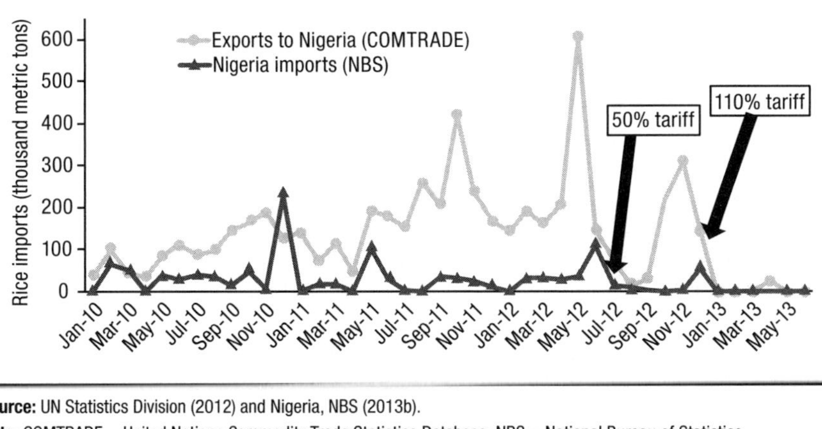

Source: UN Statistics Division (2012) and Nigeria, NBS (2013b).

Note: COMTRADE = United Nations Commodity Trade Statistics Database; NBS = National Bureau of Statistics.

following the major tariff increase in late 2012. Interviews with wholesale rice traders in Kano in October 2013 indicate that rice imports through Lagos declined precipitously, but that most imported rice was flowing from Benin through Niger, most likely by crossing into Nigeria at the Maradi-Memuja border.[11] How this might have occurred can be illustrated by a simple model showing the diversion of imports through other unofficial channels as tariffs are raised in the main Lagos port. The stylized model and results are discussed in Appendix F. From this example, the diversion of imports occurs so long as the cost of using unofficial channels is sufficiently lower than paying the tariff (inclusive of inland transportation and marketing costs). At some threshold tariff rate, all imports would be diverted from the official port of entry. Although this mechanism needs to be more formally investigated in future studies, it is consistent with the observed patterns of tariff revenues and flow of rice. The potential diversion of imported rice can pose uncertainty in tariff revenues and make them less reliable revenue sources for the government. Such uncertainty also creates challenges to rice-sector development policies that can be potentially financed by these revenues.

Import Tariffs on Other Types of Rice

The above discussion focuses on tariffs on milled rice, but Nigeria's rice tariffs vary with the degree of processing, with higher tariffs levied on products that undergo more processing (e.g., polished rice) than on unmilled rice (paddy) or brown rice. Most of Nigeria's rice imports are classified as milled and semimilled rice, which includes a wide range of qualities of packaged milled rice. In 2012, milled and semimilled rice accounted for 67 percent of the total value of recorded rice imports; husked (brown) rice and broken rice accounted for 23 and 10 percent of the total value of imports (Table F.1 in Appendix F). Average tariff rates for brown rice were only 41 percent, as compared to 64 percent for milled and semimilled rice and 74 percent for broken rice.[12] To a large extent, as discussed in Chapter 6, these differential tariff

11 COMTRADE data show that Benin imports approximately 600,000 or more metric tons (Figure 7.3) of rice per year. With a population of 8.6 million in 2009, and assuming per capita consumption of 30 to 35 kg/capita/year (FAO Food Balance Sheets report 34.4 kg/capita/year in 2009 in Benin, and 26.9 and 27 kg/capita/year in Nigeria and Ghana, respectively), Benin's total rice consumption would be 258,000–301,000 tons. Given domestic net production of 100,000 tons and imports of 897,000 tons (COMTRADE) in 2009 (based on the figures of exports to Benin, rather than imports by Benin), consumption of imports would be 158,000–201,000 tons, leaving 696,000–739,000 tons to be re-exported.

12 Note that broken rice is also milled.

rates are an attempt to support the domestic milling industry by encouraging additional supplies of imported brown (husked) rice for light milling (polishing) to white rice.

Effects of Import Tariff on the Price of Local Rice

As discussed in Chapter 2, slightly more than 50 percent of the rice consumed in Nigeria is produced and processed locally in Nigeria (local rice). By definition, import tariffs are not levied on local rice; taxing local rice would defeat the purpose of using import tariffs to protect domestic rice producers. However, it is possible that an increase in tariffs that raises the domestic price of imported rice will also lead to an increase in the price of local rice, because local and imported rice are to some extent substitutes in consumption.

The discussion of price trends in Chapter 2 suggests that the price of local rice responds to increases in the price of imported rice stemming from increased tariffs and other rice trade policies. Using the Enugu price data mentioned earlier as well as similar monthly price data from Niger State, whose time-series data are long enough to allow econometric analyses, Takeshima and Masias (2013) recently tested this formally using co-integration analysis. Their analysis tests for the relationship between changes in the price of imported rice and the price of local rice using monthly time-series data at Enugu for local rice price and the domestic market price of imported rice. In order to overcome the fact that sample sizes are still fairly small and that some series are nonstationary, they use the Autoregressive Distributed Lag (ARDL) model from Pesaran et al. (2001), which is more robust to small samples than conventional models. They obtained the following long-run relationships for Enugu between January 2006 and December 2010:

$$P_{\iota L} = 0.739 + 0.206 P^{*}_{mM} \tag{5}$$

$$P_{\iota L} = 0.477 + 0.329 P_{mM} \tag{6}$$

where $P_{\iota L}$ and P_{mM} are the observed market prices of local rice and imported rice, respectively; P^{*}_{mM} is the calculated import parity prices as discussed above and defined in equation (2) above; and 0.206 and 0.329 are estimated coefficients.[13] The estimates indicate that a 1 naira increase in the border

13 As long-run estimates, they do not signify a specific length of time. Rather, they indicate how prices converge over time to these relationships, which the ARDL model can detect. While the speed of such convergence is unknown, true price movements cannot deviate too far from this. Further detail on estimation methods and other technical matters can be found in Takeshima and Masias (2013) and are available upon request.

price of imported rice caused by increased tariff increased the price of local rice by about N 0.20. Similarly, a 1 naira change in the domestic market price of imported rice induces a 0.33 naira increase in the price of local rice. Based on the supplementary analysis through the Structural Vector Autoregression model for stationary segments of the data (Enugu, January 2003 through December 2010; Niger, January 2007 through December 2011), Takeshima and Masias (2013) find that a change in import parity price does not affect the price of local rice within the same month and only modestly affects it in the following month. In short, while import tariff may cause an increase in the price of local rice, the transmission is partial, small, and not instantaneous.[14] The implication is that tariffs and other protectionist trade policies may not provide the producer price incentives that policymakers hope for, as is also demonstrated in Chapter 4.

Conclusion

For most of the past four decades, trade and macroeconomic policies have combined to reduce incentives for domestic production of rice and lower farmer revenues from rice. Moderate tariff protection from the 1970s through the 1990s did not compensate for the macroeconomic bias against domestic rice production that occurred in periods of substantial appreciation of the real exchange rate. Moreover, the high tariff rates in the 2000s have not been totally effective in raising real prices of rice.

High tariff rates for rice did not raise substantial government revenue due to tariff evasion (most likely through underinvoicing of imports or evasion of customs altogether). The lack of a price response following the 2012 tariff increase is further evidence that, due to the high level of tariff evasion, tariff policy is a very blunt instrument for affecting domestic prices. Nonetheless, the domestic market price of imported rice is substantially higher than the import parity price without tariff, suggesting that tariffs and costs related to tariff evasion (bribes, etc.) substantially raise prices for consumers.

A number of key messages emerge from this chapter. First, rapid increases in the demand for rice and the inability of domestic production to meet this increased demand imply that rice imports in Nigeria are likely to continue to grow in the near future, despite the government's effort to decrease imports

14 Estimating the exact length of time for full transmission is left for future studies, as it requires longer data series under relatively stable international prices, where the effects of tariff changes can be clearly separated from price movements in the international market.

through trade policies. Second, increased protection through higher tariffs has not decreased rice imports to Nigeria; rather, it has encouraged tariff evasion and unofficial imports. Efforts to evade tariffs in the future can be expected, as tariffs are difficult to enforce. Finally, because of tariff evasion, not all tariff increases are transmitted to the domestic market in the form of higher domestic prices of imported rice; the divergence between the import parity price and the domestic price of imported rice increases with higher tariffs. Cointegration analysis shows that only about 20 percent of the import tariff on rice and 33 percent of changes in the domestic price of imported rice respectively are transmitted to the price of local rice. This suggests that local producers do not get the benefit of higher prices that higher tariffs are supposed to bring them.

Chapter 8

ECONOMYWIDE EFFECTS AND IMPLICATIONS OF ALTERNATIVE POLICIES

Xinshen Diao, Michael Johnson, and Hiroyuki Takeshima

The preceding chapters have focused on specific policy challenges and opportunities for transforming the domestic rice economy with regard to consumption, production, milling, marketing, and trade. This chapter considers the economywide effects of these policies on the sector's competitiveness with imports and income growth, as well as on the broader economy. Depending on how the rice sector is integrated into the rest of the agricultural sector and economy, policies targeted at the sector can significantly impact economic welfare through the effects on producer prices, food prices, rural and urban incomes, and overall economic performance. This is especially true for Nigeria, given that many staple crops may substitute for rice from both production and consumption perspectives.

The objective of this chapter is to assess the potential economywide effects of rice-sector policies on sector output growth and competitiveness with imports, changes in the sector and overall food prices, rural income growth, and overall economic performance. More specifically, we ask the following questions: (1) Which policies are likely to increase rice production and substantially reduce rice imports? (2) What are the characteristics of such production growth? (3) How will these policies affect production in the agricultural sector generally, food price inflation, food consumption, and other economywide effects? We use an economywide multimarket (EMM) model developed for Nigeria to analyze these questions. The chapter also briefly discusses the potential opportunity costs of pursuing a rice import substitution policy, in order to provide some partial insights into the cost of implementing key policies identified through the model's simulations.

The chapter begins with a brief introduction of the EMM model, its appropriateness for addressing the broad policy questions being asked, and the policy scenarios to be simulated by the model to reflect these. Three broad scenarios are considered: increased productivity from improved technology use, reduced marketing costs from market improvements, and import restrictions. Results are presented in the sections that follow and according to each scenario. We find the combination of the first two scenarios, adoption of improved technologies and improved markets, to have the highest potential impact on output growth, competitiveness with imports, and overall economic welfare. Nevertheless, because of the desire by the government to protect local rice producers through import restrictions during these early stages of transforming the domestic rice sector, we also offer a discussion of its utility as a policy instrument for promoting import substitution or price stabilization. The final section summarizes and concludes the chapter based on the model simulation results and discussion.

An Economywide Multimarket Model for Nigeria

An EMM model developed for Nigeria is particularly suitable for our purposes, as it not only captures the detailed structure of the agricultural sector but can be particularly useful for dealing with data limitations at the economywide level. Normally, a computable general equilibrium (CGE) model is more suitable for assessing economywide impacts of policy changes, as it builds on the entire structure of the economy. However, it does not break down the agricultural sector in as much detail as the EMM model or other types of agricultural sector models are able to. Moreover, the CGE model requires a detailed and accurate social accounting matrix, which is not always available due to data limitations. This was also the case for Nigeria. Therefore, for our purposes given data limitations, the EMM model proved to be the most appropriate approach. This type of model has been applied in several contexts in Africa south of the Sahara to analyze the agricultural growth options of a country or region (see, for example, Diao and Nin-Pratt 2007 for Ethiopia; Nin-Pratt et al. 2011 for the West African region; and Omamo et al. 2007 for the East and central African regions).

The generic EMM model, like other multimarket economic models, is derived from neoclassical microeconomic theory and designed to analyze the impact of a wide range of sectoral policies by measuring the interaction and interrelationships among markets in the economy. These economic relationships rely on estimates of actual demand and supply behavior in response to price and income changes. These models are easily adaptable to different specifications and level of detail within a sector or across multiple sectors with respect to production, demand, and price relationships.[1] For the Nigerian

1 For interested readers, Diao et al. (2007) provide more detail on the mathematical derivation of the generic EMM model developed at the International Food Policy Research Institute and adapted for the analysis in this chapter.

model, for example, a more detailed rice sector is specified for both demand and supply to enable a more accurate assessment of the likely responses to policy changes.

The Nigerian EMM model is designed to incorporate a detailed agricultural sector, including a more elaborate rice subsector, together with the rest of the economy. Within agriculture, there are 41 subsectors or commodities in total: 26 crops, 5 livestock products, 2 fishery products, and 8 animal products.[2] For the rest of the economy, two aggregate nonagricultural sectors are specified. An aggregate production function for each agricultural and nonagricultural sector has prices as its arguments. Production functions at the commodity level explicitly incorporate both a yield and an area of cultivation function and are defined at the state level (37 states in total). We do so in order to investigate the sources of output changes—yield increases or acreage expansion. The production function for rice is further disaggregated into four production subsystems: upland, lowland, irrigated rice, and other (a residual that represents all other types of rice production subsystems, such as deepwater or swamp rice).

Six types of demand are considered: food, seed, feed, processed, waste, and other. While food demand for a particular product is a function of prices and per capita income, other types of demand are assumed to be proportional to either own-production (for example, seed demand) or the relevant primary production (for example, feed demand proportional to livestock production and processed demand for livestock proportional to meat production). The food demand function for a particular good is defined for rural and urban population respectively, and total food demand is the sum of rural and urban demand. The domestic price for a given product is endogenously determined by the equilibrium between domestic supply and demand when that price is a departure from either import or export parities, which are defined as import or export prices adjusted for import or export tariffs and trade margins. For import (export) commodities (such as rice and wheat), prices are determined as the sum import (export) parities plus (minus) import (export) taxes and trade margins. A detailed description of an EMM model can be found in Diao and Nin-Pratt (2007).

Data used to construct the Nigerian EMM model are obtained from various sources. The agricultural production data at state level are from the Nigerian National Bureau of Statistics (NBS) (Nigeria, NBS 2013b), while FAO (2012) data are used for crops or livestock products not included in the

2 The full list of 41 agricultural sectors is provided in Appendix G.

NBS data. The two nonagricultural sectors are defined as industry and service value-added using data from the World Development Indicators (2012). With the exception of rice, the commodity trade data and data for the six types of demand are also from FAO (2012). We use COMTRADE data (UN Statistics Division 2012) to calculate rice imports. Finally, we use the Living Standards Measurement Study–Integrated Surveys on Agriculture (LSMS–ISA) (Nigeria, NBS and World Bank 2011) to calculate rural and urban food demand. We chose 2010 as the base year for the model partly because the LSMS–ISA (2011) were collected in the harvest season of 2011. Production, consumption, and trade patterns in this year are therefore likely to reflect the 2010 conditions.

As a policy simulation tool, the choice of elasticities for both the supply and demand specifications of the Nigerian EMM model will affect results. Ideally, the elasticities should be econometrically estimated if sufficient data are available. When such data are not available, the elasticities can be obtained from evidence in the literature or calibrated from the model itself. For demand, elasticities were based on model calibrations using income elasticities of demand as a basis for the calculations. The income elasticities are derived from econometric estimates using the LSMS–ISA (2011) described in Chapter 2. With more than 40 agricultural sectors included in the model and substantial data limitations, it is impossible to estimate the supply elasticity econometrically for all crops. Instead we use an estimated direct or own-price elasticity of 0.2 for rice based on the analysis presented in Chapter 4. In the supply functions for all other crops, we chose to use the same own-price elasticity value of 0.2, as this corresponds well with others in the literature despite variations around this mean (see, for example, Thiele 2000, 2003; Alemu, Oosthuizen, and van Schalkwyk 2003; Abrar, Morrissey, and Rayner 2004; Leaver 2004; and Olubode-Awosola, Oyewumi, and Jooste 2006). Cross-price elasticities were calculated for all crops by multiplying the direct or own-price elasticity by the share of the relevant sector's production in the total value of agricultural production. Given the economywide feature of the EMM model, the simulation results are generally not sensitive to the choice of the own-price supply elasticity in a range from minus 50 percent to plus 50 percent of the value we chose. We use the budget expenditure shares and income elasticities estimated in Chapter 2 to derive price elasticities of demand, assuming a linear expenditure demand system for which the budget constraint is satisfied for each demand function.

Rice is the focus of this book, and we therefore pay greater attention to the design of the rice sector in the model. We distinguish two types of rice—low-quality rice (standard) and high-quality rice (premium), as discussed in

Chapter 5. From Chapter 2, we found that many local varieties of rice are only weakly substitutable with imported rice due to their poor quality. Thus, we assume that the low-quality varieties of rice (or standard local rice) do not provide strong competition for imported rice. On the other hand, there are high-quality local rice varieties that are highly substitutable with imported rice (premium local rice). These two assumptions imply that the low-quality rice is produced and consumed domestically and its price is endogenously determined by the equilibrium condition in the domestic market, while the price for the high-quality rice (local and imported) is determined by import parity. Although low, we assume some degree of substitutability between the two types of rice, which is reflected in the model by non-zero cross-price elasticities of demand. This is a fair assumption, not only in order to be consistent with our findings in Chapter 2 of a weak substitutability between local and imported rice, but because we have assumed the premium local variety competes directly with the imported one, their interrelationship with the low-quality variety must be the same. To the extent that increases in the price of low-quality rice induce producers to increase production of low-quality rice and therefore decrease the need for imported rice, the two can be considered substitutes in production.

As noted in Chapter 6, there is very little data on the production shares of high-quality rice. The estimate of 18.3 percent was calculated for the milling model in Chapter 6 based on the maximum capacity of large-scale milling operations that existed in Nigeria; this percentage can be used as a rough indicator of the share of high-quality rice in the country because typically such rice is produced predominantly in large-scale mills (Table E.2 in Appendix E). For the EMM model, we chose to define "premium" according to the rate at which high-quality paddy varieties, and in particular certified seeds, were in use in Nigeria by 2010. Use of certified seeds improves the uniform quality of final grains processed by modern mills and therefore increases its competitiveness with higher-quality imports. A study by Bentley, Ajayi, and Adelugba (2011) suggests that between 2005 and 2009, about 4,000 tons of certified rice seeds were provided to farmers each year. According to FAO (2014), a total of 120,000 tons of rice seeds were used during this period. If these numbers are correct, the share of certified seeds in use by 2010 was about 3 percent. Further assuming, for simplicity, that yields of the two types of rice varieties are the same, then the area cultivated by certified seed is also 3 percent.

Four alternative policy scenarios are considered in the model for expanding rice production. The first scenario (S1) focuses attention on increasing rice yields and expanding the area devoted to high-quality rice varieties; we define this as a technological change policy scenario. The second scenario (S2) reduces marketing margins only; we call it a market improvement policy scenario. The third scenario (S3) combines S1 and S2. The final scenario (S4) raises the tariff for imported rice with no other policy interventions; we refer to this simply as an import restriction policy. Because we are only interested in comparing and contrasting the effects of different policy scenarios, we chose to assume that any policy instruments or public-sector interventions chosen under each scenario will be implemented effectively. This means ignoring the ineffectiveness of higher tariffs on rice imports in Nigeria due to smuggling, as pointed out in Chapter 7. Incorporating implementation inefficiencies would simply reduce the overall effects of any tariffs above 70 percent.

For all four alternative scenarios, we consider two magnitudes of policy changes: an average effort and an accelerated effort. As abstract terms, they simply refer to the extent or intensity to which a combination of policies and/or public-sector investments under each policy scenario is implemented. The "efforts" here correspond to the changes in key parameters in the model. Because we assume that much of the focus of needed public interventions will be for the production of premium local rice if the objective of reducing rice imports is to be achieved, we assume higher yield and area increases for the superior paddy variety as a result of policy changes.[3] Because superior paddy attracts a premium price, as discussed in Chapter 4, it also encourages farmers to adopt more intensive production practices and make technological improvements for high-quality rice relative to low-quality rice.

Table 8.1 summarizes these different magnitudes of policy change for each of the scenarios in terms of their effect on particular exogenous shocks in the model. For example, the table shows an "Average" policy effect under the S1 policy scenario as raising yields of high-quality paddy from 1.9 metric tons per hectare (MT/ha) in the "Base" to 2.2 MT/ha (about a 16 percent change from the base), as well as expanding the area under cultivation from 3 to 5 percent. An "Accelerated" policy effect of the same policy change is assumed to increase yields even further to 2.5 MT/ha (about a 32 percent increase from the base), and the area under cultivation also rises substantially more, from 3 to 8 percent as a share of total rice area. Such higher yields are achievable if government policy focuses on providing improved seeds, increased application of fertilizer, expanding irrigation, and improving farming practices, as we argued in Chapter 4. The "Base" case characterizes current conditions of low productivity, low adoptions

3 We refer to "superior paddy varieties" here to be consistent with its reference as a high-quality and high-yielding seed variety for paddy rice production in Chapter 4. "Premium rice varieties," on the other hand, refers to the higher-quality milled rice that relies on modern milling methods but also to the use of the superior paddy variety.

TABLE 8.1 Assumptions and targets of alternative scenarios applied in the Nigerian economywide multimarket model simulations

	Alternative Scenario	Exogenous parameters shocked in the model	Targeted key endogenous variables in the model	Magnitude of policy change and effect for each relevant scenario: Base	Average	Accelerated
S1	Technology change—standard rice (low-quality paddy)	Yield	Level of yield (MT/ha)	1.9	2.0	2.1
			% change from base		*5*	*11*
		Area	% of total rice area	97	95	92
			% change from base		*–2*	*–5*
	Technology change—premium rice (high-quality paddy)	Yield	Level of yield (MT/ha)	1.9	2.2	2.5
			% change from base		*16*	*32*
		Area	% of total rice area	3	5	8
			% change from base		*67*	*167*
S2	Market improvement	Marketing margins		70	60	51
S3	S1 + S2	Improving rice competitiveness		*All changes and effects above*		
S4	Import restrictions	Change in tariff rate only (%)		50	100	200

Source: Author's calculations for Nigerian economywide multimarket model simulations.

Note: MT = metric tons. A tariff rate of 200 percent is selected for S4. Nigeria's current average bound tariff levels for the agricultural sector notified to the World Trade Organization are 150 percent, with slightly higher rates for some commodities (Kreinin and Plummer 2012).

of high-quality paddy, and high marketing margins due to poor road infrastructure and inefficient processing sectors, as described in the previous chapters.

While the scenarios are designed to affect changes in rice production, the economywide nature of the EMM model allows us to simultaneously solve for the effects on all other economic indicators and across all sectors as a new multimarket equilibrium sets in, such as changes in prices, import volumes, and consumption. After all, production and consumption of non-rice sectors as well as the aggregated food price index can be affected when household incomes change following a change in rice production and rice prices. Model results under each scenario are reported in terms of their effect on changes in rice volumes (production, consumption, and imports), prices (food price index), and incomes (total agricultural real income and household income) to enable us to weigh their overall effect on the government's goals of expanding rice output and improving economic welfare. Due to data limitations, we do not assess the feasibility or costs associated with each intervention but address this later in the chapter when we consider policy tradeoffs.

Finally, it is important to highlight that the EMM model and selected policy scenarios introduced in this chapter—including model parameter values, underlying assumptions, and model structure—are all intended to help illustrate the broader economywide implications of alternative policy instruments on overall domestic rice production, consumption, imports, and household welfare (as measured by changes in prices and incomes). The simulation results reported here, then, are not designed to predict actual future values but rather to project the potential magnitude and direction of change that can occur among economic indicators as a result of the policy change. For this reason, we do not consider modeling the potential effects of ineffective implementation for any of the alternative policies being considered, despite evidence to the contrary (e.g., the problems of tariff evasion highlighted in Chapter 7). Instead, we discuss some of the implications of poor implementation in drawing any final conclusions.

Increasing Rice Production

The EMM model simulation results underline the importance of technological change as the key for improving rice competitiveness in Nigeria. Figure 8.1 presents the resulting changes in total rice production under each of the four scenarios. For each scenario, results are reported under the two different magnitudes of policy change: average and accelerated. For example, for the S1 scenario, the results denote the combined effect of raising rice yields and area cultivated for both standard and premium rice types.

Referring to Figure 8.1, results for the first scenario (S1) show that rice production increases by 6 percent and 18 percent under average and accelerated

FIGURE 8.1 Change in total rice production under alternative scenarios (percent change from base year)

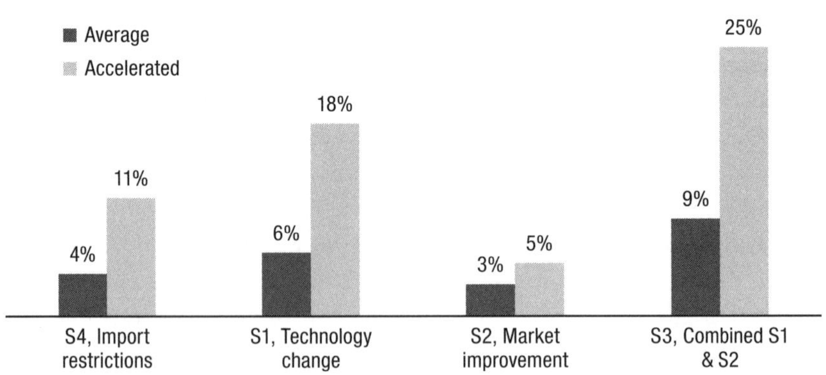

Source: Nigerian economywide multimarket model simulations.

Note: Because of synergies, the accelerated percentage in S3 is greater than the sum of the accelerated percentages for S1 and S2.

FIGURE 8.2 Change in cultivated rice area under alternative scenarios (percent change from base year)

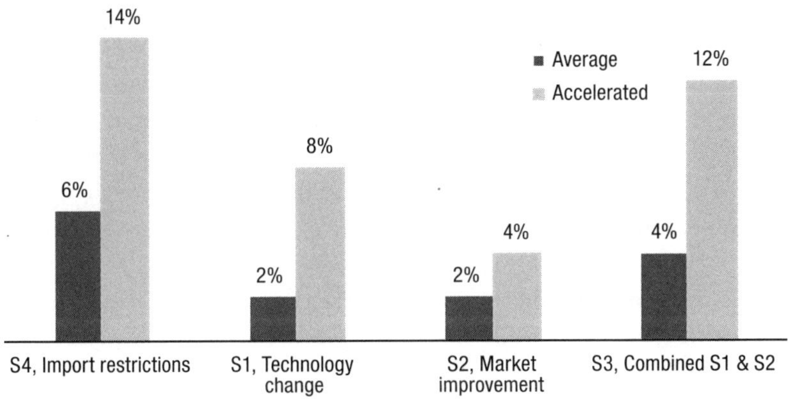

Source: Nigerian economywide multimarket model simulations.

levels of effort, respectively. The figure also shows the importance of market improvement in combination with the technology intervention. Without technological change or import restrictions, market improvements (S2) alone increase domestic supply by 3 percent and 5 percent, respectively, for average and accelerated efforts. Combining technical change and market improvement strategy (S3) increases local rice production by 9 percent and 25 percent under the two effort levels. Under the policy of import restrictions (with no technical change and market improvement), domestic rice production increases by 4 percent and 11 percent, respectively, under average and accelerated efforts.

Figure 8.2 shows the response of areas devoted to rice cultivation under different policies. It shows that under S1, acreage increases by 2 percent and 8 percent for the average and the accelerated efforts, respectively. Under S2, acreage under rice cultivation increases by 2 percent and 4 percent for the average and the accelerated efforts, respectively, while under S3 it increases by 4 percent and 12 percent under the two effort levels, respectively. Under the policy of increased import tariff (S4), the acreage devoted to rice production increases by 6 percent and 14 percent under average and accelerated efforts, respectively. Production increases under S2 and S4 are mostly due to increases in the farmgate price. Production increases under S1 and S3 are driven by an outward shift of the supply function. While production increases induced by price changes are limited if supply is inelastic, supply shifts can be sizable if new technologies help to lower the unit production costs substantially.

When rice growth strategies are centered on technology change, increased rice production will not be at the cost of reducing production of other crops. This can be assessed by changes in rice production areas under alternative growth strategies simulated in the model. As shown in Figure 8.2, the cultivated rice area rises modestly under S1, the technology change scenario. The maximum increase in rice area is 12 percent when technology is combined with the market improvement in S3 under accelerated efforts of intervention. Area increases only modestly because of the production structure changes due to the new technology that allow each hectare of land to produce more output than the old technology. This means that this policy is unlikely to result in an expansion of rice production that replaces other crops.

The results from Figures 8.1 and 8.2 suggest that most of the increase in total rice production under S1 to S3 stems from increased yield/ha (increased land productivity), while under S4 most of the increased rice production comes from acreage expansion, as there is no technical change. Here there is potential for decreased yields per hectare of land; acreage expansion is more likely to displace the production of other local food staples.

Raising Import Tariffs

We now turn to the discussion of raising import tariffs. Increasing the tariff rate is intended to reduce rice imports by making rice imports more expensive relative to domestically produced rice and hence induce consumers to substitute locally produced for imported rice in consumption and producers to increase production of local rice. This has been referred to as the "infant industry argument"—the idea of using tariffs to protect an "infant industry" until it is able to compete on its own. In Nigeria, this seems to be the goal as well, to promote local production and eventually achieve rice self-sufficiency.[4] However, the prospect of achieving self-sufficiency in rice production is unclear if supply response is weak, as Chapter 4 suggests. Moreover, we also argued in Chapter 7 that it is very difficult to enforce tariff policies in Nigeria due to tariff evasion, either through underreporting or smuggling. Therefore, assessing whether higher tariffs can significantly reduce rice imports would seem to be impractical for Nigeria in this case. However, we chose to ignore this for our own illustrative purposes by assuming that policymakers will

4 This can be inferred from public pronouncements of Nigerian government officials. For example, see Nnabuife (2014) and the February 6, 2013, circular released by the Central Bank of Nigeria (CBN) on the subject (*Nation* 2013). Note that the degree of protection is made up of the 10 percent duty + 100 percent levy, for a total tax rate of 110 percent.

FIGURE 8.3 Ratio of rice imports to consumption under alternative scenarios (percent)

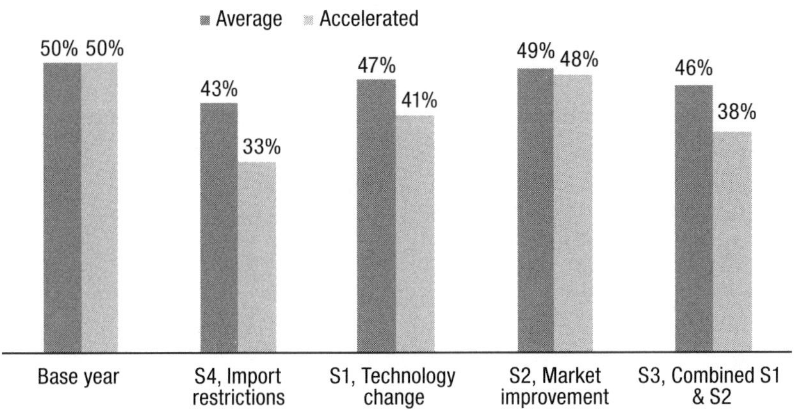

Source: Nigerian economywide multimarket model simulations.

become more effective in implementing these tariff policies and that higher tariff rates will consequently be binding. Additionally, given that rice imports tend not to be very sensitive to small changes in tariff rates, we chose to introduce larger discrete changes in tariffs to test whether tariff increases can help achieve the desired result of significantly reducing rice imports, irrespective of the political feasibility for introducing them.

Figure 8.3 presents the model results of rice import-consumption ratios under alternative scenarios. In the base year (2012, with a tariff rate of 50 percent) of the model, imports account for about half of total rice consumption in Nigeria.[5] When the rice tariff rate increases to as high as 200 percent (S4 scenario) from the current 50 percent, the import-consumption ratio falls to about 33 percent of total consumption.[6] When the tariff rate increases from 50 to 100 percent (average level in 2013), the import-consumption ratio only decreases to 43 percent. This shows that tariff increases alone may be ineffective in reducing rice imports. Because we also assumed that the tariff policy is implemented effectively in these scenarios—that is, with no smuggling or underreporting—its overall effect on imports would be less, and thus even more ineffective, if we remove this assumption.

5 We chose 2012 as the base year when the tariff rate was 50 percent and a policy of increasing the rate to a 100 percent was being debated in parliament. The tariff rate on imported milled polished rice was increased to 110 percent effective January 1, 2013.

6 Given the political economy of tariff policy, we admit that this high tariff rate may be politically and practically impossible to implement. Indeed, the 110 percent tariff rate on milled rice imposed in January 2013 was eventually reduced to 60 percent in 2014.

FIGURE 8.4 Change in rice demand under alternative scenarios (percent change from base year)

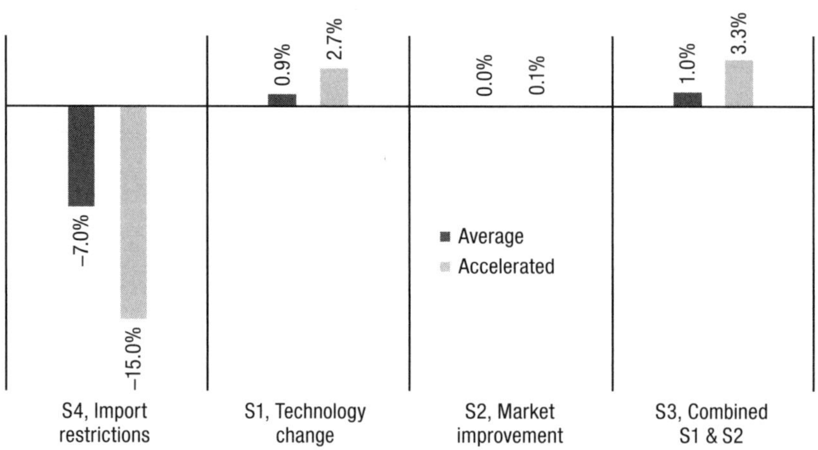

Source: Nigerian economywide multimarket model simulations.

As shown in Figure 8.4 (and Figure 8.1), the decline in the rice import dependency ratio due to increased tariff is a combined result of increasing domestic rice production and reducing rice consumption due to higher prices of rice; that is, the decline in the import–consumption ratio comes partly at the expense of the consumers. For example, rice consumption is reduced by 15 percent with the accelerated tariff rate of 200 percent. In contrast, consumption increases by about 3 percent under the scenario of both increases in productivity and reduced marketing margins (S3), thanks to reduced production costs and thus prices of premium- and standard-quality rice. To further understand the relationship between rice production and consumption at different rice tariff rates, we run a series of additional scenarios with different rice tariff rates, ranging from 50 to 200 percent. Figure 8.5 presents these results.

As shown in Figure 8.5, in the base year, with a 50 percent tariff rate, Nigeria consumes about 5.2 million MT of rice, of which 2.8 million MT are produced domestically, with about 2.4 million MT imported. When the tariff rate rises to 100 percent (2013 tariff rate) from the base rate of 50 percent, domestic rice production increases only modestly, to 2.9 million MT, while demand falls to below 4.8 million MT, which is a 4:1 consumption decline/production increase ratio. This is consistent with the findings from Chapters 2 and 4; while the own-price elasticity of demand for rice is generally low as in Chapter 2 (Table 2.8), the elasticity of supply is even lower at 0.2. At a tariff rate of 200 percent, rice

FIGURE 8.5 Rice production and demand under different tariff rates

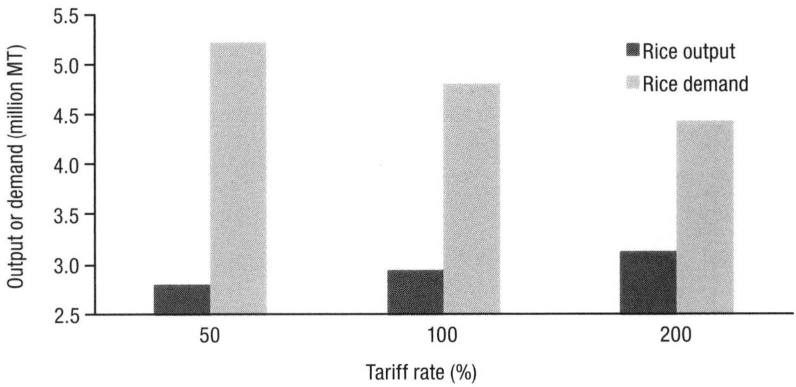

Source: Nigerian economywide multimarket model simulations.
Note: MT = metric tons.

production increases to 3.1 million MT but is still far below the level of rice consumption (4.4 million MT). Thus, under a high tariff policy, the gap between domestic supply and rice consumption would remain, albeit smaller. But even more importantly, the reductions in the size of the gap would be driven mainly by the decline in rice consumption as consumers face big price hikes, a reduction of up to 15 percent. The political feasibility of such a policy, therefore, would need to be seriously considered. Even if this policy option is feasible, there are very large consumer welfare losses associated with it.

The overall effect of the high tariff policy on the economy might be negative. As shown in Figure 8.6, with high tariff rates on rice, the consumer price index (CPI) rises. This increase in CPI is not due to increased demand for goods but to the higher rice prices. The increased food price is driven by two reinforcing factors: a higher domestic price of rice due to the higher tariff and increased prices of domestic staples whose production declines as a result of shifting domestic agricultural resources to produce rice inefficiently. Demand for nontraded products (such as staples) may fall, but increased resource allocation to rice production induced by higher rice prices also raises production costs for these nontraded products. Real gross domestic product (GDP) would also be negatively affected by the rising CPI and misallocation of agricultural resources to the production of rice. Figure 8.6 presents these results. At the tariff rate of 200 percent, nominal GDP falls by more than 0.7 percent while CPI rises by almost 4 percent. Given that rice production increases and its contribution to agriculture—and hence to GDP—is positive with high tariff rates, the decline

FIGURE 8.6 Changes in real GDP and consumer price index under different tariff rates

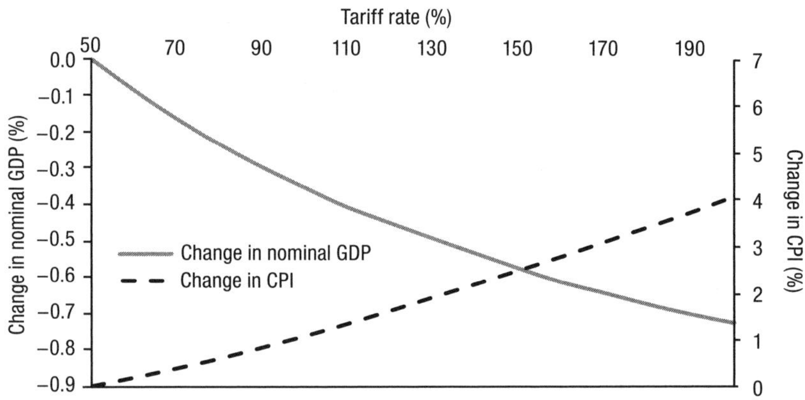

Source: Nigerian economywide multimarket model simulations.
Note: GDP = gross domestic product; CPI = consumer price index.

in GDP implies that other agricultural sectors (that is, aside from rice production) and the nonagricultural sector will be negatively affected by the rice tariff policy. Although one may argue that increased tariff revenues could be used to offset some of these welfare costs, this will depend on how well the government can effectively collect tariff revenues and ensure the transfers. As is shown in Chapter 7, very little tariff revenue is effectively collected in reality.

In Figure 8.7, we assess the impact of alternative rice growth strategies by examining the impact on the incomes of rural and urban households. This provides an indication of the overall welfare effect of different policies. Changes in households' real incomes are measured as their per capita nominal incomes deflated by the CPI in each simulation against the corresponding base-year levels. As shown in Figure 8.7, both rural and urban households are hurt by the import restriction policy, while both gain significantly under the policies that promote technology change and improved market efficiency.

For both urban and rural households, the negative welfare effect under a restrictive import policy is due to declines in nominal income and increases in CPI. For example, at a tariff rate of 200 percent, rural and urban nominal household incomes fall by about 2.7 and 2.0 percent (not shown in the tables), respectively, while CPI rises by 4 percent (Figure 8.6). Declines in nominal incomes are the result of a fall in agricultural and nonagricultural production. Given the low price elasticity of demand, consumers are likely to consume nearly the same amount of rice even at a higher price, which reduces the demand for other agricultural and nonagricultural products. The prices of

FIGURE 8.7 Changes in rural and urban per capita real income under alternative scenarios

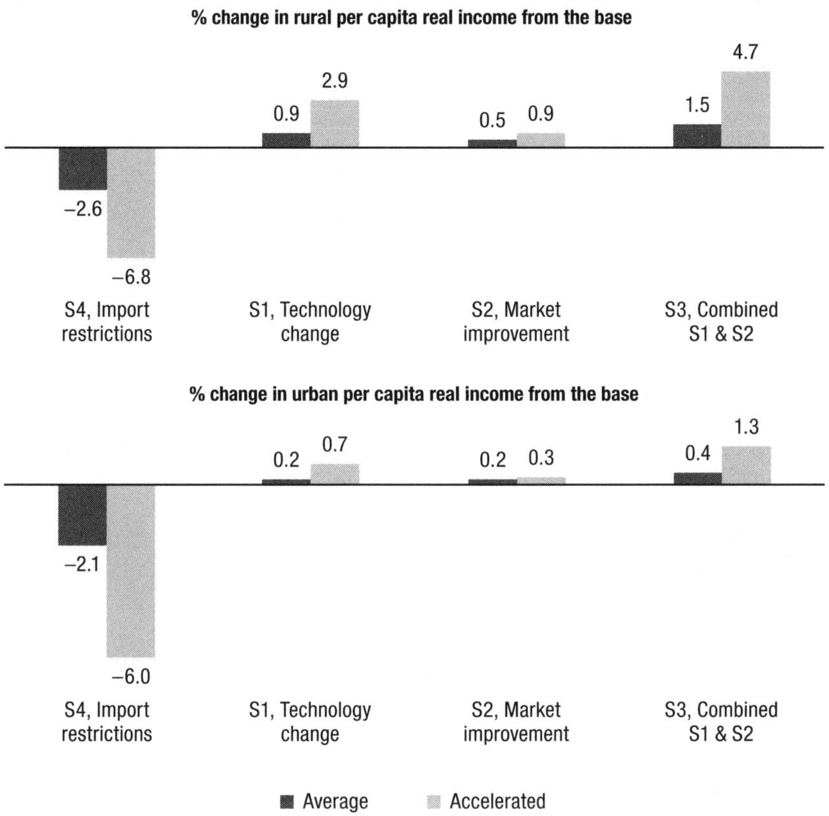

Source: Nigerian economywide multimarket model simulations.

Note: Because of synergies, the percentages in S3 might be greater than the sum of the percentages for S1 and S2.

those commodities produced mainly for domestic markets fall, generating a negative supply response.

Policies that promote technological change and/or improved market efficiency, such as those under scenarios S1 to S3, benefit both rural and urban households, with rural households benefiting more. Under these scenarios, while CPI also rises, nominal incomes of rural and urban households rise even more. For example, under S3, with the accelerated level of productivity change and reductions in marketing margins, rural and urban households' nominal incomes increase by 6.8 and 3.4 percent, respectively, while CPI increases by 2.1 percent. As a result, increases in real income turn out to be 4.7 and 1.3 percent for rural and urban households, respectively (Figure 8.7).

FIGURE 8.8 Changes in relative rice prices under alternative scenarios (percent change from base year)

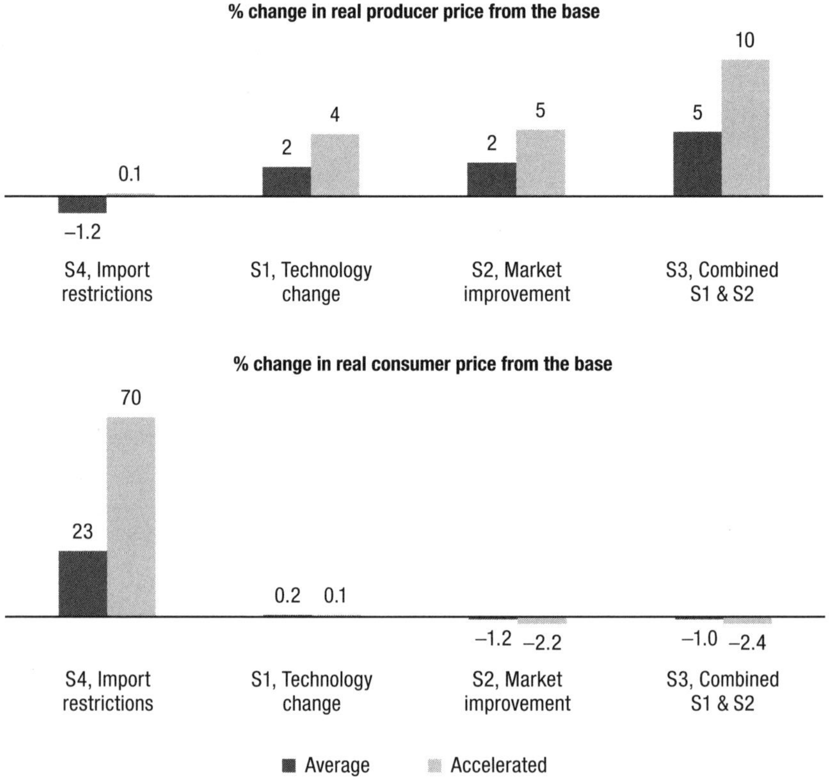

Source: Nigerian economywide multimarket model simulations.

Note: Real prices are defined here as the nominal rice price deflated by the consumer price index. Because of synergies, the percentages in S3 might be greater than the sum of the percentages for S1 and S2.

Figure 8.8 also reports this phenomenon for changes in real producer and consumer prices of rice. Technical changes in rice production and increased income from rice production further increase demand for local rice relative to other crops, which raises producer incomes of rice relative to CPI. The combined technology changes and reduction in marketing margins increase the incomes of rice producers while marginally reducing the price consumers pay (both deflated by the CPI).

If the tariff policy (S4) is maintained together with the combined technology change and reduction in marketing margins (S3), we would expect the prices consumers pay to be higher but still significantly lower than under the tariff regime scenario alone. In this case, a rice import tariff may be justifiable under the "infant industry" argument—especially if such a combination helps to transform the country's domestic rice sector through productivity and output growth. The challenge for the Nigerian government under such a scenario would be as follows: (1) ensuring consistency in tariff policies (for example, to avoid using tariffs simply as an instrument for domestic price stabilization); (2) effectively enforcing the tariff (given the leakages observed in Chapter 7); (3) committing sufficient resources to transform the sector; and (4) committing to a credible exit strategy to avoid a budgetary crisis on account of increasing an unsustainable support of the industry. More important, it is necessary for the government to provide infrastructure in order to attract private-sector investment.

An Import Tariff for Price Stabilization or Self-Sufficiency

A combined strategy of improving rice competitiveness and protecting the sector through import tariffs may have greater effects on rice production than when each one is pursued individually. As noted earlier, such a dual strategy is often adopted using the "infant industry" argument, as the Nigerian government was trying to do under the Agricultural Transformation Agenda strategy of President Jonathan. Aside from introducing tariffs, the government also focused attention on raising the productivity of rice production and processing. However, as the evidence from Chapters 4 and 6 show, much of the focus has been on expanding the large-scale milling sector, with less attention being given to the development and diffusion of improved and superior-quality paddy varieties, as well as improved access to technologies and infrastructure for the more dominant small milling sector. As we found in Chapter 6, while the large-scale milling sector has a greater potential to compete with imports, it will always more likely require a protective tariff policy in place unless significant and sustainable improvements in productivity occur. This is a real challenge, considering the high costs of capital and the current state of paddy production in Nigeria. The small milling sector, on the other hand, is shown to be far more resilient to shifts in policy than the large-scale sector and thus more sustainable in the long run. This becomes important if prices become volatile either because of a sudden change in global markets or in the domestic supply of paddy or stocks of milled rice.

In the long run, it is important to maintain a transparent and predictable tariff policy. The challenge is not so much about justifying a tariff policy per se but more about determining the optimal level of protection needed to maintain sufficiently stable prices without overly distorting the economy. The "infant industry" argument can be justified if policies are consistent with efficient resource allocation over time and across sectors. Overall, welfare costs and benefits to the economy that are likely to occur as a result of the policy must be weighed. As demonstrated in Chapters 2 and 7, the costs of managing a high import tariff rate in Nigeria are likely to be considerable, both with respect to their effects on prices and on consumer welfare. The benefits are also very low, considering the current poor record of tariff implementation and revenue collection. As discussed in Chapter 7, beyond some level of high tariff rate, a very large proportion of rice imports to Nigeria flows through unofficial channels; similarly, low tariff rates may not be enough to provide farmers and millers the protection they need under the "infant industry" argument. It is very likely that there is a threshold at which the cost of a certain tariff just equals its benefits. More research may be needed to determine such an optimal tariff rate.

Opportunity Costs of Pursuing Rice Self-Sufficiency through Technological Change

Technological change in rice production and market improvement both lead not only to more efficient domestic rice production growth, but they also have positive effects on the economy even without the import tariff. Investments required for these policies may, however, involve substantial opportunity costs. While a rigorous assessment of such costs is beyond the scope of this book, we highlight the importance of weighing in these costs in this section.

As discussed in Chapter 4, some of the key investments needed are in rice research and development (R&D) and irrigation infrastructure development. Rice R&D in Sri Lanka provides a useful insight. While starting from 50 percent import dependency on rice in the 1960s, Sri Lanka had achieved self-sufficiency by the mid-1980s. In the late 1990s, the country had invested approximately five to seven times more into rice R&D per rice area than Nigeria. As presented in Table D.3 in Appendix D, the rice R&D expenditure in Nigeria in 1998 was 0.3–0.5 (2010 US$ purchasing power parity) per rice hectare, while it was 2.1 in Sri Lanka. In addition, Nigeria had only six rice scientists for 1 million ha of rice land in 1999, while many Asian countries (including Sri Lanka) had approximately 10–20 rice scientists per 1 million ha in 1983 and 1999 (Takeshima 2014). Roughly speaking, R&D in rice production and processing in Nigeria must increase by a multiple of several times from existing levels to realize the varietal seed development and productivity growth needed to achieve the Nigerian government's goal of rice self-sufficiency. We will come back to this in the next chapter when discussing some of the costs and benefits associated with the rice self-sufficiency goal.

Similarly, as with R&D investments, if Nigeria were to expand the irrigated area under rice from the current 5 percent to 50 percent (approximately 1 million ha), which is closer to the ratio in many South and Southeast Asian countries, it will need to rehabilitate 63,000 ha of existing irrigation schemes and construct new ones of up to 900,000 ha, according to one estimate (see Enplan Group 2004). This can be very expensive, as costs of building irrigation schemes in Africa are generally very high. Estimates by Inocencio et al. (2007) suggest a cost of US$13 billion (nominal). In addition, operations and maintenance (O&M) costs of about US$50 million will be needed annually.[7] At these cost levels, it could take approximately 80 years to meet these costs if current total spending levels of US$160 million per year for irrigation schemes are maintained (Foster and Pushak 2011). While substantially more public funding is needed for rice R&D and irrigation expansion, this should not come at the cost of ignoring investments in agricultural extension and rural infrastructure, particularly rural feeder roads (Fan 2008).

Given scarce public-sector resources, the decision to allocate more resources to the development of the rice sector must be made carefully as it will likely affect investments in other sectors. It may require, for example, reducing spending for fertilizer subsidies, which until recently accounted for a significant share of the agricultural budget. It may also require reducing expenditures on other crops for which Nigeria may have greater competitive advantages, such as cassava and sorghum. Spending in other major sectors such as education, defense, and health could also be affected, all of which have traditionally accounted for a significant share of the federal government's budget (Mogues et al. 2008). Although resources could come from increased taxation, this is hardly an option considering current low incomes, difficulties of income tax collection, and its effects on reducing private investment. Alternative sources, such as revenues from petroleum exports and any import tariffs in place, offer the best option. Ultimately, the push for achieving self-sufficiency in rice is not going to be cheap; indeed, the opportunity costs of such a policy, such as the returns from alternative investments, may exceed its benefits. The Nigerian government must carefully weigh all these costs and benefits.

7 Applying O&M costs in Asian countries in the 1980s (Gupta, Miranda, and Parry 1995) and converting them to the current US dollar, typical O&M costs in these countries were $50 per hectare.

Conclusion

This chapter assesses how various policies may affect rice production, import dependency, and overall economic welfare. Our analysis suggests that, given the importance of rice as a major food staple in Nigeria, rice policies will have significant effects on GDP, consumer prices, and real incomes. This indicates that appropriate rice policies will need to be carefully identified.

Historical trade data analyzed in Chapter 7 indicate that import tariffs may not be effective in reducing rice imports and increasing rice production in Nigeria. The modeling analysis in the current chapter further suggests that even if the tariff were effective, its impact may be limited due to inelastic supply and demand for rice, while welfare effects can be substantially negative. The tariff-induced supply response occurs through rice area expansion without yield growth, which implies that production of other crops can be negatively affected, including overall effects on welfare. Both rural and urban households can be hurt by high import tariffs. Finally, despite the reduced consumption that occurs after introducing very large tariffs (e.g., at 200 percent and assuming tariffs are effective), imports would still account for a significant share (up to 33 percent) of total consumption. Evidently, even with no tariff evasion, achieving self-sufficiency through increased tariffs is unlikely to be successful.

According to the model results, alternative rice growth strategies that focus on technological change and market improvement offer the best options to transform the domestic rice sector. With a modest increase in rice yield of about 16 percent per hectare—so long as it is in conjunction with the expansion of superior-quality varieties that replace low-quality ones—the competitiveness of local rice can increase substantially. In fact, rice output can potentially rise to levels close to those that can be achieved under a highly protectionist policy. Similarly, the import dependency ratios also fall under both scenarios, in the range of 33 to 38 percent from the current 50 percent. The difference between the two, however, is their effects on rice consumption and incomes among both rural and urban households. With high import restrictions, overall consumption and real per capita urban and rural incomes could fall by 15 percent and 6.8 percent and 6.0 percent, respectively, hurting consumers the most. In sharp contrast, they could rise with technology change and market improvement: for consumption, by 3.3 percent and for real incomes by 4.7 and 1.3 percent for rural and urban incomes, respectively.

Aside from the positive effects of technology change and market improvement on production, consumption, incomes, and trade, the economywide impact effects are also positive. Other crop production will not be hurt, and some crops even benefit from the expansion of rice production. With income gains going to both rural and urban populations, including rice producers, as the rice growth is not accompanied by domestic food price inflation (of an overall rise in food prices), consumers will have more disposable income to spend on other goods and services and contributing to overall GDP growth.

The challenge, however, is how to go about investing in technology change and market improvement, as these also come with significant costs. No matter what policies and investments are pursued, a careful cost-benefit analysis will be needed to select the best mix to produce the greatest return to economic and social welfare in the long run. In the short run, if the government still considers the tariff policy as politically imperative using the "infant industry" argument, it will be essential to determine an optimal and consistent tariff that can be effectively managed. In the meantime, public investments in the rice sector must be maintained. As part of this, it will be important also to attract private-sector investments, not only in partnership with the public sector, but to help ensure productivity growth and improved competitiveness over the long term so that tariffs would no longer be needed in the future. This is important, as history shows that it is not so easy for governments to stop supporting an "infant industry": powerful interests within the affected industry tend to fight against ending the protection they have come to rely on.

Chapter 9

TRANSFORMING THE RICE SECTOR

Kwabena Gyimah-Brempong, Michael Johnson, and Hiroyuki Takeshima

This book presents analyses of primary and secondary data on the production, domestic markets, processing, consumption, and trade for the rice economy in Nigeria using a variety of economic models. The goal has been to identify constraints and opportunities for policy in order to transform and improve the domestic rice sector's competitiveness with imports. In this chapter, we summarize the key results and findings from this analysis and discuss the broad policy implications.

Increasing Rice Yields through Technology and Market Improvement

A number of key messages emerge from our analysis. First, rice has grown in importance and is now one of the most important staples in the Nigerian diet. The trend has been driven not only by income and population growth, rapid urbanization, and lifestyle changes but also by past policies that implicitly reduced the ability of local rice producers to respond to the growing demand and compete with cheaper imports.

The demand for rice will likely continue to grow. Moreover, given that a large component of rice in Nigeria is now being consumed by low-income households as well, rice has become an important component of household food security in the country. Our analysis shows that local rice and imported rice are consumed for different purposes. However, they are not easily substitutable for each other. This is confirmed by econometric estimates of the cross-price elasticity of demand between the two types of rice.

Therefore, a change in the price of imported rice alone is not likely to be sufficient to affect local rice consumption (which currently accounts for more than half of total rice consumption in rural areas). In addition, low-income groups—especially low-income urban households—are likely to be hurt the most by increases in the price of imported rice, which the government tries to achieve through import tariffs. Policies that focus on increasing the supply of rice and thereby decrease the price of rice will generally be pro-poor, improving the welfare of low-income groups. Policies that restrict supply and thereby increase the price will generally be regressive, as rice accounts for a relatively large share of poor households' rice expenditures.

Thus, existing rice consumption patterns suggest that rice imports in Nigeria will continue to grow in the near future despite the government's effort to decrease rice imports through trade policies, unless domestic production increases substantially.

However, growing demand alone does not seem to lead to substantial rice paddy production growth in Nigeria. Our analyses indicate that while Nigeria has the potential to increase the competitiveness of local rice through increasing yield growth, there are many challenges to overcome. For example, given highly diverse and fragile soil types and a number of pests, diseases, and weeds that are relatively unknown outside Africa south of the Sahara (SSA), improved and higher-yield rice varieties imported from other countries are not as well adapted to local conditions.

The challenges and prospects for increasing yields vary widely. Northern Nigeria is relatively free of many of the most prevalent pests, diseases, and weeds, but areas with suitable soils are limited to the major river basins of *fadamas* (narrow inland valley bottoms along the two major rivers, the Benue and Niger). Supplementary investments will be needed to exploit fully these narrow suitable areas.

The potential to raise yields from expanding areas under irrigation also has challenges. For example, fertilizer responses of current irrigated varieties appear to be low, and irrigated rice production is often less profitable than rainfed rice production. These challenges indicate that many production technologies, including improved varieties, may be outdated, particularly for lowland and irrigated ecologies, leading to low adoption rates of irrigated rice. Empirical household analyses provide evidence that is consistent with this view: neither rice output nor irrigated rice area is very responsive to rice price increases. Various socioeconomic factors, such as underdeveloped input markets, weak extension systems, and high marketing transaction costs, have led to low levels of rice intensification and productivity.

Historical and international experiences are consistent with these findings. On the technology front, governments in many Asian and Latin American countries invested heavily in rice research and development (R&D), and as a result many of their domestically bred varieties became popular with a majority of producers. Additionally, earlier private investments in irrigation facilitated the successful adoption of high-yield varieties. Governments also invested heavily in public irrigation schemes at a much higher intensity than Nigeria (given the size of arable land compared to Nigeria's). This ultimately affects a whole range of rice-sector activities, including efficiency of both input and output markets or capacity of R&D and extension delivery.

The growing demand has also not translated into improved postharvest processing of domestic rice for the market. Although the bulk of rice milling is still dominated by a cottage industry of small-scale operators located in villages and small town markets who produce an inferior rice variety relative to imports, the sector should not be ignored in favor of promoting large integrated mills for the sole purpose of ensuring locally produced premium-quality rice, as the government has been doing. While meeting the demand for premium-quality rice in the short run is feasible only through the use of the larger industrial mills, there is a real danger that in years to come the sector will still have to rely heavily on government support to remain competitive.

Model simulations of the rice milling sector highlight the vulnerability of this sector to changes in import tariffs and global rice markets, showing that the sector is unable to compete effectively without tariff protection. Experiences elsewhere in West Africa and Asia illustrate this same problem. Meanwhile, by not paying sufficient attention to improving the small-to-medium-scale sector, the demand for standard-quality rice would remain unmet.

Promoting (or at least not discriminating against) the small-to-medium-scale sector makes sense, as this sector is more economically efficient and more resilient to tariff changes or volatility in global markets. While this type of rice is less substitutable with imported brands, a majority of poorer consumers continue to rely on it (about half of total rice consumption), and the potential to upgrade the sector exists, as our results show. The sector can easily absorb a 25 to 33 percent increase in costs to further process and improve the quality of milled rice through de-stoning, polishing, and colorization. Returns can be raised if the current high trader and marketing cost margins can be reduced. One possible way of doing this is through the development of milling clusters, wherein economies of scale and location can be taken advantage of by a greater number of small-to-medium-scale millers choosing to locate within or in proximity to these clusters.

Experiences in Asia and elsewhere in West Africa also suggest that the growth and resilience of the small-to-medium-scale mills can play an important role in contributing to the employment growth and transformation of rural economies. Over time, as incomes have risen and consumer preferences have shifted more toward higher-quality rice, the sector has been able to adapt and attract investors along its value chain in order to upgrade equipment and expand milling capacities. Government support has been critical, but it has mostly been in the following areas: providing access to technology and training; subsidizing credit and sometimes even the price of paddy; and investing in infrastructure to lower costs for energy, transportation, and access to markets and information.

With these supports in place, subagents for paddy producers and other new trader entrants appear on the scene, including those specializing in post-milling operations. Eventually, this route has led to the commercialization of specific rice brands, together with a system of quality grades and standards. While this process can take time, the Nigerian small-to-medium milling sector is poised to respond, given the right enabling environment.

Weak institutional capacity often constrains effective policy implementation and regulatory enforcement in SSA. In the Nigerian rice sector, this problem is manifested in its rice trade policy. Increased tariffs and import bans have not succeeded in substantially reducing rice imports. In part, this is potentially because high import tariffs can encourage unofficial imports (smuggling), both through Lagos port and other cross-border trading posts (including the border with Niger). Efforts to evade higher tariffs are difficult to enforce given the size of the market in Nigeria, the difficulties of collecting tariffs on imports coming through Nigeria's ports, and the expanse of Nigeria's porous borders with neighboring countries. In fact, at a high enough tariff level, imports could simply cease to pass through official channels, as they are diverted through other channels.

Thus, only a small percentage of tariff increases may be transmitted to the domestic market in the form of higher domestic prices of imported rice; moreover, the divergence between import parity price and domestic price of imported rice increases with higher tariffs. Empirical analyses show that only about 20 percent of increases in rice import tariffs and 33 percent of changes in the domestic price of imported rice are transmitted to the price of local rice. Producer prices, therefore, may not rise substantially either.

Simulations using an economywide multimarket model developed for Nigeria show that a combination of interventions related to production, markets, and trade may be necessary, although the feasibility of such an approach is unknown at this point. Trade and price policies alone, even if they are effectively enforced, will not result in stimulating a large supply response. If rice prices rise, consumers—especially the poor, who spend a larger share of their income on food—will be significantly hurt, offsetting any benefits to producers.

Alternatively, a rice-sector development strategy that focuses more attention on technological change and market improvement is likely to be more successful in benefiting both producers and consumers, while also contributing to stable food prices and overall GDP growth. If an import tariff is warranted, maintaining a consistent tariff policy together with increased public investments in technological change and market improvements will be required to spur productivity growth. This is something that will not be easy to do, considering the government's record for implementing similar initiatives in the past. In addition, public investment in the rice sector can incur substantial opportunity costs, such as investments in other sectors, that must be carefully assessed.

Policies to Encourage Domestic Rice Production

Several factors may explain Nigeria's pursuit of transforming the rice economy. Recent food price spikes, growing demand for cereals like rice in Asia, including China, and slowing growth in rice production in major exporting regions all point toward increasing costs for importing rice for Nigeria. There is still considerable scope for developing the domestic rice sector. Unlike in other tropical West African countries, a substantial share of rice production in Nigeria is in the lowland ecologies, where much of the Green Revolution took place in Asia. Given its market size and the potential impact it has in international rice markets and West African food security, how the rice economy in Nigeria can be transformed is a question that merits serious examination.

Compared with its counterparts outside Africa south of the Sahara, Nigeria not only faces the dual challenges of catching up to Asia's Green Revolution and transforming the rice postharvest sector, it also faces the additional task of reversing decades of neglect in agriculture. Doing so in a fashion that will lead to a complete transformation of the rice economy in a short period of time is a formidable challenge. Therefore, many of the issues that have been addressed in this book have focused more on a medium-to-long-run perspective. A successful strategy will depend partly, but crucially, on stocks of knowledge and capital accumulated through past investments and long-term future investments in rice R&D and irrigation, supplemented by investment in physical infrastructure.

Though it is beyond the scope of this book to discuss the political economy aspects of long-term public investments, these investments need to increase substantially if Nigeria is to improve the competitiveness of its rice sector. According to recent estimates, the share of agriculture in total government expenditures was only 5.3 and 5.7 percent in 2009 and 2010, respectively (Benin and Yu 2013). The 2013 budget allocation for agriculture remained at

5 percent as a share of the total projected budget (Ukaoha and Ngene 2013). This is not very different from the 1970s, when the government was flush with petrodollars. Moreover, it is well below the 10 percent target agreed upon by African countries, including Nigeria, under the Comprehensive Africa Agriculture Development Programme, which is an initiative launched by the African Union to raise awareness of the importance of increasing government support of agriculture in Africa. It is critical for the Nigerian government to assess how effective support for the rice sector can be enhanced as part of broader efforts in raising agricultural-sector support.

The national government's commitment to the agricultural sector is encouraging and should be maintained, even though short-term responses may be limited, in part because of the small share of public expenditure still devoted to agriculture. Under its recent strategy for agriculture, the Nigerian government laid out an ambitious plan to achieve a transformation in the rice economy through policy reforms and fiscal investments. Aside from the use of import tariffs, the government also introduced reforms that include the deregulation of the fertilizer sector, marketing reforms that involve promoting the setup of private market corporations to help coordinate the market and set grades and standards, and innovative financing mechanisms for supplying credit. Additionally, interstate barriers to trade in paddy rice, such as taxes, were expected to be eliminated to reduce market transaction costs.

The policy and investment framework for transforming the rice economy is promising if it can be successfully implemented, kept consistent, and on course. Asia's own Green Revolution decades ago relied heavily on protective policies against imports and strong state involvement in providing subsidies and intervening in the market to help coordinate access to input and output markets and to stabilize prices (Johnson, Hazell, and Gulati 2003). In Asia, governments played important roles in various aspects throughout the course of the Green Revolution. Many of their roles turned out to be effective because they invested heavily in rice R&D, irrigation, and physical infrastructure. Moreover, many of these countries had established strong monitoring and evaluation systems by the 1970s, so that adequate time-series statistics on irrigated area, number of tractors, fertilizer use, and so forth are readily available. This has proved crucial for these countries to evaluate future policy options and monitor progress over time. It is now critical for Nigeria to do the same.

Finally, there is always the danger that a rice development strategy that focuses too much attention on displacing imports of higher-end quality rice will always have a strong urban bias, as Moseley, Carney, and Becker (2010) caution. Unfortunately, this is already the case, considering that much of the focus on displacing rice imports across many countries in West Africa is being driven more by concerns of increased vulnerability to volatile global markets, threats of depleting foreign-exchange reserves, and political risks associated with not having sufficient supplies of affordable rice, especially in urban areas. In other words, policymakers' concerns about reducing the dependency on imported rice (especially higher-quality premium rice) are strongly urban biased. Concerns about improving the livelihoods of rice farmers appear to be secondary. Identifying the right balance between these is likely to remain an important policy question as Nigeria tries to define its rice-sector development strategy.

Appendix A

A BRIEF CHRONOLOGY OF NIGERIA'S POLITICAL HISTORY

Kwabena Gyimah-Brempong, Michael Johnson, and Hiroyuki Takeshima

TABLE A.1 A brief chronology of Nigeria's political history, 1960–2013

1960	Independence from Great Britain, with Prime Minister Sir Abubakar Tafawa Balewa leading a coalition government.
1963–1966	**First Republic (3 years)**
1963	Nigeria is proclaimed as the Federal Republic of Nigeria, and former Governor General Nnamdi Azikiwe becomes the country's first president.
1966–1978	**First period of military rule (12 years)**
1966	Prime Minister Balewa is killed in a coup in January. Major General Johnson Aguiyi-Ironsi heads up a military administration. Ironsi is subsequently killed in a countercoup in July and replaced by Lieutenant Colonel Yakubu Gowon.
1967–1970	**Biafra Civil War:** Three eastern states secede as the Republic of Biafra, sparking bloody civil war. The Biafran leaders surrender in 1970, and former Biafran regions are reintegrated into the country.
1975	Gowon is overthrown and flees to Britain. He is replaced by Brigadier General Murtala Ramat Mohammed, who begins the process of moving the federal capital to Abuja.
1976	Mohammed is assassinated in a failed coup attempt. He is replaced by his deputy, Lieutenant General Olusegun Obasanjo, who helps introduce an American-style presidential constitution.
1978–1983	**Second Republic (5 years)**
1978	A ban on political activity is lifted and a new constitution is drafted and published.
1979	Elections bring Alhaji Shehu Shagari to power.
1983	August–September: Shagari is reelected amid accusations of irregularities. By December, Major General Muhammad Buhari seizes power in a bloodless coup.
1984–1993	**Second period of military rule (9 years)**
1985	Ibrahim Babangida seizes power in a bloodless coup and curtails political activity.
1993	Third Republic (0 years)
1993	In June, the military annuls elections when the preliminary results show victory by Chief Moshood Abiola. By August, power is transferred to an Interim National Government.
1993–1999	**Third period of military rule (6 years)**
1993	November: General Sani Abacha seizes power from the Interim National Government and suppresses any opposition.
1994	Abiola is arrested after proclaiming himself the president of the republic.
1995	Writer and campaigner Ken Saro-Wiwa is executed following a hasty trial for protesting against the petroleum industry for the damages to his Ogoni homeland.
1998	Abacha dies and is succeeded by Major General Abdulsalami Abubakar. Chief Abiola dies in custody.

1999–present	**Fourth Republic (16+ years)**
1999	Parliamentary and presidential elections are held under Abubakar. Olusegun Obasanjo is sworn in as president.
2003	In April, the first civilian-run presidential elections are held since the end of military rule. Olusegun Obasanjo is reelected for a second term. Also first legislative elections are held, with Obasanjo's People's Democratic Party winning the parliamentary majority.
2006	Nigeria becomes the first African nation to pay off its debt to the Paris Club of rich lenders, helped by record petroleum prices.
2007	Umaru Yar'Adua of the ruling People's Democratic Party is proclaimed the winner of the presidential election.
2010	President Umaru Yar'Adua dies after a long period of illness. Vice President Goodluck Jonathan, already acting in Yar'Adua's stead, succeeds him.
2011	Goodluck Jonathan wins the presidential elections.

Source: Nigeria country profile, BBC (2013), and other sources (e.g., Ehigiamusoe 2012).

Appendix B

THE LINEAR EXPENDITURE SYSTEM MODEL

Kwabena Gyimah-Brempong and Oluyemisi Kuku-Shittu

There are several approaches to the systems demand for commodities by households, each with strengths and weaknesses. All these estimate how households allocate their expenditures on food (total household expenditure) among broad classes of goods. In this sense, these models estimate household budget shares rather than the demand for each individual commodity. Among the most popular models are Deaton's Almost Ideal Demand Systems (Deaton and Muellbauer 1980) and the Direct Addilog Consumer Demand Systems popularized by Houthakker and Taylor (1970, 2010). Another convenient model used to study household expenditures is the Linear Expenditure System (LES) popularized by Stone (1954). Each of these models is derived from a plausible demand function based on a utility maximization model.

We have chosen to use the LES for a number of reasons. Besides the ease of estimation and interpretation of results, it allows for "subsistence consumption," after which additional budget increases are allocated to "supernumerary" consumption as marginal budget shares (MBS). Subsistence consumption as used here is different from the way it is used in the development literature, where one thinks of "subsistence farming" as farmers growing just enough food to feed their families without generating a surplus for sales in the market. As used here, subsistence consumption refers to the level of consumption expenditures that the household is "committed" to without regard to its income level. Subsistence demand as used here is akin to demand to satisfy "basic needs"; the household will purchase this level (quantity) of consumption of all commodities in its consumption basket first before it purchases additional quantities of any good with its remaining income. It reflects the minimum quantity (expenditure) of a commodity that enters the household's consumption basket. In a low-income country like Nigeria, such subsistence consumption may be an important component of the welfare of households. MBS may also be important for ensuring food security and may therefore be of interest to policymakers. In addition, concerns about increasing food imports in Nigeria resulting from increased incomes and population growth as well as concerns with high rates of food insecurity make the MBS approach an attractive feature of the LES for modeling the demand for rice in Nigeria.

We note, however, that the LES is not without its weaknesses. In particular, the assumption of strong separability among broad classes of commodities may imply that the utility function is additive and that the marginal utility from commodity group *i* does not depend on the consumption of commodity group *j*. Second, it assumes a strict proportionality between income elasticities and price effects, indicating that own-price elasticities of demand for a particular product depend on expenditure elasticity of demand for that good and the sum of cross-price elasticities of demand for that good and all other goods in the consumer's consumption basket. Finally, it does not allow for inferior goods in the system. Despite these weaknesses, Taylor and Houthakker (2010) show that the direction of change in budget shares is very similar among the most popular approaches to household demand systems.

We estimate household consumption of rice and its two components—local and imported—in Nigeria as part of the demand for a large number of broad categories of food, including rice, yam, millet, sorghum, maize, meat, fish, poultry, fruits, and vegetables, among others. The demand system we estimate is derived from maximizing a Stone-Geary utility function, subject to a household budget constraint. Formally, the household maximizes a utility function:

$$Max\ U = \prod_i^I (Q_i - \gamma_i)^{\alpha_i}$$
$$S.T. \sum_i^I P_i * Q_i \leq Y, \tag{1}$$

where Q_i is the quantity demanded for good i, γ_i is the subsistence level of i, α_i is the MBS of good i, Y is household income, and P_i is the price for Q_i.

Solving this constrained maximization problem, one can then derive the demand function for Q_i, which takes the following form:

$$Q_i = \frac{\alpha_i (Y - \sum_{j \neq i}^I p_j * y_j)}{p_i} + \gamma_i, \tag{2}$$

where all variables are as defined in (1) above. Equation (2) is written for food group (category) i and can be converted to budget shares by multiplying both sides by p_i / Y to get the share of the household's budget that is spent on that particular food item, so that

$$S_i = \frac{p_i Q_i}{Y} = p_i \frac{\alpha_i (Y - \sum_{j \neq i}^I p_j * y_j)}{p_i Y} + \frac{\gamma_i}{Y}.$$

In this specification, the share equation is a function of prices and quantities of all goods consumed by the household and the total expenditure (income) of the household. Note that α_i and γ_i may be different for each food category. One estimates the share equations as a system of linear equations with the appropriate restrictions (linear homogeneity, homotheticity, etc.). There are as many share equations as there are food groups. The error vector of all the share equations is singular, and therefore the systems cannot be estimated. To make it estimable, we drop one of the share equations and recover the parameters of that equation from the parameters of other share equations using the restrictions imposed for estimation. Taylor and Houthakker (2010) show that parameter estimates are invariant as to which share equation is dropped from the estimation. For our estimation, we dropped the other grain category and recovered the parameters of that equation from the estimated coefficients.

If we use the average budget share (ABS) of food group i, that is, $s_i = \frac{p_i Q_i}{Y}$, with $\sum_i^I P_i * Q_i = Y$, as the budget constraint, the income elasticity of demand for food group i can be derived from the share equation as

$$\varepsilon_{iY} = \frac{\partial Q_i}{\partial Y} * \frac{Y}{Q_i} = \frac{\alpha_i}{s_i}, \tag{3}$$

showing that the expenditure (income) elasticity of demand for food group i is the ratio of MBS to ABS. The cross-price elasticity of the demand is derived as

$$\varepsilon_{ij;\, i \neq j} = \frac{\partial Q_i}{\partial p_j} * \frac{p_j}{Q_i} = -\frac{\alpha_i p_j \gamma_j}{p_i Q_i} = -\frac{\alpha_i}{s_i} * \frac{p_j \gamma_j}{Y}, i \neq j. \tag{4}$$

This suggests that the cross-price elasticity of demand in the LES depends on the income elasticity of demand for that food item as well as the expenditure shares of the other food items. With the imposition of homogeneity restriction, that is, $\left(\varepsilon_i^Y + \sum_{j \neq i}^I \varepsilon_{ij}^p + \varepsilon_{ii}^p\right) = 0$, the own-price elasticity of demand can be derived from the known income elasticity, and cross-price elasticity as

$$\varepsilon_{ii} = \frac{\partial Q_i}{\partial p_j} * \frac{p_i}{Q_i} = -\frac{\alpha_i}{s_i} + \frac{\alpha_i}{s_i} \sum_{j \neq i}^I \frac{p_j \gamma_j}{Y} = -\left(\varepsilon_i^Y + \sum_{j \neq i}^I \varepsilon_{ij}^p\right). \tag{5}$$

From equations (4) and (5), it is clear that once we know income elasticity, ε_i^Y, subsistence parameter, γ_i, and price, P_i, we can calculate price elasticities of demand. Box B.1 uses a three-good utility function to illustrate this.

The derived price elasticities of demand are estimated based on the average price at the national level, estimated income elasticity for national, urban,

BOX B.1 Calculating price elasticities of demand for a three-good example

$$Q_1 = \frac{\alpha_1(Y - \gamma_2 P_2 - \gamma_3 P_3) \; + \gamma_1}{P_1} + \gamma_1,$$

$$\frac{aQ_1}{ap_1} = \frac{\alpha_1 Y}{P_1{}^2} + \frac{\alpha_1(\gamma_2 P_2 - \gamma_3 P_3)}{P_1^2}$$

$$\frac{aQ_1}{ap_2} = \frac{\alpha_1 \gamma_2}{p_1}$$

$$\frac{aQ_1}{ap_3} = \frac{\alpha_1 \gamma_3}{p_1}$$

$$\frac{aQ_1}{aY} = \frac{\alpha_1}{p_1}$$

$$\varepsilon_{iy} = \frac{aQ_1}{aY} * \frac{Y}{Q} = \frac{\alpha_1 Y}{p_1 Q_1} = \frac{\alpha_1}{s_1} = \alpha_1 = \varepsilon_{iy} s$$

$$\frac{aQ_1}{ap_2} * \frac{p_2}{Q_1} = -\frac{\alpha_1 \gamma_2 p_2}{p_1 Q_1}\left(s_1 = \frac{p_1 Q_1}{Y}, p_1 Q_1 = s_1 Y\right) = \frac{\alpha_1}{s_1 Y} * p_2 \gamma_2$$

$$\frac{aQ_1}{ap_2} = -\frac{\alpha_1}{s_1 Y} * p_3 \gamma_3 \frac{aQ_1}{ap_1} * \frac{p_1}{Q_1}$$

and rural average households, and crucially on a subsistence parameter, γ_i, discussed above. In our analysis, we calibrated the subsistence parameter for national, urban, rural, and income quintile groups in rural and urban areas. Although we are interested in estimating the demand for rice in both the imported and local forms, we recognize that households spend their food budgets—and indeed, their entire household budgets—on more than these two food items. We therefore model and estimate a household demand system for 18 food items and one non-food item that cover all the household expenditures.[1] Because we are interested not only in the demand for rice as a whole but also in the demand for imported rice and domestically produced rice and the relationship among them, we treat imported rice and local rice as separate food items in our estimation. The analysis is carried on at the household level.

1 The food items and the non-food item covered in the model are presented in the following tables. Multiple items are covered by the terms "Other grains" and "Other roots."

TABLE B.1A Demand for major commodities by zone (North Central and North East)

	Urban			Rural		
Group	Expenditure (million naira/year)	Quantity (thousand tons/year)	Quantity (kg/capita)	Expenditure (million naira/year)	Quantity (thousand tons/year)	Quantity (kg/capita)
North Central						
Rice	26,660	190	28.6	54,605	495	28.8
Local rice	10,758	96	14.4	37,126	389	22.6
Imported rice	15,902	95	14.2	17,479	106	6.2
Maize	9,795	167	25.0	22,277	466	27.1
Wheat	10,016	52	7.8	9,463	52	3.0
Sorghum	5,325	107	16.1	19,141	520	30.2
Millet	928	22	3.2	7,243	170	9.9
Other grains	—	—	—	168	1	0.1
Cassava root	319	6	0.3	13,282	282	5.5
Garri	15,693	231	34.7	26,562	396	23.0
Yam	34,826	619	31.0	80,481	1,573	30.5
Other roots	2,444	47	2.3	4,755	103	2.0
Total	**106,006**			**237,977**		
North East						
Rice	19,690	138	39.7	61,477	543	29.3
Local rice	12,580	99	28.5	51,237	482	26.0
Imported rice	7,109	39	11.2	10,240	61	3.3
Maize	10,043	174	50.2	36,325	760	41.1
Wheat	3,563	22	6.4	13,428	85	4.6
Sorghum	2,889	86	24.9	29,507	1,109	60.0
Millet	2,285	66	19.0	24,233	677	36.6
Other grains	197	1	0.4	56	0	0.0
Cassava root	122	2	0.2	2,633	48	0.9
Garri	563	7	2.0	4,898	73	3.9
Yam	4,881	65	6.2	15,778	230	4.1
Other roots	1,490	32	3.1	7,131	146	2.6
Total	**45,723**			**195,466**		

Source: Authors' calculations based on National Bureau of Statistics and World Bank (2011).

TABLE B.1B Demand for major commodities by zone (North West and South East)

	Urban			Rural		
Group	Expenditure (million naira/year)	Quantity (thousand tons/year)	Quantity (kg/capita)	Expenditure (million naira/year)	Quantity (thousand tons/year)	Quantity (kg/capita)
North West						
Rice	62,462	513	53.0	142,428	1,259	39.6
Local rice	43,401	389	40.1	124,084	1,144	35.9
Imported rice	19,061	125	12.9	18,344	116	3.6
Maize	29,360	549	56.7	70,541	1,490	46.8
Wheat	12,147	56	5.8	26,827	120	3.8
Sorghum	13,627	291	30.1	84,013	2,428	76.3
Millet	19,216	407	42.0	72,926	1,855	58.3
Other grains	939	8	0.8	4,727	46	1.4
Cassava root	659	19	0.7	3,062	62	0.6
Garri	4,004	42	4.3	3,876	41	1.3
Yam	13,391	226	7.8	15,508	280	2.9
Other roots	6,560	100	3.4	12,971	234	2.4
Total	**162,365**			**436,879**		
South East						
Rice	13,022	192	36.3	29,447	363	26.2
Local rice	2,225	25	4.7	12,059	111	8.1
Imported rice	10,797	167	31.5	17,387	252	18.2
Maize	674	118	22.3	727	92	6.6
Wheat	5,309	76	14.4	9,543	79	5.7
Sorghum	127	20	3.8	14	0	0.0
Millet	13	2	0.4	—	—	—
Other grains	111	22	4.1	419	32	2.3
Cassava root	585	83	5.2	22,011	700	16.9
Garri	8,441	427	80.6	21,511	794	57.4
Yam	15,029	599	37.7	39,597	1,156	27.9
Other roots	10,328	357	7.5	16,542	478	11.5
Total	**53,639**			**139,811**		

Source: Authors' calculations based on National Bureau of Statistics and World Bank (2011).

TABLE B.1C Demand for major commodities by zone (South South and South West)

	Urban			Rural		
Group	Expenditure (million naira/year)	Quantity (thousand tons/year)	Quantity (kg/capita)	Expenditure (million naira/year)	Quantity (thousand tons/year)	Quantity (kg/capita)
South South						
Rice	41,773	234	24.4	53,440	314	20.9
Local rice	2,117	19	2.0	7,780	78	5.2
Imported rice	39,655	215	22.4	45,660	236	15.7
Maize	193	2	0.2	166	1	0.1
Wheat	17,280	71	7.4	22,402	80	5.3
Sorghum	—	—	—	58	2	0.1
Millet	—	—	—	5	0	0.0
Other grains	75	0	0.0	857	8	0.5
Cassava root	7,254	161	16.8	30,405	634	14.1
Garri	38,994	451	47.0	71,933	846	56.4
Yam	26,992	432	45.0	53,617	814	18.1
Other roots	22,121	282	29.4	40,199	506	11.2
Total	**154,682**			**273,082**		
South West						
Rice	101,058	774	32.8	30,576	184	26.4
Local rice	1,171	9	0.4	1,339	10	1.5
Imported rice	99,888	765	32.4	29,236	174	24.9
Maize	4,978	53	2.2	2,505	30	4.3
Wheat	44,757	249	10.6	9,201	50	7.1
Sorghum	9	0	0.0	231	7	1.0
Millet	—	—	—	107	3	0.5
Other grains	7,599	53	2.2	140	1	0.1
Cassava root	2,753	37	1.6	1,017	31	1.5
Garri	64,538	932	39.6	19,686	232	33.2
Yam	113,149	1,375	58.3	38,345	547	26.1
Other roots	19,409	212	9.0	7,145	99	4.7
Total	**358,250**			**108,953**		

Source: Authors' calculations based on National Bureau of Statistics and World Bank (2011).

TABLE B.2 Own- and cross-price elasticities of demand for food in Nigeria

Commodity	Local rice	Imported rice	Maize	Wheat flour	Bread	All cassava	Sugar	Yam	Pulses	Oil and fat	Fruits and vegetables	Milk	Poultry	Fish	Other grains	Other roots	Other meat	Other food	Dining out	Non-food
Local rice	−0.309	−0.003	−0.006	−0.001	−0.002	−0.004	0.000	−0.016	−0.008	−0.015	−0.015	−0.001	−0.001	−0.012	−0.017	−0.005	−0.014	−0.004	−0.001	−0.002
Imported rice	−0.019	−0.614	−0.012	−0.003	−0.003	−0.008	−0.001	−0.032	−0.017	−0.030	−0.030	−0.002	−0.002	−0.025	−0.035	−0.011	−0.028	−0.009	−0.001	−0.003
Maize	−0.013	−0.004	−0.427	−0.002	−0.002	−0.006	−0.001	−0.022	−0.012	−0.020	−0.021	−0.001	−0.001	−0.017	−0.024	−0.008	−0.019	−0.006	−0.001	−0.002
Wheat flour	−0.021	−0.005	−0.013	−0.659	−0.004	−0.009	−0.001	−0.035	−0.018	−0.032	−0.033	−0.002	−0.002	−0.027	−0.037	−0.012	−0.030	−0.009	−0.001	−0.003
Bread	−0.022	−0.006	−0.014	−0.003	−0.696	−0.010	−0.001	−0.037	−0.019	−0.034	−0.035	−0.002	−0.002	−0.028	−0.039	−0.012	−0.031	−0.010	−0.001	−0.004
All cassava	−0.008	−0.002	−0.005	−0.001	−0.001	−0.269	0.000	−0.014	−0.007	−0.013	−0.013	−0.001	−0.001	−0.011	−0.015	−0.005	−0.012	−0.004	0.000	−0.001
Sugar	−0.013	−0.003	−0.008	−0.002	−0.002	−0.006	−0.416	−0.022	−0.011	−0.020	−0.021	−0.001	−0.001	−0.017	−0.024	−0.007	−0.019	−0.006	−0.001	−0.002
Yam	−0.012	−0.003	−0.008	−0.002	−0.002	−0.005	−0.001	−0.400	−0.011	−0.019	−0.019	−0.001	−0.001	−0.015	−0.022	−0.007	−0.017	−0.005	−0.001	−0.002
Pulses	−0.010	−0.003	−0.006	−0.001	−0.002	−0.004	0.000	−0.017	−0.324	−0.015	−0.016	−0.001	−0.001	−0.013	−0.018	−0.006	−0.014	−0.004	−0.001	−0.002
Oil and fat	−0.010	−0.003	−0.006	−0.001	−0.002	−0.004	0.000	−0.016	−0.009	−0.327	−0.016	−0.001	−0.001	−0.013	−0.018	−0.006	−0.014	−0.004	−0.001	−0.002
Fruits and vegetables	−0.011	−0.003	−0.007	−0.002	−0.002	−0.005	−0.001	−0.019	−0.010	−0.017	−0.375	−0.001	−0.001	−0.015	−0.020	−0.006	−0.016	−0.005	−0.001	−0.002
Milk	−0.023	−0.006	−0.015	−0.003	−0.004	−0.010	−0.001	−0.038	−0.020	−0.035	−0.036	−0.715	−0.002	−0.029	−0.041	−0.013	−0.032	−0.010	−0.001	−0.004
Poultry	−0.041	−0.011	−0.026	−0.005	−0.007	−0.018	−0.002	−0.067	−0.035	−0.062	−0.064	−0.004	−1.279	−0.052	−0.072	−0.023	−0.058	−0.018	−0.002	−0.007
Fish	−0.014	−0.004	−0.009	−0.002	−0.002	−0.006	−0.001	−0.023	−0.012	−0.021	−0.022	−0.001	−0.001	−0.449	−0.025	−0.008	−0.020	−0.006	−0.001	−0.002
Other grains	−0.001	0.000	−0.001	0.000	0.000	−0.001	0.000	−0.002	−0.001	−0.002	−0.002	0.000	0.000	−0.002	−0.048	−0.001	−0.002	−0.001	0.000	0.000
Other roots	−0.015	−0.004	−0.009	−0.002	−0.002	−0.006	−0.001	−0.025	−0.013	−0.023	−0.023	−0.001	−0.001	−0.019	−0.026	−0.473	−0.021	−0.007	−0.001	−0.002
Other meat	−0.024	−0.006	−0.015	−0.003	−0.004	−0.010	−0.001	−0.039	−0.021	−0.036	−0.037	−0.002	−0.002	−0.030	−0.042	−0.013	−0.777	−0.010	−0.001	−0.004
Other food	−0.023	−0.006	−0.014	−0.003	−0.004	−0.010	−0.001	−0.038	−0.020	−0.035	−0.035	−0.002	−0.002	−0.029	−0.040	−0.013	−0.032	−0.721	−0.001	−0.004
Dining out	−0.035	−0.009	−0.023	−0.005	−0.006	−0.015	−0.002	−0.059	−0.031	−0.054	−0.055	−0.003	−0.003	−0.045	−0.063	−0.020	−0.050	−0.016	−1.112	−0.006
Non-food	−0.033	−0.009	−0.021	−0.004	−0.006	−0.014	−0.002	−0.055	−0.029	−0.051	−0.052	−0.003	−0.003	−0.043	−0.060	−0.019	−0.048	−0.015	−0.002	−1.055

Source: Linear Expenditure System model results.

Note: Own- and cross-price elasticities measure how responsive quantity demanded is to changes in a good's own-price and the price of other goods, respectively.

TABLE B.3 Estimates of own-price elasticity

	National			Urban			Rural		
Commodity	High	Medium	Low	High	Medium	Low	High	Medium	Low
Local rice	–0.37	–0.31	–0.28	–0.18	–0.15	–0.15	–0.59	–0.47	–0.42
Imported rice	–0.75	–0.61	–0.55	–0.47	–0.40	–0.38	–0.85	–0.66	–0.58
Maize	–0.52	–0.43	–0.39	–0.57	–0.50	–0.46	–0.59	–0.47	–0.41
Wheat flour	–0.81	–0.66	–0.59	–0.53	–0.46	–0.43	–1.38	–1.08	–0.94
Bread	–0.85	–0.70	–0.63	–0.62	–0.54	–0.50	–0.91	–0.71	–0.62
All cassava	–0.33	–0.27	–0.24	–0.20	–0.17	–0.16	–0.39	–0.31	–0.27
Sugar	–0.51	–0.42	–0.37	–0.54	–0.47	–0.44	–0.58	–0.45	–0.39
Yam	–0.48	–0.40	–0.37	–0.41	–0.36	–0.34	–0.55	–0.44	–0.40
Pulses	–0.39	–0.32	–0.29	–0.36	–0.31	–0.29	–0.48	–0.38	–0.33
Oil and fat	–0.39	–0.33	–0.30	–0.33	–0.29	–0.28	–0.46	–0.37	–0.33
Fruits and vegetables	–0.45	–0.38	–0.34	–0.44	–0.39	–0.37	–0.47	–0.38	–0.34
Milk	–0.88	–0.72	–0.64	–0.91	–0.79	–0.74	–0.84	–0.66	–0.57
Poultry	–1.57	–1.28	–1.15	–2.02	–1.75	–1.63	–1.57	–1.22	–1.07
Fish	–0.54	–0.45	–0.41	–0.37	–0.33	–0.31	–0.60	–0.48	–0.43
Other grains	–0.06	–0.05	–0.04	–0.14	–0.12	–0.11	–0.12	–0.09	–0.08
Other roots	–0.57	–0.47	–0.43	–0.53	–0.46	–0.43	–0.61	–0.48	–0.42
Other meat	–0.93	–0.78	–0.71	–0.73	–0.64	–0.60	–1.05	–0.85	–0.75
Other food	–0.88	–0.72	–0.65	–0.86	–0.75	–0.70	–0.86	–0.67	–0.59
Dining out	–1.36	–1.11	–1.00	–1.40	–1.21	–1.13	–1.23	–0.96	–0.84
Non-food	–1.29	–1.05	–0.95	–1.20	–1.04	–0.97	–1.26	–0.98	–0.86

Source: Linear Expenditure System model results.

TABLE B.4 Average and marginal budget shares and income elasticities

	National			Urban			Rural		
	ABS (%)*	MBS (%)**	Income elasticity	ABS (%)	MBS (%)	Income elasticity	ABS (%)	MBS (%)	Income elasticity
Rice	6.60	4.2	0.63	5.70	2.4	0.43	7.05	5.4	0.77
Local rice	3.79	1.6	0.44	1.84	0.4	0.20	4.69	3.0	0.64
Imported rice	2.82	2.5	0.88	3.82	2.0	0.53	2.36	2.4	1.03
Maize	2.21	1.3	0.61	1.17	0.7	0.64	2.69	1.9	0.72
Wheat flour	0.14	0.1	0.95	0.13	0.1	0.60	0.15	0.3	1.69
Bread	1.43	1.4	1.01	1.60	1.1	0.70	1.35	1.5	1.12
All cassava	4.28	1.6	0.38	3.21	0.7	0.22	4.78	2.3	0.48
Sugar	1.14	0.7	0.60	0.69	0.4	0.61	1.35	0.9	0.71
Yam	5.05	2.8	0.55	4.59	2.1	0.46	5.26	3.4	0.65
Pulses	3.59	1.6	0.46	2.99	1.2	0.40	3.87	2.2	0.57
Oil and fat	6.09	2.8	0.45	4.70	1.7	0.37	6.73	3.7	0.55
Fruits and vegetables	6.48	3.4	0.52	5.22	2.6	0.49	7.07	4.0	0.56
Milk	1.05	1.1	1.04	1.07	1.1	1.03	1.04	1.1	1.03
Poultry	1.01	1.9	1.85	0.86	2.0	2.28	1.07	2.1	1.92
Fish	5.76	3.6	0.63	5.47	2.3	0.42	5.89	4.3	0.72
Other grains	4.39	0.3	0.07	1.37	0.2	0.15	5.78	0.8	0.14
Other roots	1.54	1.0	0.68	1.30	0.8	0.60	1.64	1.2	0.74
Other meat	5.86	6.3	1.08	5.30	4.3	0.81	6.12	7.7	1.26
Other food	2.41	2.5	1.03	2.39	2.3	0.97	2.42	2.5	1.04
Dining out	10.42	16.8	1.61	12.77	20.2	1.58	9.34	14.1	1.51
Non-food	30.53	46.5	1.52	39.50	53.7	1.36	26.41	40.6	1.54
Total	**100**	**100**		**100**	**100**		**100**	**100**	

Source: Linear Expenditure System model results.

Note: *ABS = average budget shares. **MBS = marginal budget shares. Because of rounding, column entries might not exactly total 100.

TABLE B.5 Sample statistics of consumption (national, urban, and rural)

Commodity	Expenditure (million naira/year)	Total quantity (thousand tons/year)	Expenditure (naira/per capita)	Quantity (kg/per capita)
a. Sample statistics of consumption (national)				
Rice	636,638	5,200	3,938	32.2
Local rice	305,878	2,851	1,892	17.6
Imported rice	330,760	2,349	2,046	14.5
Maize	187,585	3,900	1,160	24.1
Wheat	183,936	993	1,138	6.1
Bread	161,203	837	997	5.2
Wheat flour	22,733	156	141	1.0
Sorghum	154,941	4,572	958	28.3
Millet	126,956	3,202	785	19.8
Other grains	15,289	173	95	1.1
Cassava	364,798	5,161	2,257	31.9
Yam	451,595	2,946	2,794	18.2
Other roots	151,096	865	935	5.4
Total	**2,272,834**	**27,012**	**14,060**	**167.0**
b. Sample statistics of consumption (urban)				
Rice	264,665	2,041	4,541	35.0
Local rice	72,252	636	1,240	10.9
Imported rice	192,413	1,405	3,301	24.1
Maize	55,044	1,062	944	18.2
Wheat	93,072	527	1,597	9.0
Bread	83,071	443	1,425	7.6
Wheat flour	10,001	84	172	1.4
Sorghum	21,977	505	377	8.7
Millet	22,442	496	385	8.5
Other grains	8,922	84	153	1.4
Cassava	143,923	2,193	2,469	37.6
Yam	208,269	1,324	3,573	22.7
Other roots	62,353	343	1,070	5.9
Total	**880,665**	**8,576**	**15,110**	**147.0**

Commodity	Expenditure (million naira/year)	Total quantity (thousand tons/year)	Expenditure (naira/per capita)	Quantity (kg/per capita)
c. Sample statistics of consumption (rural)				
Rice	371,973	3,159	3,599	30.6
Local rice	233,626	2,215	2,260	21.4
Imported rice	138,347	944	1,338	9.1
Maize	132,541	2,838	1,282	27.5
Wheat	90,864	466	879	4.5
Bread	78,132	394	756	3.8
Wheat flour	12,732	72	123	0.7
Sorghum	132,964	4,067	1,286	39.3
Millet	104,515	2,706	1,011	26.2
Other grains	6,367	89	62	0.9
Cassava	220,875	2,968	2,137	28.7
Yam	243,326	1,622	2,354	15.7
Other roots	88,743	522	859	5.1
Total	**1,392,169**	**18,437**	**13,468**	**178.0**

Source: Author's calculations based on Nigeria, NBS and World Bank (2011).

Appendix C

ADDITIONAL TABLES TO CHAPTER 3

Hiroyuki Takeshima and Oladele Samuel Bakare

TABLE C.1 Major soils and rice ecologies

(1) Order	(2) Lowland	(3) Transitional	(4) Upland / dryland	(5) Not in rice growing areas
Inceptisols	Cambisols Eutric, Calcaric, Dystric, Humic Gleysols	Dystric, Ferralic Cambisol Andosols		Eutric, Calcic, Vertic Cambisols, Humic Cambisols
Alfisols	Gleyic Luvisols		Gleyic-Solonetz, Luvisols, Eutric Nitosols	
Ultisols	Gleyic Acrisols, Plinthic Acrisols, Dystric Planosols (swamp type)	Humic Acrisols, Humic Nitosols, Dystric Nitosols (steep terrace)		Arenosols (little) Regosols (little)
Entisols	Fluvisols Gleysols			
Vertisols	Vertisols			
Mollisols				Mollic Gleysols pt., Kastanozems, Chernozems, Rendzinas, Greyzems, Planosols
Oxisols			—	Ferralsols Gleysols
Aridisols			—	
Histosols			—	
Spodosols			—	
Others				Lithosols
				Podsols

Source: Authors' analysis based on Moormann and van Breemen (1978).

TABLE C.2 Major known rice pests in Nigeria

(1) Insect pest	(2) Stage of attack on rice	(3) Control
(a) Stem borer: Striped borer (*Chilo zacconius* Bleszynski), white borer (*Maliarpha separatelia* Ragonot), stalk-eyed flies (*Diopsis longicornis* Macquart)	Vegetative/reproductive	Plough residue into the soil, burn stubble after harvest. Spray Miral at 0.75 kg a.i./ha (NCRI 2012).
(b) African rice gall midge (AfRGM) (*Orseolia oryzivora* Harris and Gagné)	Vegetative	Use of tolerant variety; e.g., FARO 51 (Ukwungwu and Joshi 1992b). Use clean seed; Spray Miral at 0.75 kg a.i./ha (NCRI 2012). Avoid late planting (Wada et al. 2012).
(c) Stem and leaf suckers (white flies; leaf hoppers and plant hoppers)	Vegetative/reproductive	Good plant density and good field sanitation (NCRI 2012).
(d) Leaf/stem feeder (lady bird; cricket and grasshoppers)	Vegetative/reproductive	Use of insecticide (NCRI 2012).
(e) Vertebrate pests (rat, squirrel, and grass cutter)	Seeding Vegetative/reproductive	Good sanitation, trapping (NCRI 2012). Rodenticide.
(f) Birds (weaver, quelea, bishop, sparrow)	Reproductive/maturity	Human scarer, flash tapes, resistant varieties, synchronized planting (NCRI 2012).
(g) Nematodes	Vegetative/reproductive	Nematicides (NCRI 2012).

Source: Authors' compilations from various studies cited in the table.

Note: NCRI = National Cereals Research Institute. kg a.i./ha = kilograms of active ingredient per hectare.

TABLE C.3 Major pathogens causing diseases of rice in Nigeria

(1) Pathogen	(2) Symptoms of attack on rice	(3) Control
(a) Blast (*Pyricularia oryzae*)	Spots/lesion; stunted growth	Chemical spray (NCRI 2012).
(b) Brown spot (*Helminthosporium oryzae*)	Leaves lesion, spotted grains	Seed treatment (NCRI 2012), healthy seeds.
(c) Leaf scald (*Gerlachia oryzae* Hashioka and Yokogi)	Zonated brown lesions	Avoid high use of nitrogen (NCRI 2012).
(d) Sheath blight (*Rhizochonia solani* Kuhn)	Reddish-gray spots on leaf sheath	Avoid overpopulation (NCRI 2012).
(e) Sheath rot (*Sarocladium oryzae*)	Lesions on uppermost leaf sheath	Use fully exerted panicle rice variety (NCRI 2012).
(f) Rice yellow mottle virus (RYMV)	Yellow leaves, stunted growth	Tolerant variety, good sanitation, seed treatment (NCRI 2012).

Source: Authors' compilations based on NCRI (2012, 2013).

Note: NCRI = National Cereals Research Institute.

TABLE C.4 Some weed species among lowland rice in Nigeria

(1) Weed species	(2) Family	(3) Common name	(4) Types	(5) Known outside Africa
Ipomoea spp.	Convolvulaceae	Wild potato		Known
Ludwigia spp.	Convolvulaceae	Water primrose		Known
Sphenoclea zeylanica (Gaertn)	Campanulaceae	Chicken spike	Annual (RJ)	Known
Nymphaea lotus Linn	Nymphaeaceae	Water lily		Unknown
Echinochloa spp.	Poaceae	Jungle rice	Annual (RJ)	Known
Oryza barthii A. chev.	Poaceae	Barth's rice	Annual (RJ)	NA
Cyperus rotundus Linn	Cyperaceae	Purple nut sedge	Perennial (RJ)	NA
Kyllinga spp.	Cyperaceae	Kyllinga		Sri Lanka (B)
Cyperus difformis Linn	Cyperaceae	Small-flowered umbrella sedge	Annual (RJ) (infest rice fields as a result of farmers' inability to maintain permanent flooding in poorly leveled fields) (I)	NA
Cyperus haspan Linn	Cyperaceae	Sharp edge sedge		NA
Cyperus iria Linn	Cyperaceae	Rice flat sedge	Annual (RJ)	Known
Fimbristylis littoralis Gaudet	Cyperaceae	Fimbry	Annual (RJ)	Known
Mariscus longibraceteatus Cherm	Cyperaceae	—		NA

Source: Authors' compilations based on Ismaila et al. (2013), Wada et al. (2012) (B); Imeokparia, Lagoke, and Olunuga (1992) (I); Rodenburg and Johnson (2009) (RJ).

Note: NA = information not available.

TABLE C.5 Weed species status at some lowland experimental sites of the National Cereals Research Institute, 2008 and 2009

		Years of experiment	
(1) Grasses	(2) Family of grasses	(3) 2008	(4) 2009
Echinochloa obtusiflora Stapt	Poaceae	+	++
Leersia hexandra sw	Poaceae	+	—
Paspalum polystachylum R. Br	Poaceae	+	+
Bracharia deflexa Schumach	Poaceae	+	+
Cynodon dactylon (Linn) pers	Poaceae	++	+
Broadleaved			
Ipomoea aquatic Forsk	Convolvulaceae	+	++
Nymphaea maculata Schum & Thorn	Nymphaeaceae	+	+
Nymphaea lotus Linn	Nymphaeaceae	—	++
Sphenoclea zeylanica Gaertn	Sphenocleaceae	++	++
Sedges			
Cyperus difformis L	Cyperaceae	+	+
Fimbristylis littoralis Gaudet	Cyperaceae	+	+
Kylinga bulbosa Beauv	Cyperaceae	+	—
Cyperus esculentus L	Cyperaceae	+	+
Sceleria naumanniana Boeck	Cyperaceae	—	+

Source: Authors' compilation based on NCRI (2011), Bakare, Ukwungwu, and Aliyu (2008a, 2008b).

Note: The degree of prevalence of weeds is indicated by the symbols + = low, ++ = moderate; — = information not available.

Appendix D

SUPPLY RESPONSE ANALYSIS

Hiroyuki Takeshima

We estimate that

$$I_{it}^{*} = f(p_{it}, w_{it}, H_i, \bar{z}_i, AEZ_i, Access_i, D_y) + \varepsilon_{it},$$

$$I_{it} = 1 \text{ if } I_{it}^{*} > 0, \text{ else } I_{it} = 0 \quad (1)$$

$$y_{it} = f(p_{it}, w_{it}, H_i, \bar{z}_i, AEZ_i, Access_i, D_y) + u_{it} \text{ if } I_{it} = 1. \quad (2)$$

The variable $I_{it} = 1$ if a farmer i grew cassava in year t and 0 otherwise. The variable y_{it} is the production of rice (kg). Both are determined by the vectors of price ratios between rice and other major crops (p_{it}), input prices and access to technologies (w_{it}), household characteristics (H_i), agroecological factors (AEZ_i), access to infrastructure, market, and research and development (*Access*), and year dummy (D_y) that controls for year-specific shocks. We also include $\bar{z}_i$, which are time averages of all the time-varying variables. This extends Chamberlain's (1984) method to a pseudopanel specification in order to control partly for cohort-specific effects that are unobserved.

For p_{it}, maize, cassava, and yam are included as crops potentially competing with rice. Prices of local rice and maize are calculated from farm-level sales information. Other price data were taken from community-level surveys included in the Living Standards Measurement Study–Integrated Surveys on Agriculture (LSMS–ISA) (Nigeria, NBS and World Bank 2011, 2013), where one community was randomly selected for each enumeration area (EA). Some communities did not report some of the price information. In order to achieve sufficient degrees of freedom and obtain efficient estimators, we calculated the local government area (LGA)–level, state-level, and zonal-level averages of these prices and replaced missing observations with these prices. Importantly, the community survey for LSMS–ISA (2013) or price information were not available at the time of the analyses. We therefore used the figures from the LSMS–ISA (2011) postharvesting survey, which were reported around March 2011, as a proxy for the prices farmers might have faced at the planting time in 2012. Although this is not the perfect proxy, given the generally low level of market integration for certain crops, a greater increase (or lesser decrease) of price between August 2010 (post-planting) and March 2011 (postharvesting) could have sustaining effects, keeping the price levels high all throughout 2011 and up to the planting time of August 2012.

w_{it} includes the price of commercial fertilizer, nominal farm wage for land clearing (adult male), and whether tractor owners are in the sample within the EA (to control for the access to tractor service). Due to the data limitations, fertilizer price and nominal farm wage from LSMS–ISA (2011) are used for 2012 as well. H_i includes household size, age, and gender of the household head, total farm size, whether the household head had any formal education or was literate, whether any family member was employed by the government institution, combined value of livestock and nonproductive household assets excluding land, and incomes from remittance, retirements, or dividend, which are exogenous to the cassava production decision.

AEZ_i are represented by the dominant soils of the area (FAO/IIASA/ISRIC/ISSCAS/JRC 2012), the Euclidean distance to the nearest major rivers (FAO 2000) and dams (AQUASTAT 2012), and the historical average and standard deviation of rainfall (University of East Anglia 2012). Because Nigeria has diverse soil types, we also include variables measuring various characteristics of the topsoil (0–5 cm) of interviewees' locations (ISRIC 2013). They include soil organic carbon, pH, sand content, silt content, clay content, cation exchange capacity, and bulk density of soils.[1]

$Access_i$ consists of the distance to the nearest road, population center, market, and whether the EA is in an urban or rural area to capture the effect of market and urban access. $Access_i$ also includes the average Euclidean distances to the two rice-breeding institutions in Nigeria (the National Cereals Research Institute in Bida and the International Institute of Tropical Agriculture in Ibadan) and seven screening stations of rice in Nigeria (Amakama, Birnin Kebbi, Ibadan, Mokwa, Uyo, Warri, Yandev) to partly control for the level of exposure and suitability of improved rice varieties. The levels are expected to be higher for areas closer to these institutions because of the similarity of the production environment to the research environment and the frequency of interactions with researchers. Note that the above method weighs distance from Ibadan: Ibadan is of particular importance because it has both breeding and screening stations.

1 The cation exchange capacity is the soils' ability to store cation, which is a group of a type of positively charged ions.

In the double-hurdle model, (1) and (2) are estimated by Tobit and truncated regression, respectively, with a possible correlation between error terms. We estimate (1) and (2) separately instead of jointly, assuming errors are uncorrelated. Such separate estimation makes the model free from potential misspecification in the other model, which is important in our case because reliable variables are limited. Using the estimated coefficients, we estimate the partial effects (*PEs*) of the change in rice price on the probability of deciding to grow rice (PE_P), rice production upon making such decision (PE_T), and combined effects of both (*PE*). Specifically, they are calculated by the formula

PE_P:
$$PE_P = \phi(\bar{X}\hat{\beta}) * \hat{\beta}$$

$$PE_T = \hat{\gamma} * \left[1 - \hat{\lambda} * \left(\frac{\bar{X}\hat{\gamma}}{\hat{\sigma}} + \hat{\lambda}\right)\right]$$

where
$$\hat{\lambda} = \frac{\phi\left(\frac{\bar{X}\hat{\gamma}}{\hat{\sigma}}\right)}{\Phi\left(\frac{\bar{X}\hat{\gamma}}{\hat{\sigma}}\right)}$$

$$PE = \left(\bar{X}\hat{\gamma} + \hat{\sigma}\hat{\lambda}\right) * PE_P + \Phi\left(\bar{X}\hat{\beta}\right) * PE_T,$$

in which $\bar{X}$ is a vector of the sample means of all variables; $\hat{\beta}$, $\hat{\gamma}$, and $\hat{\sigma}$ are estimated parameters; and ϕ and Φ are density and distribution functions of standard normal distribution. From *PE*, we obtain the elasticity evaluated at the sample mean.

While the LSMS–ISA (2011, 2013) is by far one of the best datasets in Nigeria, it is not free from outliers and some questionable data. Partly due to the high inflation rates in Nigeria in 2010 and 2011 (more than 10 percent each), growth rates of crop prices were somewhat volatile across locations. We assumed, however, that prices were unlikely to have either more than halved or doubled between 2010 and 2012. For households reporting such price movements, we replaced the prices in both 2010 and 2012 with LGA median prices. In order to include as many observations as possible, we also calculated median prices at state- and geopolitical-zone levels and substituted them for missing observations on prices.

Results are shown in Table D.1 in marginal effects on corresponding outcomes, and standard errors of *PEs* are estimated using delta methods, adjusted for EA–level cluster effects. For irrigated rice area, we present the marginal effects based on McDonald and Moffitt's (1980) decomposition. We present only the coefficients of time-variant variables because time-invariant variables only control for unobserved EA specific effects, and their coefficients have no direct meaning.

TABLE D.1 Rice production and irrigated rice area responses in Nigeria (pseudopanel double-hurdle model/Tobit)

	Rice production				Irrigated rice area (ha)			
	Models							
	Probit		Truncated		Tobit			
	Dependent variable/outcomes							
	Probability of growing rice		Rice output (ton)		Probability of growing rice with irrigation		Irrigated rice area (ha)	
Variables	(1) Marginal effect	(2) Std. err	(3) Marginal effect	(4) Std. err	(5) Marginal effect	(6) Std. err	(7) Marginal effect	(8) Std. err
Price ratio (natural log)								
Local rice/yam	.022*	.010	.052	.043	.001	.001	.018†	.009
Local rice/cassava	.042*	.020	.496*	.200	.001	.002	.013	.039
Local rice/maize	.006	.005	.074	.084	.000	.000	.001	.007
Whether tractor owners are in the EA (1 if yes)	–.006	.023	–.047	.117	.001	.002	.026	.027
Own tractor (1 if yes)	.019	.019	–.137	.149	.019	.019	.007	.015
Total farm size (ha)	.000	.000	.029**	.008	.000	.000	–.001	.001
Time-invariant factors								
Fertilizer price	Included		Included		Included		Included	
Nominal farm wage	Included		Included		Included		Included	
Household size	Included		Included		Included		Included	
Household head age	Included		Included		Included		Included	
Household head gender	Included		Included		Included		Included	
Household head has no formal education	Included		Included		Included		Included	
Household head is literate	Included		Included		Included		Included	
Household asset excluding land	Included		Included		Included		Included	
Livestock	Included		Included		Included		Included	
Exogenous income	Included		Included		Included		Included	
Whether employed in the government	Included		Included		Included		Included	
Average historical rainfall	Included		Included		Included		Included	

Standard deviation of historical rainfall	Included	Included	Included	Included
Distance to the nearest road	Included	Included	Included	Included
Distance to the nearest population center	Included	Included	Included	Included
Distance to the nearest market	Included	Included	Included	Included
Distance (Euclidean) to the nearest river	Included	Included	Included	Included
Distance (Euclidean) to the nearest dam	Included	Included	Included	Included
Average distance to the rice research stations	Included	Included	Included	Included
Dominant type of soil	Included	Included	Included	Included
Topsoil cation exchange	Included	Included	Included	Included
Topsoil pH	Included	Included	Included	Included
Topsoil clay content	Included	Included	Included	Included
Topsoil organic contents	Included	Included	Included	Included
Topsoil silt	Included	Included	Included	Included
Topsoil bulk density	Included	Included	Included	Included
Time EA averages	Included	Included	Included	Included
Sector dummy (1 if rural)	Included	Included	Included	Included
State dummies	Included	Included	Included	Included
Year dummy (1 if 2012)	Included	Included	Included	Included
Constant	Included	Included	Included	Included
Number of obs	4,325	482	4,325	4,325
p-value (H0: insignificant overall fit)	.000	.000	.000	.000

Source: National Bureau of Statistics and World Bank (2011, 2013).

Note: Symbols indicate the statistical significance: **1%; *5%; †10%. Standard errors are adjusted for EA cluster effects. Marginal effects are evaluated at the means of all covariates.

TABLE D.2 Reported production costs for rice (US$ per hectare)

Upland	Lowland	Irrigated	Variety	State	Year	Source
1,010			*Ofada*	Ogun	2010	Omonona et al. (2012)
540–656			*Ofada*	Ekiti	2004–07	PrOpCom (2007c)
543			Unspecified	Anambra	Unspecified	Anozie, Okereke, and Anozie (2012)
	886		Unspecified	Ebonyi	2009	Nwaobiala (2010)
	296–1,512		Unspecified	Niger, Kaduna, Ondo	Unspecified	Kolawole, Oladele, and Wakatsuki (2011)
	416		Unspecified	Kogi	2011	Ibitoye, Orebiyi, and Shaibu (2012)
	2,370 (Deepwater)		Local	Kebbi	2013	IFPRI fieldwork in 2013
		773	Improved	Kano	2003–04	Enplan Group (2004)
		1,156	Improved	Adamawa	2011	Jamala, Shehu, and Garba (2011)
		1,124	Improved	Kwara	2006	Akanbi, Omotesho, and Ayinde (2011)
		597	Unspecified	Anambra	Unspecified	Anozie, Okereke, and Anozie (2012)
		2,235	Local	Zamfara	2013	IFPRI fieldwork in 2013
		2,461	FARO 44	Kebbi	2013	IFPRI fieldwork in 2013

Source: Author's compilation based on cited studies and fieldwork in Nigeria.

Note: Figures were converted from naira to US dollars using annual nominal exchange rates from USDA (2014). When years are unspecified, exchange rates of published years were used. IFPRI = International Food Policy Research Institute.

TABLE D.3 Rice research expenditure in selected countries

(1) Country	(2) Amount (thousand, 2010 US$)		(3) PPP adjusted		(4) Amount per rice area (2010 US$/ha, PPP)		(5) Amount per 1,000 people (2010 US$, PPP)	
	1974	1998	1974	1998	1974	1998	1974	1998
Bangladesh[a]	646	4,548	922	9,097	0.1	0.9	13	73
Indonesia	2,959	NA	3,699	NA	0.4	NA	28	NA
India	20,982	16,455	34,970	54,848	0.9	1.2	58	54
Japan	247,480	NA	224,982	NA	82.6	NA	2,055	NA
Nepal	538	NA	1,076	NA	0.9	NA	83	NA
Philippines[a]	2,690	4,682	5,380	11,705	1.5	3.2	135	158
Pakistan	1,130	NA	2,260	NA	1.4	NA	34	NA
South Korea	1,345	NA	1,921	NA	1.6	NA	56	NA
Sri Lanka[b]	NA	547	NA	1,824	NA	2.1	NA	98
Thailand	1,614	NA	2,690	NA	0.4	NA	65	NA
Thailand[c]	6,690	14,000	11,149	26,923	1.5	2.9	270	449
Taiwan	9,146	NA	11,433	NA	14.7	NA	NA	NA
Vietnam	861	NA	1,722	NA	0.3	NA	35	NA
Latin America (Argentina, Brazil, Chile, Colombia, Mexico, Peru, and Venezuela) (1980)	22,200	NA	33,783	NA	4.5	NA	123	NA
Nigeria	NA	197	NA	538–985	NA	0.3–0.5	NA	5–8
IRRI	12,912	45,484	NA	NA	NA	NA	NA	NA
WARDA	NA	2,200	NA	3,300	NA	NA	NA	NA

Source: Author's modifications based on Hossain et al. (2003, 7ff) for Asian countries and IRRI; Ojehomon et al. (1999, cited in Dalton and Guei 2003) for Nigeria; unpublished WARDA 1999 "Workplans and Budget" paper, cited in Dalton and Guei (2003) for WARDA, and de Janvry, Runsten, and Sadoulet (1987, Table 13) for Latin America. All 1974 figures are from Evenson and Flores (1978) and Barker, Herdt, and Rose (1985, Table 14.9).

Note: IRRI = International Rice Research Institute. PPP = purchasing power parity. WARDA = West Africa Rice Development Association (now Africa Rice Center). WDI = World Development Indicators of the World Bank. NA = information not available. Figures are converted into 2010 US$. PPP for Asian countries are estimated by authors using "PPP Conversion Factor (GDP) to Market Exchange Rate" by the World Bank WDI. For Latin America, we took the average across seven countries. For Nigeria, PPP adjusted figures have some ranges due to a conflict between Dalton and Guei (2003) figures and WDI-based figures. For figures in 1974, we applied PPP conversion rates for 1980 (except Vietnam, for which we used 1985) due to the lack of information. [a] Figures for 1998 are for the total funding for Bangladesh Rice Research Institute and the Philippines Rice Research Institute. [b] Figures are for 1999, calculated based on Niranjan, Jogaratnam, and Weerahewa (2001). [c] Figures are the averages of 1970–1979 and 1990–2000, respectively, with the author's recalculation into 2010 US$ based on Jaroensathapornkul (2007). Alternative estimates for Thailand are included as well, due to the somewhat large variations in estimates across sources. For this reason, two "Thailand" rows are included in the table.

TABLE D.4 Producer price of rice and fertilizer price in selected countries and Nigeria

										Nigeria[a]				
										Official exchange rate		Parallel market exchange rate		
(1) Items/ years	(2) Bang-ladesh	(3) Colombia	(4) India	(5) Indonesia	(6) Philip-pines	(7) Sri Lanka	(8) Thailand	(9) Côte d'Ivoire	(10) Benin	(11) Sub-sidized	(12) Unsub-sidized	(13) Sub-sidized	(14) Unsub-sidized	
Price of urea per ton of nutrients paid by farmers (current US$ per ton of nutrients)														
66–70		261	279		263	178		202						
71–75	251	542	385	277	177	332		554	235					
76–80	291	594	429	315	529	222	359	613	359	162	610	162	610	
81–85	383	705	475	242	646	254	599	639	385	262	1,055	129	493	
86–90	332	329	355	214	386	284	375	646	570	130	529	87	329	
91–95	304	548	235	256	504	434	574	894	738	206	857	103	429	
96–00	255	446	212	273	447	262	434	744		1,324	1,560	410	469	
01–02	220	463	214	212	367	220	327							
Farm harvest price of paddy (current US$ per ton)														
66–70		122	95		82	123		74		175		175		Nigeria rice import restriction[b]
71–75	159	112	142	82	109	200	79	163	89	291		291		
76–80	146	180	164	175	141	161	118	273	192	452		452		Tariff (10–20%), partial restrictions
81–85	160	213	184	179	157	153	118	194	254	647		345		
86–90	172	202	147	138	161	151	132	226	326	498		357		Import ban
91–95	161	202	149	172	213	168	151	212	329	710		409		
96–00	128	260	134	153	236	173	159	199		874		284		Tariff (50–100%)
01–02	111	208	124	123	166	142	113	182						

							Fertilizer-paddy price ratio						
66–70		2.1	2.9		3.2	1.4		2.7					
71–75	1.6	4.9	2.7	3.4	1.6	1.7		3.4	2.6				
76–80	2.0	3.3	2.6	1.8	3.8	1.4	3.0	2.2	1.9	0.4	1.5	0.4	1.5
81–85	2.4	3.3	2.6	1.3	4.1	1.7	5.1	3.3	1.5	0.4	1.6	0.8	3.5
86–90	1.9	1.6	2.4	1.5	2.4	1.9	2.8	2.9	1.7	0.3	1.0	0.4	1.4
91–95	1.9	2.7	1.6	1.5	2.4	2.6	3.8	4.2	2.2	0.3	1.3	0.8	3.2
96–00	2.0	1.7	1.6	1.8	1.9	1.5	2.7	3.7		1.5	1.7	4.5	5.3
01–02	2.0	2.2	1.7	1.7	2.2	1.6	2.9						

Source: Authors modifying Estudillo and Otsuka (2012, Table 2.2), World Rice Statistics (farm harvest price of paddy), and FAOSTAT (urea price).

Note: Blank spaces indicate information is not available. [a]Unsubsidized figures for Nigeria are calculated using the historical fertilizer subsidy rates summarized in Liverpool-Tasie and Takeshima (2013, Table 3). Approximate parallel market rates were assessed from figures or tables in various sources: Ashworth, Evans, and Teriba (1999) and Kuijs (1998) for the 1980s and 1990–1994, Hashim and Meagher (1999) for 1995 and 1996, Falola and Adebayo (2000) for 1997 and 1998, and Nnanna (2002) for 1999–2001. While the exact parallel market rates may vary across time and location within each year, the figures in the table should provide good approximations. [b]More descriptions are provided in Chapter 7.

Appendix E

A MATHEMATICAL DERIVATION OF THE RICE MILLING MODEL FOR NIGERIA

Michael Johnson

The rice milling model (RMM) for Nigeria can be considered as a two-commodity and spatially oriented optimization model with an objective of maximizing industry profits across three miller sectors in all 37 states. These sectors have technologies, operating costs, and production capacities of differing scales. Prices are assumed to be given (i.e., they are exogenous to the model), while quantities of paddy and milled rice are treated as decision variables. Results can be reported by milling sector, zone or state, and at the national level.[1]

Mathematically, the RMM can be described as a simple linear programming optimization model with the following industrywide objective function:

$$Maximize\text{: } \pi\left(Q_{ikt}^{pm}, Q_{ikt}^{bm}, Q_{iktj}^{pi}, Q_{iktj}^{pt}, Q_{iktj}^{b}, Q_{ikt}^{ps}, Q_{ikt}^{bs}\right)$$

$$= \sum_{i=1}^{N}\sum_{t=1}^{T}\sum_{k=1}^{K}\left[P_{ikt}^{m}\left(\theta_k^p Q_{ikt}^{pm} + \theta_k^b Q_{ikt}^{bm}\right) - f_{ik}^{p}\left(Q_{ikt}^{pm} + Q_{ikt}^{bm}\right) - c_{ik}^{p}\left(Q_{ikt}^{pm} + \delta_k^b Q_{ikt}^{bm}\right)\right]$$

$$-\left(p_{it}^{pt} + tc_{ii} + mkt_i\right)Q_{ikt}^{pt} - \sum_{j=1}^{N}\left[\left(p_{it}^{pi} + tc_{ij} + mkt_j\right)Q_{iktj}^{pi} + \left(p_{it}^{b} + tc_{ij} + mkt_i\right)Q_{iktj}^{b}\right]$$

$$- s_k\left(Q_{ikt}^{ps} + Q_{ikt}^{bs}\right) \qquad (1)$$

[Revenue from milled rice] – [Fixed costs] – [Milling costs] –
[Cost of traditional common paddy] – [Cost of improved superior paddy + Cost of transporting paddy from zone *j*] – [Cost of storage],

where $i, j = 1, \ldots, N(N = 37)$; $k = 1, \ldots, K(K = 3)$; $T = 1,2$. N here represents all 37 states in Nigeria, K the three miller sectors (small, medium, and large), and T for two rice-production seasons ("on" and "off"). Table E.1 describes in more detail the annotations and symbols used in the equations above.

The objective function in equation (1) states that profits of the industry are maximized given the choice of rice milled, stored, and sold during the year by

1 A list of the 37 states in Nigeria and corresponding geopolitical zones is provided in Table E.7.

TABLE E.1 Annotations and symbols for rice milling model equation (1)

Symbol		Description
Variables		
Q_{ikt}^{pm}	=	Quantity of rice in MT milled from local paddy by miller sector k in zone i and in period t.
Q_{ikt}^{bm}	=	Quantity of rice in MT milled from imported brown rice by miller sector k in zone i and in period t.
Q_{iktj}^{pi}	=	Quantity of improved superior paddy purchased by miller sector k in zone i, from zone j, and in period t.
Q_{iktj}^{pt}	=	Quantity of traditional common paddy purchased by miller sector k in zone i, from zone j, and in period t.
Q_{iktj}^{b}	=	Quantity of quality brown rice imported by miller sector k in zone i, from zone j (where is restricted to Lagos), and in period t.
Q_{ikt}^{ps}	=	Quantity of local paddy stored (either common or superior) by miller sector k in zone i and in period t.
Q_{ikt}^{bs}	=	Quantity of imported brown rice stored by miller sector k in zone i and in period t.
P_{ikt}^{m}	=	Price of milled rice from miller sector k in zone i and in period t.
Parameters		
f_{ik}^{p}	=	Annualized fixed costs for milling operations for miller sector k in zone i, in $ per MT of paddy rice (p).
c_{ik}^{p}	=	Variable costs for milling operations for miller sector k in zone i, in $ per MT of paddy rice (p).
p_{it}^{pt}	=	Observed prices of traditional common paddy (pt) in zone i during period t.
p_{it}^{pi}	=	Prices of improved superior paddy (pi) in zone i and in period t as a proportionate share above p_{it}^{pi} (assumed here as 25%, although higher than the reported 6% in Chapter 5).
tc_{ij}	=	Transportation costs for shipping paddy or rice between zone i and zone j, in $/MT.
mkt_i	=	Marketing costs for paddy or rice trade within zone, in $/MT.
τ	=	Import tariff rate (percent).
Coefficients		
θ_k^p, θ_k^b	=	Technical efficiency coefficients of miller sector k for converting paddy (brown) into milled rice, as a weight ratio of milled rice output to paddy (brown) input, respectively.
δ_k^b	=	Coefficient of brown rice milling costs as a proportion of paddy milling costs for miller sector k (restricted to k = large).
β	=	Coefficient in supply response function in equations 4 and 5 (fixed at 0.4).

Source: Authors' rice milling model.

Note: MT = metric tons. The zones refer to the six geopolitical zones.

each milling sector, including their fixed and variable operating costs, cost of paddy purchased from any state, transportation and storage costs, and given output prices for milled rice. The costs of paddy are distinguished by two quality varieties: improved superior and traditional common. Superior paddy varieties refer to the most commonly preferred quality seed varieties for milling premium-quality rice among medium- and large-scale millers, while common varieties refer to both traditional and other older modern varieties for milling standard-quality rice by the small milling sector.

Although paddy prices are treated as exogenous in the model, the price of milled rice $\left(P^m_{ikt}\right)$ is explicitly linked to the rice import tariff (τ), the world price for premium imported rice $\left(p^w_{it}\right)$, and the price for standard domestic rice $\left(p^d_{it}\right)$[2]. The relationship is described in the model as follows:

$$P^m_{ikt}(k = small) = p^d_{it}(1 + \tau)^{\mu_k}\left[\frac{1 + (2 - mkt_i)}{2} - tc_{ii}\right], \forall i \in N, \forall t \in T \tag{2}$$

$$P^m_{ikt}(k = large) = p^w_{it}(1 + \tau)^{\mu_k} + tc_{ij} + mkt_i, \forall i \in N, \forall t \in T, \forall j = Lagos \tag{3}$$

$$P^m_{ikt}(k = large) = \begin{cases} P^m_{ikt}(large) & \text{if } \alpha P^m_{ikt}(large) > \varpi P^m_{ikt}(small) \\ \varpi P^m_{ikt}(small) & \text{if } \alpha P^m_{ikt}(large) < \varpi P^m_{ikt}(small) \\ \forall i \in N, \forall t \in T \end{cases} \tag{4}$$

$$P^m_{ikt}(k = medium) = \begin{cases} \tau P^m_{ikt}(large) & \text{if } \alpha P^m_{ikt}(large) > \varpi P^m_{ikt}(small) \\ \varpi P^m_{ikt}(small) & \text{if } \alpha P^m_{ikt}(large) < \varpi P^m_{ikt}(small) \\ \forall i \in N, \forall t \in T \end{cases} \tag{5}$$

where μ_k in equation (3) is a parameter that represents the extent to which changes in the global price and/or import tariffs are transmitted to domestic prices within each of the milling sectors and type of rice market (standard versus premium rice). α is a parameter representing a constant ratio of medium miller rice prices to large miller rice prices (in the model, α is set at 0.85). ϖ is a parameter representing a minimum floor price by which medium millers would still enjoy a price premium over small millers (in the model, ϖ is set at 1.05). τ is the rice import tariff. The right-hand terms in equations (2) and (3) explicitly account for transportation and marketing costs according to

2 Note that p^{di}_t is an observed price for domestic rice in zone i. Because the price already includes local transportation and marketing costs in zone i, these costs are removed to avoid double counting when linking the price to an import parity price that already includes these costs.

each location. Equations (4) and (5) simply ensure that the prices faced by medium and large millers can never fall below the prices small millers face—a reasonable assumption considering the inferiority of rice from the small milling sector.

The industry's profits, as expressed in the objective function in equation (1), are subject to a number of resource and location constraints. First, the operations of each milling sector are subject to capacity and storage constraints and are expressed in the model as

$$Q_{ikt}^{pm} + Q_{ikt}^{bm} \leq \overline{q}_{ikt}^{m}, \forall i \in N, \forall k \in K, \forall t \in T \tag{6}$$

[Amount of paddy milled] + [Amount of brown milled] ≤ [Milling capacity]

$$Q_{ikt}^{ps}, Q_{ikt}^{bs} \leq \overline{q}_{ikt}^{s}, \forall i \in N, \forall k \in K, \forall t \in T \tag{7}$$

[Amount of paddy stored] + [Amount of brown stored] ≤ [Storage capacity],

where $\overline{q}_{ikt}^{m}$ is the total milling capacity of miller sector k in geopolitical zone i during period t, in total MT of paddy per period, and $\overline{q}_{ikt}^{s}$ is the total storage capacity of miller sector k in zone i during period t, in total MT of paddy per period. Equations (6) and (7) constrain milling and storage according to their maximum capacities at each location, by the miller scale type, and according to each seasonal period. Rice milling occurs using either improved (superior-type) paddy varieties or traditional (common-type) paddy varieties. The former produces better-quality premium rice after milling, while the latter produces a standard- and lower-quality rice. Although local paddy can be purchased from any state, this is limited to the premium paddy varieties that are the preferred type for milling by both the medium- and large-scale operators. However, purchasing from outside a state can incur higher transport and marketing costs, as the next set of constraints will specify.

In addition to capacity and storage constraints, millers across all three sectors are also constrained to purchase not more than the supply of improved (superior) or traditional (common) local paddy available anywhere in Nigeria. Additionally, they purchase from further afield only if the relative price of paddy in that location, inclusive of any transportation and marketing costs, is lower than any other location during each seasonal period. To allow for a positive supply response to price changes, we assume positive own-price elasticities of 0.4 for superior varieties and 0.2 for common varieties in the inequality constraints of excess supply modeled as inverse supply functions (f) with respect to the price relationships in equations (2) and (3) as follows:

$$\sum_{k=1}^{K}\sum_{i=1}^{N} Q_{iktj}^{pt} \leq \overline{q}_{jt}^{pi}[p_{jt}^{pi}(1+\tau)^{\beta}[2-mkt_j]+tc_{ij}]^{0.4}, \forall i \in N, \forall t \in T \quad (8)$$

[Supply of improved superior paddy] ≤ [f(Price of improved superior paddy)]

$$\sum_{k=1}^{K} Q_{ikt}^{pt} \leq \overline{q}_{it}^{pt}[p_{it}^{pt}(1+\tau)^{\frac{\beta}{2}}[1+(2-mkt_i)]+tc_{ij}]^{0.2}, \forall i \in N, \forall t \in T \quad (9)$$

[Supply of traditional common paddy] ≤
[f(Price of traditional common paddy)].

The inverse supply response functions (f) in the right-hand side of equations (8) and (9) for each paddy variety type are intended to account for changes in output as prices are made to adjust, either due to a change in the world price or the rice import tariff (τ). The value of $\overline{q}_{jt}^{pi}$ is calculated using currently observed paddy prices and production levels for both superior and common paddy varieties, respectively, and in each zone.

In addition to considering the purchase of paddy across different locations in the country, millers can also access stored paddy from the previous season. Storage of paddy rice is permitted in the model between an *on* season (November to March) and an *off* season (April to October) during each rice-production year. This is illustrated in equations (10) and (11), which allow for the storage of paddy or brown rice, respectively, between the seasons.

$$Q_{ikt}^{pm} + Q_{ikt}^{ps} = \sum_{j=1}^{N} Q_{iktj}^{pi} + Q_{ikt}^{pt} + Q_{ikt-1}^{ps}, \forall i,j \in N, \forall k \in K, \forall t \in T \quad (10)$$

[Paddy milled and stored in current season] +
[Paddy purchased in current and stored from previous seasons]

$$Q_{ikt}^{bm} + Q_{ikt}^{bs} = \sum_{j=1}^{N} Q_{iktj}^{b} + Q_{ikt-1}^{bs}, \forall i \in N, \forall k \in K, \forall t \in T \quad (11)$$

[Brown milled and stored in current season] =
[Brown purchased in current and stored from previous seasons].

For simplicity, there are no carryover stocks of stored paddy or brown rice between years in equations (10) and (11) based on the assumption that all stocks are milled and sold by the end of each production year to meet domestic demand.

Finally, the model places some restrictions on the type of paddy variety traded between states and by whom, as well as for brown rice, which is restricted for use by the large milling sector only as follows:

$$Q^{pi}_{iktj} + Q^{pt}_{iktj} = 0, \forall i \neq j \in N, \forall k = small, \forall t \in T \tag{12}$$

[Small millers cannot purchase any paddy from other zones]

$$Q^{pt}_{iktj} = 0, \forall i, j \in N, \forall k \neq small, \forall t \in T \tag{13}$$

[Large and medium millers cannot purchase traditional common paddy varieties]

$$Q^{b}_{iktj}, Q^{bm}_{ikt}, Q^{bs}_{ikt} = 0, \forall i \in N, \forall j \neq lagos, \forall k \neq large, \forall t \in T \tag{14}$$

[Brown rice can only be purchased from Lagos by large millers]

$$Q^{pm}_{ikt}, Q^{bm}_{ikt}, Q^{pi}_{iktj}, Q^{pt}_{iktj}, Q^{b}_{iktj}, Q^{ps}_{ikt}, Q^{bs}_{ikt} \geq 0, \forall i, j \in N, \forall k \in K, \forall t \in T \tag{15}$$

[Paddy or brown purchased, stored, and milled cannot be negative].

The restrictions on paddy trade only occur for the small milling sector, equation (12), as this is dominated by small millers located within village or small town market centers, which operate mostly on a fee basis for local producers. The rest are variable restrictions in (13) through (15) and based on our assumptions of the milling industry structure and type of products. For example, we restrict medium and large millers to purchase superior paddy varieties only.

Model Assumptions and Validation

The model is validated to 2010 base-year quantities of paddy production and prices, as discussed in the main text. Production of paddy occurs seasonally (unless there is irrigation) and serves as the major input into production of milled rice. Table E.2 summarizes the baseline production levels used in the model by major zones (see Total column). Almost 90 percent of paddy production comes from the North Central, North East, and North West zones. The type of milling technologies and hourly capacities considered are as follows: Small (about 0.1 MT/hr, unconstrained)—about 70–75 percent of total rice milled; Medium (0.5–2 MT/hr)—unknown (about 9 percent from 2002/2003 surveys); Large (12 MT/hr)—potentially 15–21 percent of total rice milled (excluding brown rice). In the baseline, a majority of paddy is assumed to be milled by the dominant small cottage industrial sector in rural areas (totaling about 73 percent nationally), although this

TABLE E.2 Milling output by miller sector and zones

	Milling output ('000 MT)				Share of zonal output (%)		
State	Large	Medium	Small	Total	Large	Medium	Small
North Central	172.8	77.1	957.4	1,207.3	14.3	6.4	79.3
North East	0.0	164.5	738.0	902.5	0.0	18.2	81.8
North West	345.6	76.6	792.2	1,214.4	28.5	6.3	65.2
South East	172.8	5.4	150.3	328.5	52.6	1.6	45.8
South South	0.0	0.0	22.1	22.1	0.0	0.0	100.0
South West	0.0	2.2	98.0	100.2	0.0	2.2	97.8
Total	*691.2*	*325.8*	*2,758.0*	*3,775.0*	*18.3*	*8.6*	*73.1*

Source: Nigeria, NAERLS (2009) and author's calculations based on milling capacities.

Note: The total also reflects total supply of paddy in each zone, as the small miller sector was treated as a residual in calculating milling capacities (i.e., total paddy production less large and medium milling capacities).

varies by zone, as shown in Table E.2 (e.g., 81.8 percent in North East, and 65.2 percent in North West). Total monthly and annual capacities for large mills are binding for each state (later relaxed with the introduction of new mills); medium mills have nonbinding capacity except in low-producing states with no capacity (< 30,000 MT); and small mills serve as a residual for all states.

Supply of paddy is constrained by current production levels of the two types of paddy—traditional common and improved superior varieties. Paddy output from superior varieties averages about 29 percent nationally in the baseline case (summarized in Table E.3). This is quite conservative, as estimates in the NAERLS (2009) puts this closer to 35 percent on average, while NTBS data shows an average of 51 percent.[3]

Storage capacity varies by miller sizes and is measured in terms of storable months (Small = 1 month, Medium = 2 months, and Large = 3 months). Prices and all costs are assumed given (i.e., exogenous), and baseline includes import tariffs of 50 percent for both milled rice and brown rice. The price for superior paddy is assumed to command a 20 percent premium over more common paddy rice varieties. Rice from small-scale mills versus medium and large is assumed to be non-substitutable. Rice prices for large-scale millers are the same as for imported rice. The latter are a function of the cost, insurance, and freight price of imported rice at the port of Lagos plus any transportation and

3 The lower conservative estimates were purposefully selected to introduce a binding constraint in the model in the baseline scenario for the medium- and large-scale millers, who are assumed to strictly prefer superior paddy varieties.

TABLE E.3 Use of superior and common paddy varieties by major zone in the baseline scenario

	Improved superior varieties	Common varieties	Total	All improved varieties (NBS) (1)	Improved superior varieties (2)	Common varieties: Older improved varieties (3)	Common varieties: Traditional varieties (4)
	Output (metric tons)				*Percent of area*		
North Central	333.3	874	1,207.3	50	27.6	22.4	50
North East	247.1	655.5	902.6	58	27.4	30.6	42
North West	418.3	796.1	1,214.4	55	34.4	20.6	45
South East	70.4	258.2	328.5	30	21.4	8.6	70
South South	6.0	16.1	22.1	55	27.2	27.8	45
South West	21.5	78.7	100.2	53	21.4	31.6	47
Total	1,096.5	2,678.5	3,775.1	51	29.0	22.0	49

Source: Nigeria, NAERLS (2009), Nigeria, NBS (2013b), and author's calculations.

Note: "Superior varieties" here are limited to the most commonly preferred paddy rice varieties for milling premium-quality rice and hence are assumed to be less than the adoption of all improved varieties. The lower, more conservative estimates were purposefully selected to introduce a binding constraint in the model's baseline scenario for the medium- and the large-scale millers, who are assumed to strictly prefer superior paddy varieties. "Common varieties" refers to both traditional and other older improved varieties that are considered inferior for milling purposes. Adding columns 2 and 3 equals column 1. Column 4 is 100 less column 1.

marketing costs to local markets in each state, and therefore by definition it is the import parity price observed in that market. For medium scale, the price is set at a premium above the smaller rice millers that is 50 percent of the price gap with the large-scale and imported rice prices. Such a price assumption was necessary given the advantage of medium over small millers with respect to the quality of rice produced. The opposite is true when compared with the larger industrial mills.

Figure E.1 shows the cross-sectional levels of original price data used in the model and sorted by distance from Lagos port. The sorting was necessary to compare the resulting calculations of the import parity price in each market (which here only accounts for transportation in estimating inland costs) with local price data. As the figure shows, the calculated import parity price of imported rice (excluding marketing costs) lies below the actual observed price of imported rice in local markets. Assuming the difference measures additional marketing costs, we use this information to estimate such costs for the model. From the same figure, it is worth noting that prices for local rice appear more variable, while prices for paddy (more generally) are more consistent across all the states.

FIGURE E.1 **Baseline data of average state prices for paddy, local, and imported rice in Nigeria, 2012**

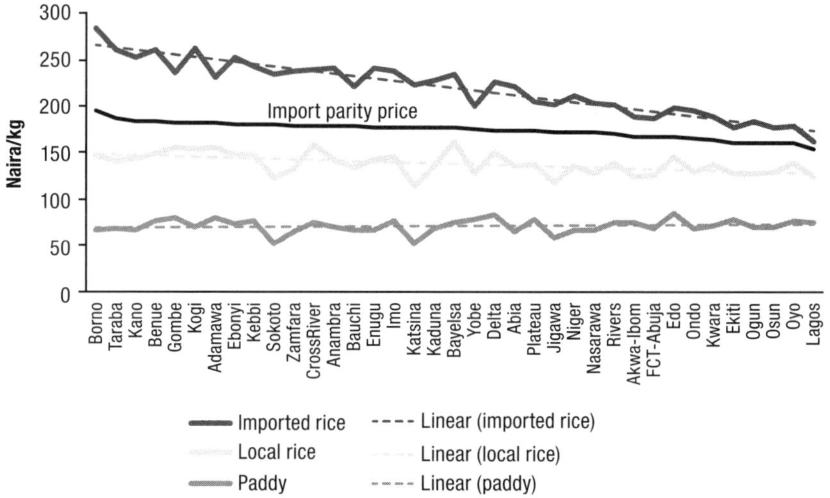

Source: Data are from Nigeria, NBS (2012); Nigeria, NBS and World Bank (2013).

Note: Import parity price is calculated as Lagos FOB plus unloading costs, port fees, and transportation and marketing costs to each state. FOB = free on board.

Paddy can be bought from any state by medium and large mills, with the option to import brown rice for large mills only. Large- and medium-scale mills demand premium-quality paddy only, while small mills can use either. Marketing and transaction costs are determined by analyzing actual price differentials from NBS state average price data between local and imported rice. Rice can be sold by large or medium mills in any state (inclusive of transportation cost). A small sample of trailer costs for hauling rice from Lagos to a number of cities in Nigeria was collected to estimate average per-unit transportation costs (naira/MT/km) and then multiplied by the actual distances between cities (Table E.4). The final per-unit (MT/km) transportation costs (naira/MT/km) within each zone are presented in Table E.5.

From Table E.5, not surprisingly, the highest cost of transporting a metric ton of rice is from southern to northern regions: the cost can range from 13.2 to 17.3 naira per metric ton. With regard to marketing costs, the highest are reportedly in the South East. This is a region that showed wider gaps in the data between actual and import parity rice prices. Here we present only 30 percent of the difference, which we have assumed in the model to represent genuine marketing costs.

TABLE E.4 Estimation of average transportation costs from Lagos port

Zone	To [city, state]	Total kms[a]	Naira/trailer[b]	Naira/MT	Naira/MT/km
South West	Ibadan, Oyo	122	200,000	6,667	55
South West	Osogbo, Osun	234	220,000	7,333	31
South East	Enugu, Enugu	617	280,000	9,333	15
South South	Calabar, Cross River	737	400,000	13,333	18
North East	Jalingo, Taraba	1,153	450,000	15,000	13
North East	Yola, Adamawa	1,304	450,000	15,000	12
North West	Kano, Kano	984	380,000	12,667	13
North West	Sokoto, Sokoto	959	380,000	12,667	13
North West	Katsina, Katsina	940	380,000	12,667	13
North Central	Abuja, FCT	722	320,000	10,667	15
North Central	Ilorin, Kwara	280	220,000	7,333	26

Source: Authors' calculations.

Note: Naira/trailer = naira per trailer load, N/MT = naira per metric ton, N/MT/km = naira per metric ton and per km. [a] Based on Google Maps in actual distances along major highways. [b] One trailer load = 30 metric tons, 600 bags of 50 kg each.

Technology and unit cost data are summarized in Table E.6. They include paddy-to-milled rice conversion factors, annualized fixed costs, milling operational costs, and storage costs (all in naira per kg). Transportation costs are calculated based on actual distances, per-unit costs (naira/MT/km), and include 20 percent unobservable marketing (or transaction) costs. Cost values are calculated for each miller sector and based on field visits and interviews with milling operators (Figure E.2). The values were cross-checked with other estimates in the literature.

Fixed costs for the proposed integrated large mills were taken directly from the Agricultural Transformation Agenda, a document laying out a proposal for the new mills. The fixed costs are annualized over the life of any up-front capital investments, including building, milling facilities, storage space, vehicles, and so on. Notice that the up-front investment costs for the newer integrated mills are much higher.

Milling costs are variable and based on operations at maximum capacity. We assume the per-unit output milling costs to be similar between newer and older large integrated mills. However, we expect the newer ones to be more technologically efficient by proving a higher paddy-to-milled rice conversion factor. Medium-scale millers experience much higher milling costs per unit of output. This is primarily because of high labor and energy costs for milling

TABLE E.5 Average transportation and marketing costs between and within states, respectively (naira/MT)

	North Central	North East	North West	South East	South South	South West	From Lagos
a. Between states zonal average transportation costs (naira/MT)							*Naira/MT/km*
North Central	6.4	11.9	10.5	9.2	10.4	9.9	18.0
North East	11.9	5.9	12.1	13.8	15.3	17.3	12.2
North West	10.5	12.1	6.4	14.3	15.3	13.2	13.2
South East	9.2	13.8	14.3	2.8	4.6	9.8	15.1
South South	10.4	15.3	15.3	4.6	4.5	9.4	18.1
South West	9.9	17.3	13.2	9.8	9.4	3.5	22.5
b. Regional average marketing costs (naira/MT)							
In state	15.4	16.6	17.4	20.9	15.7	10.2	—

Source: Author's calculations from actual distances, based on Google Maps, along major highways between states and zones.
Note: Unit costs were recalculated as a matrix measuring the costs of transporting a metric ton between capitals of each state. Only averages are reported here. These reflect the total transportation cost given distance and the per-unit (naira/MT/km) estimated cost in the last column. MT = metric tons.

TABLE E.6 Parameter estimates per unit of output production costs at maximum capacity and transportation costs by miller type (naira/kg)

Miller scale	Paddy/rice CF[1]	Fixed costs[2]	Milling costs[3]	Storage costs[4]
Large (new)	0.62	18.00	6.08	0.10
Large (current)	0.60	9.83	6.08	0.10
Medium	0.64	1.30	15.26	0.20
Small	0.66	0.07	4.65	0.20

Source: Author field data.
Note: [1] CF = conversion factor. [2] Fixed costs are annualized over the life of any up-front capital investments, including building, milling facilities, storage space, vehicles, etc. The up-front investment costs are higher for the newer integrated mills. [3] Milling costs are variable and based on operations at maximum capacity. We assume costs to be similar between newer and older large integrated mills. However, we expect the newer ones to be more technologically efficient by proving a higher paddy-to-milled rice conversion factor. [4] Storage costs are minimal but higher for the small- and medium-scale operators.

premium-quality rice. Smaller operators are less concerned with quality and typically mill on a fee basis, hence the lower costs. Large-scale millers enjoy economies of scale and are able to keep their milling costs per unit of output at much lower rates than medium-scale operators. However, they have the highest up-front costs in capital investments and hence higher discounted annualized fixed costs over their life. With combined fixed and operating costs, the medium- and large-scale operators end up with similar per-unit costs.

FIGURE E.2 Average baseline cost data for rice milling model by miller type (naira/kg)

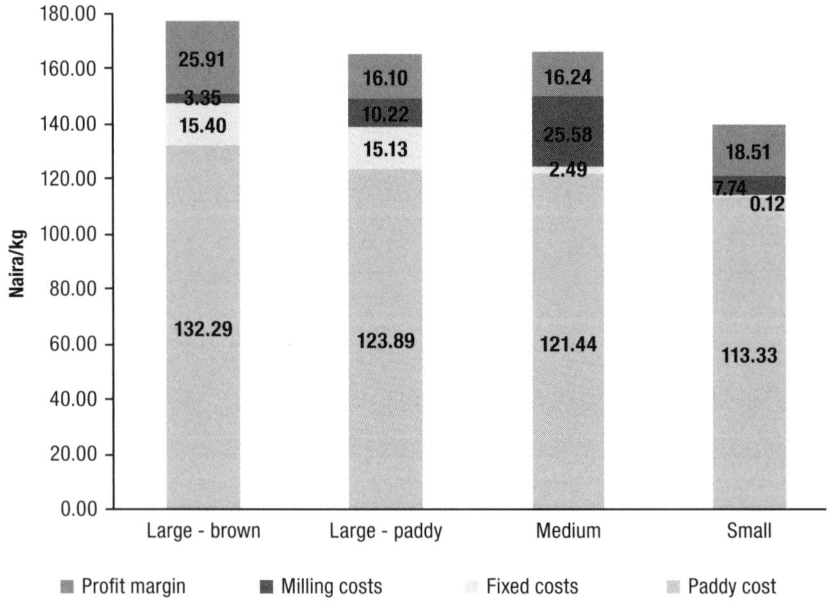

Source: Marketing costs are authors' calculations based on 30 percent of the price differential between actual imported rice prices and the import parity price in each state (inclusive of transportation costs from Lagos port).

TABLE E.7 **Geopolitical zones, states, and state capital cities in Nigeria**

Zone	State	Capital city	Zone	State	Capital city
North Central	Benue	Makurdi	South East	Abia	Umuahia
	FCT	Abuja		Anambra	Awka
	Kogi	Lokoja		Ebonyi	Abakaliki
	Kwara	Ilorin		Enugu	Enugu
	Nasarawa	Lafia		Imo	Owerri
	Niger	Minna	South South	Akwa Ibom	Uyo
	Plateau	Jos		Bayelsa	Yenagoa
North East	Adamawa	Yola		Cross River	Calabar
	Bauchi	Bauchi		Delta	Asaba
	Borno	Maiduguri		Edo	Benin City
	Gombe	Gombe		Rivers	Port Harcourt
	Taraba	Jalingo	South West	Ekiti	Ado Ekiti
	Yobe	Damaturu		Lagos	Ikeja
North West	Jigawa	Dutse		Ogun	Abeokuta
	Kaduna	Kaduna		Ondo	Akure
	Kano	Kano		Osun	Osogbo
	Katsina	Katsina		Oyo	Ibadan
	Kebbi	Birnin Kebbi			
	Sokoto	Sokoto			
	Zamfara	Gusau			

Source: Nigeria, NBS (2013a).

Note: FCT = Federal Capital Territory, home of the national capital of Nigeria, Abuja.

Appendix F

A STYLIZED RICE TARIFF MODEL

Paul A. Dorosh and Mehrab Malek

A simple model of rice imports is developed as a stylized representation of what could be occurring in the Nigerian example in order to provide a plausible economic explanation. The model is intended to help illustrate how increases in rice import tariffs beyond a certain threshold may become ineffective under conditions when the tariff is evaded by diverting imports through unofficial channels. Two ports of entry for rice imports are assumed on the outset: the first port (1) is assumed to represent imports through official channels (as would be the case at the Lagos port), and the second port (2) is through other channels that evade the import tariff either through underinvoicing or smuggling (as could be occurring through northern Nigeria).

For rice imported through customs at the first port (1) at Lagos, the import parity price of rice in wholesale markets in southern Nigeria $\left(P_{m1}^{S}\right)$ is given by

$$P_{m1}^{S} = EP_w(1 + \tau) + mktg_1,$$

where $mktg_1$ is the margin for port handling and marketing costs to wholesale markets. For imported rice that is transported to the north, the import parity price is

$$P_{m1}^{N} = EP_w(1 + \tau) + mktg_1 + \delta,$$

where δ is the margin for additional marketing costs between wholesale markets in southern Nigeria and those in northern Nigeria.

For rice imported through the northern border with Niger,[1] the import parity price of rice in wholesale markets in northern Nigeria $\left(P_{m2}^{N}\right)$ is given by

$$P_{m2}^{N} = EP_w + mktg_2,$$

1 This simplified model abstracts from imports through Benin. While generally small, COMTRADE data (UN Statistics Division 2012) for Benin exports to Nigeria may indicate exports that flow through Niger as well.

where $mktg_2$ is the margin for port handling in Cotonou, Benin, and marketing costs to wholesale markets in northern Nigeria (note that there are no import tariffs on rice in Benin or Niger and that rice flowing from Niger to Nigeria evades customs).[2] For imported rice through this channel that is further transported to southern Nigeria, the import parity price is

$$= EP_w + mktg_2 + \delta.$$

The prevailing market price for imports in each region will be the minimum of the import parity prices for rice coming through Lagos or from the northern border:

$$P^S_m = min\{P^S_{m1}, P^S_{m2}\}P^N_m$$
$$= min\{P^N_{m1}, P^N_{m2}\}.$$

Thus, the import parity price of rice from Lagos determines the market price in the south as long as $P^S_{m1} < P^S_{m2}$; that is,

$$EP_w(1 + \tau) + mktg_1 < EP_w + mktg_2$$
$$+ \delta\ EP_w\tau + mktg_1 < mktg_2 + \delta.$$

Similarly, the import parity price of rice from the northern border determines the market price in northern Nigeria as long as $P^N_{m2} < P^N_{m1}$; that is,

$$EP_w + mktg_2 < EP_w(1 + \tau) + mktg_1$$
$$+ \delta\ mktg_2 < EP_w\tau + mktg_1 + \delta.$$

Figure F.1 shows the results of simulated trade flows under alternative import tariff rates using this simple stylized model. (These calculations assume $mktg_1 = 15$ naira/kg; $mktg_2 = 50$ naira/kg; $\delta = 25$ naira/kg; $P_w = \$600$/ton; ER = 160 naira/\$.) At low tariff rates, all imports will come through Lagos port and imported rice will flow from southern to northern Nigeria. At higher tariff rates (30–70 percent in this example), traders supplying northern Nigerian markets will find it cheaper to import rice through the northern border (and evade the tariff), in spite of higher transport and other marketing costs through Benin and Niger. At very high tariff rates (> 70 percent in this example), it is cheaper to import rice through Niger, even for sales in southern Nigerian markets.

Implicitly, the model assumes that the domestic price of imports in the south is not directly linked to the domestic price of imports in the north

2 If there is a parallel market for exchange, E is replaced by E^* in this equation.

FIGURE F.1 Simulated effects of rice tariff increases

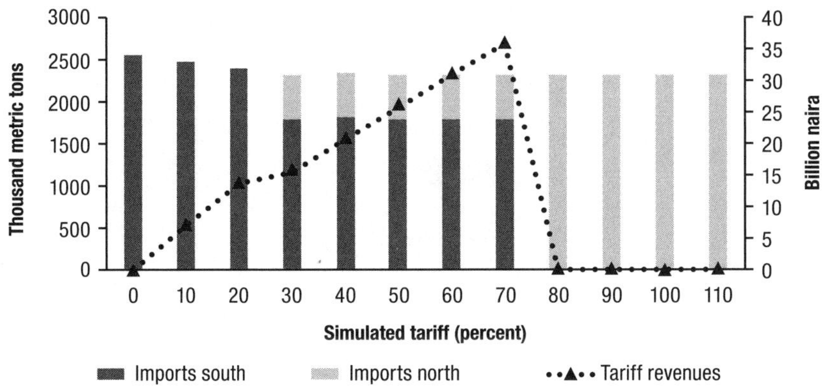

Source: Model simulations.

Note: These preliminary simulations are based on rough approximations of marketing costs and do not necessarily indicate traders' actual incentives.

(except when all import trade flows exclusively through Lagos or all import trade flows exclusively through Niger). If the markets for local rice in northern and southern Nigeria are integrated, then the ratio of P_m/P_l (price ratio of imported to local rice) varies between south and north. Supply and demand response for changes in P_m are overstated compared to a model that takes into account these variations across space and imperfect substitution. To take this into account, the current model uses conservative elasticities. Further refinements of the model could include explicit equations for imperfect substitution and variations in the price of P_l.

For this stylized model, we assume that demand and supply for rice can be specified as a function of the price of imported rice only, either because the price ratio of imported to local rice is constant or because the share of local rice in total consumption is small (here assumed to be only 8.8 percent in the south). Additional assumptions include the following: total milled rice production is 2.9 million metric tons, total rice demand is 5.2 million metric tons, and total rice imports are 2.3 million metric tons; the country has two subnational regions (north and south); about 9 percent of local rice consumption is in the south (the rest is in the north); 40 percent of imported rice demand is in the south (the rest is in the north).[3] In this case, price changes for imported

3 Calculated from Living Standards Measurement Study–Integrated Surveys on Agriculture (2011), with downward adjustments in quantities for three states in the South East zone.

TABLE F.1 Nigeria: Official data on rice imports and tariffs by type of rice, 2007–2012

	Imports: net weight ('000 MT)	Imports: value (billion naira)	Gross tariff revenue (billion naira)	Average price (naira/kg)	Average tariff collected (%)
Broken rice					
2007	100	10	22	104	208
2008	19	1	2	43	205
2009	64	13	6	207	42
2010	2	2	1	827	39
2011	8	3	2	413	55
2012	40	25	19	632	74
Husked (brown) rice					
2007	41	1	1	13	210
2008	10	< 1	1	40	202
2009	3	4	1	1,211	22
2010	32	12	2	370	15
2011	7	10	1	1,335	10
2012	117	57	24	488	41
Other milled rice*					
2007	2,314	50	82	22	162
2008	318	10	20	31	203
2009	330	53	23	160	43
2010	551	61	35	110	58
2011	270	140	75	517	53
2012	261	169	108	646	64
All rice**					
2007	2,455	61	105	25	171
2008	348	11	23	32	203
2009	398	70	29	176	42
2010	585	74	38	127	51
2011	286	153	78	536	51
2012	418	251	150	600	60

Source: All figures are based on unedited monthly data from Nigerian customs.
Note: MT = metric tons. The actual amounts of imported rice are likely higher than the official figures. Other milled rice imports in 2007 may be substantially overstated. The alternative estimate is 1.014 million tons of rice imports using approximate average prices to calculate the quantity of imports in June and July 2007.
* "Milled and semimilled rice." ** Includes very small amounts of rice seed, rice in husk (paddy or rough rice), and "other rice." Figures may also not match exactly the sum of rice in the three categories due to rounding.

rice determine changes in production, consumption, and imports. Import tax revenues are then calculated as the quantity of imports multiplied by the C&F price and effective tariff rate (here we assume that 30 percent of the official rate is captured to take into account under-invoicing).

As shown in Figure F.1, import tariff revenues rise with increases in tariff rates until the tariff exceeds a threshold (in our stylized model this is at the 70 percent tariff rate), at which point tariff revenues fall to zero as imports through Lagos stop and all imports flow through the northern border (and evade tariffs). Thus, due to the assumption of a high level of tariff evasion, raising tariffs beyond a certain threshold has no effect on domestic prices, making tariff policy an ineffective instrument for further raising revenues or domestic prices in this example.

Appendix G

AGRICULTURAL SECTORS INCLUDED IN THE ECONOMYWIDE MULTIMARKET MODEL

Xinshen Diao, Michael Johnson, and Hiroyuki Takeshima

TABLE G.1 List of agricultural sectors (products) included in economywide multimarket model

Crops		Livestock and fisheries
Cereals	*Cash crops*	*Livestock*
1. Maize	18. Cotton	27. Cattle
2. Rice	19. Sugar	28. Chickens
3. Millet	20. Cocoa	29. Sheep and goats
4. Sorghum	21. Coffee	30. Pigs
5. Wheat		31. Other livestock
6. Other cereals		
Root crops and tubers	*Fruits and vegetables*	*Animal products*
7. Cassava	22. Total vegetables	32. Beef
8. Yam	23. Total plantains	33. Poultry
9. Sweet potato	24. Other fruits	34. Lamb and goat
10. Potato		35. Pork
11. Cocoyam		36. Other meat
		37. Milk
Oilseeds and pulses	*Tree nuts and spices*	38. Eggs
12. Groundnuts	25. Total spices	39. Skin
13. Soybeans	26. Nuts	
14. Oil palm		*Fisheries*
15. Sesame seed		40. Sea fish
16. Other oilseed		41. Fresh fish
17. Total pulses		

Source: Authors, from economywide multimarket model.

References

Abbott, P. 2007. "Distortions to Agricultural Incentives in Côte d'Ivoire." World Bank Agricultural Distortions Working Paper 46, Washington, DC. Accessed July 30, 2015. www.researchgate.net/profile/John_Baffes/publication/228666994_Distortions_to_Agricultural_Incentives_in_Cote_d'Ivoire/links/0deec5331aa4a486da000000.pdf.

Abo, M. E., A. A. Sy, and M. D. Alegbejo. 1997. "Rice Yellow Mottle Virus (RYMV) in Africa: Evolution, Distribution, Economic Significance on Sustainable Rice Production and Management Strategies." *Journal of Sustainable Agriculture* 11 (2-3): 85–111.

Abo, M. E., M. N. Ukwungwu, and A. Onasanya. 2002. "The Distribution, Incidence, Natural Reservoir Hosts and Insect Vectors of Rice Yellow Mottle Virus (RYMV), Genus *Sobemovirus* in Northern Nigeria." *Tropicultura* 20 (4): 198–202.

Abrar, S., O. Morrissey, and T. Rayner. 2004. "Crop-Level Supply Response by Agro-Climatic Region in Ethiopia." *Journal of Agricultural Economics* 55 (2): 289–311.

Adejumo-Ayibiowu, D. 2010. "Sector Strategies and Policies Related to Rice Development in Nigeria." Mapping of Poverty Reduction Strategies Papers (PRSP), Coalition for African Rice Development (CARD), AGRA, Nairobi, Kenya. Accessed on July 31, 2015. www.riceforafrica.org/downloads/CARD%20DOCS/NIGERIA_FINAL_REPORT_revised.pdf.

Adeniran, K. A., M. F. Amolu, M. O. Amolu, and F. A. Adeniji. 2010. "Water Requirements of Some Selected Crops in Kampe Dam Irrigation Project." *Australian Journal of Agricultural Engineering* 1 (4): 119–125.

Adesina, A. 2012. "Unlocking the Potential of Agriculture in Sub-Saharan Africa: Nigeria's Transformation Agenda for Agriculture." PowerPoint presentation, Ministry of Agriculture and Rural Development, Federal Republic of Nigeria, Abuja, Nigeria.

Adeyeye, J. A., E. P. Navesero, O. J. Ariyo, and S. A. Adeyeye. 2010. "Consumer Preference for Rice Consumption in Nigeria." *Journal of Humanities, Social Sciences and Creative Arts* 5 (1): 26–36.

Agrifood Consulting International. 2005. "Northeast Thailand Rice Value Chain Study." Project Brief Series, Agrifood Consulting International, Bethesda, MD, US. Accessed on July 31, 2015. www.value-chains.org/dyn/bds/docs/553/Agrifood%20Consulting%20Rice%20VC%20 Thailand%202005.pdf.

Agwu, N., and J. Ibeabuchi. 2011. "Socio-Economic Analysis of Wholesale Rice Marketers in Abia State, Nigeria." *International Journal of Social Science and Humanity* 1: 285–288.

Ahonsi, M. O., A. A. Adeoti, I. D. Erinle, M. D. Alegbejo, B. N. Singh, and A. A. Sy. 2000. "Effect of Variety and Sowing Date on False Smut Incidence in Upland Rice in Edo State, Nigeria." *International Rice Research Notes* 25 (1): 14–21.

Akaeze, H. O. 2010. "Consumer Preferences for Imported Rice in Nigeria: Perceived Quality Differences or Habit Persistence?" PhD dissertation, Michigan State University, East Lansing, MI, US.

Akanbi, O. A., and C. B. Du Toit. 2011. "Macro-Econometric Modelling for the Nigerian Economy: A Growth Poverty Gap Analysis." *Economic Modelling* 28: 335–350.

Akanbi, U. O., O. Omotesho, and O. E. Ayinde. 2011. "Analysis of Technical Efficiency of Rice Farms in Duku Irrigation Scheme Kwara State, Nigeria." *Nigerian Journal of Agriculture, Food and Environment* 7 (3): 65–72.

Akande, S., and G. Akpokodje. 2003. "Rice Prices and Market Integration in Selected Areas in Nigeria: A Study Report." WARDA/NISER, Cotonou, Benin. Accessed on July 31, 2015. http://pdf.usaid.gov/pdf_docs/PNADB907.pdf.

Akande, T. 2001. "An Overview of the Nigerian Rice Economy." Monograph of the Nigerian Institute of Social and Economic Research (NISER), Ibadan, Nigeria.

———. 2002. "An Overview of the Nigerian Rice Economy." Monograph of the Nigerian Institute of Social and Economic Research (NISER), Ibadan, Nigeria.

———. 2005. "The Role of the State in the Nigerian Green Revolution." In *The African Food Crisis: Lessons from the Asian Green Revolution*, edited by G. Djurfeldt, H. Holmen, M. Jirstrom, and R. Larsson, 161–179. Wallingford, UK: CABI Publishing.

Akighir, D. T., A. S. Ngutsav, and S. Asom. 2007. "Assessment of Poverty Level among Rice Millers in Kwande Local Government Area of Benue State, Nigeria." *International Journal of Humanities and Social Science* 1: 89–96.

Akinbile, L. A. 2007. "Determinants of Productivity Level among Rice Farmers in Ogun State, Nigeria." In African Crop Science Conference proceedings, 8: 1339–1344.

Akintayo, O. I., M. A. Y. Rahji, T. T. Awoyemi, and A. I. Adeoti. 2011. "Determination of Yield Gap in Lowland Rice Production in North Central Nigeria." *Agrosearch* 11 (1 and 2).

Akobundu, I. O. 1987. *Weed Science in the Tropics: Principles and Practices.* John Wiley: Chichester, UK.

Akpokodje, G., F. Lançon, and O. Erenstein. 2001. "Nigeria's Rice Economy: State of the Art." West Africa Rice Development Association (WARDA), Bouake, Côte d'Ivoire. Accessed on July 31, 2015. http://pdf.usaid.gov/pdf_docs/Pnadb851.pdf.

Alam, M. S., K. Alluri, T. M. Masajo, K. Zan, and V. T. John. 1986. "Upland Rice Improvement in Humid and Subhumid Tropics of West Africa." In *Progress in Upland Rice Research: Proceedings of the 1985 Jakarta Conference.* Manila: International Rice Research Institute.

Alemu, Z. G., K. Oosthuizen, and H. D. van Schalkwyk. 2003. "Grain-Supply Response in Ethiopia: An Error-Correction Approach." *Agrekon* 42 (4): 389–404.

Aluwihare, P. B., and M. Kikuchi. 1991. "Irrigation Investment Trends in Sri Lanka: New Construction and Beyond." International Irrigation Management Institute, Colombo, Sri Lanka. Accessed on July 31, 2015. http://publications.iwmi.org/pdf/H_9282i.pdf.

Anderson, K., ed. 2009. *Distortions to Agricultural Incentives: A Global Perspective, 1955–2007.* Washington, DC: World Bank.

Andriesse, W., and L. O. Fresco. 1991. "A Characterization of Rice-Growing Environments in West Africa." *Agriculture, Ecosystems and Environment* 33: 377–395.

Anozie, R. O., S. N. Okereke, and S. N. Anozie. 2012. "Comparative Profitability Analysis of Small Holder Upland and Irrigated Rice Productions in Anyamelum, Anambra State, Nigeria." *Journal of Agriculture and Food Technology* 2 (9): 159–163.

Apland, J., and H. Andersson. 1996. "Optimal Locations of Processing Plants: Sector Modeling Considerations and an Example." *Review of Agricultural Economics* 18: 491–504.

AQUASTAT. 2012. Geo-Referenced Database on Dams. Rome: Food and Agriculture Organization of the United Nations. Computer disk.

Arndt, C., and R. Schiller. 2000. "Grain Transport and Rural Credit in Mozambique: Solving the Space-Time Problem." *Agricultural Economics* 25: 59–70.

Ashworth, J., L. Evans, and A. Teriba. 1999. "Structural Breaks in Parallel Markets? The Case of Nigeria, 1980–1993." *Journal of Development Economics* 58 (1): 255–264.

ASTI (Agricultural Science and Technology Indicators). 2013. *Diffusion and Impact of Improved Varieties in Africa (DIIVA).* Washington, DC: ASTI.

Awotide, B. A., A. Karimov, A. Diagne, and T. Nakelse. 2013. "The Impact of Seed Vouchers on Poverty Reduction among Smallholder Rice Farmers in Nigeria." *Agricultural Economics* 44 (6): 647–658.

Ayanwale, A., V. Akinyosoye, S. Yusuf, and A. Oni. 2011. "Rice Supply Response in Nigeria: Whither Changing Policies and Climate." *World Rural Observations* 3: 78–84.

Ayoola, J. B., T. Kudi, C. Dangbegnon, C. Daudu, A. Mando, I. Amapu, J. Adeosun, and K. Ezui. 2011. "Gender Perspectives of Action Research for Improved Rice Value Chain in Northern Guinea Savanna, Nigeria." *Journal of Agricultural Science* 4: 211–219.

Bakare, S. O. 2007. "Participatory Use of Force Field Analysis in Evaluating Farmers' Response to Weed Control Methods." *Global Approaches to Extension Practice (GAEP)* 3 (1): 95–104.

Bakare, S. O., M. N. Ukwungwu, and A. Aliyu. 2008a. "Evaluation of Garil Efficacy as Post-Emergence Herbicide in Lowland Rice." *Africa Journal of General Agriculture* 4 (4): 213–217.

———. 2008b. "Determination of Rainbow Efficacy as Post-emergence Herbicide in Lowland Rice." *Africa Journal of General Agriculture* 4 (4): 219–223.

Baland, J. M., and J. P. Platteau. 1996. *Halting Degradation of Natural Resources: Is There a Role for Rural Communities?* Oxford, UK: Oxford University Press.

Balasubramanian, V., R. Rabesan, L. Razafinjara, and J. Ratsimandresy. 1995. "Rice Soil Constraints and Fertility Management in the Highlands of Madagascar." In *Fragile Lives in Fragile Ecosystems: Proceedings of the International Rice Research Conference*, 313–324. Los Baños, Philippines: IRRI.

Balasubramanian, V., M. Sie, R. J. Hijmans, and K. Otsuka. 2007. "Increasing Rice Production in Sub-Saharan Africa: Challenges and Opportunities." *Advances in Agronomy* 94: 55–133.

Bamidele, F., O. Abayomi, and O. Esther. 2010. "Economic Analysis of Rice Consumption Patterns in Nigeria." *Journal of Agricultural Science Technology* 12: 1–11.

Barker, R., R. W. Herdt, and B. Rose. 1985. *The Rice Economy of Asia.* Washington, DC: Resources for the Future.

Barrett, C. B., and P. Dorosh. 1996. "Farmers' Welfare and Changing Food Prices: Nonparametric Evidence from Rice in Madagascar." *American Journal of Agricultural Economics* 78 (3): 656–669.

Basorun, J. 2008. "A Bivariate Analysis of Factors Affecting Rice Processing in Igbemo-Ekiti, Nigeria." *Agricultural Journal* 3: 442–446.

BBC (British Broadcasting Corporation). 2013. "Nigeria Profile." Accessed December 16, 2013. www.bbc.co.uk/news/world-africa-13951696.

Behrman, J. R. 1968. *Supply Response in Underdeveloped Agriculture: A Case Study of Four Major Annual Crops in Thailand 1937–1963.* Amsterdam: North-Holland.

Beintema, N. M., and G. B. Ayoola. 2004. *Nigeria.* ASTI Country Brief No. 10. Washington, DC: International Food Policy Research Institute and International Service for National Agricultural Research.

Bello, N. 1987. "An Assessment of Water Supply for Agriculture in the Niger River Basin Development Authority Area, Nigeria." *Agricultural and Forest Meteorology* 40: 109–121.

Benin, S., and B. Yu. 2013. "Complying with the Maputo Declaration Target: Trends in Public Agricultural Expenditures and Implications for Pursuit of Optimal Public Agricultural Spending." ReSAKSS Annual Trends Report. Washington, DC: International Food Policy Research Institute.

Bentley, J. W., O. Ajayi, and K. Adelugba. 2011. "Nigeria: Clustered Seed Companies." In *African Seed Enterprises: Sowing the Seeds of Food Security*, edited by P. van Mele and R. G. Guei, 38–64. Wallingford, UK: CABI Publishing.

Bierlen, R., E. J. Wailes, and G. L. Crammer. 1997. "The MERCOSUR Rice Economy." Arkansas Agricultural Experiment Station Bulletin No. 954, Fayetteville, AR, US. Accessed on July 31, 2015. http://arkansasagnews.uark.edu/954.pdf.

Binswanger, H. 1986. "Agricultural Mechanization: A Comparative Historical Perspective." *World Bank Research Observer* 1 (1): 27–56.

Brown, J. P., R. J. G. M. Florax, and K. T. Mcnamara. 2009. "Determinants of Investment Flows in U.S. Manufacturing." *Review of Regional Studies* 39 (3): 269–286.

Byerlee, D., and G. Traxler. 2001. "The Role of Technology Spillovers and Economies of Size in the Efficient Design of Agricultural Research Systems." In *Agricultural Science Policy: Changing Global Agendas,* edited by J. M. Alston, P. G. Pardey, and M. J. Taylor, 161-186. Baltimore: Johns Hopkins University Press.

Carletto, C., S. Savastano, and A. Zezza. 2013. "Fact or Artifact: The Impact of Measurement Errors on the Farm Size–Productivity Relationship." *Journal of Development Economics* 103: 254–261.

Chamberlain, G. 1984. "Panel Data." In *Handbook of Econometrics.* Vol. 2, edited by Z. Grilliches and M. D. Intriligator, 1247–1318. Amsterdam: North-Holland.

Chambers, R. 1980. "Basic Concepts in the Organization of Irrigation." In *Irrigation and Agricultural Development in Asia: Perspectives from the Social Sciences,* edited by E. W. Coward Jr., 28–50. Ithaca, NY, US: Cornell University Press.

Chandler, R. F. 1979. *Rice in the Tropics: A Guide to the Development of National Programs.* Boulder, CO, US: Westview Press.

Chemonics. 2009. "Overview of Nigerian Rice: Diamond Analysis." Working paper, PowerPoint presentation, USAID/MARKETS Project, Abuja, Nigeria.

Chern, W. S., K. Ishibashi, K. Taniguchi, and Y. Tokoyama. 2003. *Analysis of the Food Consumption of Japanese Households.* FAO Economic and Social Development Paper No. 152. Rome: Food and Agriculture Organization of the United Nations.

Choeun, H., Y. Godo, and Y. Hayami. 2006. "The Economics and Politics of Rice Export Taxation in Thailand: A Historical Simulation Analysis, 1950–1985." *Journal of Asian Economics* 17 (1): 103–125.

Cragg, J. G. 1971. "Some Statistical Models for Limited Dependent Variables with Application to the Demand for Durable Goods." *Econometrica* 39 (5): 829–844.

DAI (Development Alternatives International). 2009. "Nigeria Rice Value Chain Analysis." Unpublished working paper, DAI, USAID/MARKETS Project, Washington, DC.

Dalton, T., and R. Guei. 2003. "Ecological Diversity and Rice Varietal Improvement in West Africa." In *Crop Variety Improvement and Its Effect on Productivity: The Impact of International Agricultural Research,* edited by R. Evenson and D. Gollin, 109–134. Wallingford, UK: CABI Publishing.

Daramola, A. G. 2005. "Government Policies and Competitiveness of Nigerian Rice Economy." Paper presented at the Workshop on Rice Policy & Food Security in Sub-Saharan Africa, Cotonou, Republic of Benin, November 7–9. Accessed on July 31, 2015. www.africaricecenter.org/workshop/RicePolicy/Biyi/Biyi.D.Nigeria.Paper.pdf.

Deaton, A. 1997. *The Analysis of Household Surveys: A Microeconomic Approach to Development Policy.* Baltimore: Johns Hopkins University Press.

Deaton, A., and J. Muellbauer. 1980. *Economics and Consumer Behaviour.* New York: Cambridge University Press.

De Datta, S. K. 1981. *Principles and Practices of Rice Production.* Los Baños, Philippines: IRRI.

de Janvry, A., D. Runsten, and E. Sadoulet. 1987. *Technological Innovations in Latin American Agriculture.* San José, Costa Rica: Inter-American Institute for Cooperation on Agriculture (IICA) and Biblioteca Venezuela.

Demont, M., and A. C. Rizzotto. 2012. "Policy Sequencing and the Development of Rice Value Chains in Senegal." *Development Policy Review* 30: 451–472.

Demont, M., P. Rutsaert, M. Ndour, and W. Verbeke. 2013. "Reversing Urban Bias in African Rice Markets: Evidence from Senegal." *World Development* 45: 63–74.

Diagne, A., S. K. G. Midingoyi, M. Wopereis, and I. Akintayo. 2011. "Increasing Rice Productivity and Strengthening Food Security through New Rice for Africa (NERICA)." In *Yes Africa Can: Success Stories from a Dynamic Continent,* edited by P. Chuhan-Pole and M. Angwafo, 253–268. Washington, DC: World Bank.

Diao, X., P. Hazell, D. Resnick, and J. Thurlow. 2007. "The Role of Agriculture in Development: Implications for Sub-Saharan Africa." Research Report, International Food Policy Research Institute, Washington, DC.

Diao, X., and A. Nin-Pratt. 2007. "Growth Options and Poverty Reduction in Ethiopia: An Economy-Wide Model Analysis." *Food Policy* 32 (2): 205–228.

Diao, X., M. Nwafor, V. Alpuerto, K. Akramov, and S. Salau. 2010. "Agricultural Growth and Investment Options for Poverty Reduction in Nigeria." IFPRI Discussion Paper. Washington, DC: International Food Policy Research Institute.

Diarra, S. B., J. Staatz, R. B. James, and N. N. Dembélé. 1999. "The Reform of Rice Milling and Marketing in the Office Du Niger: Catalyst for an Agricultural Success Story in Mali." East Lansing, MI, US: Michigan State University.

Dim, C., and U. Ezenekwe. 2013. "Does Agriculture Matter for Economic Development? Empirical Evidence from Nigeria." *Journal of Economics and Finance* 1: 61–77.

Dinar, A., and A. Keck. 1997. "Private Irrigation Investment in Colombia: Effects of Violence, Macroeconomic Policy, and Environmental Conditions." *Agricultural Economics* 16 (1): 1–15.

Djurfeldt, G., E. Aryeetey, and A. C. Isinika, eds. 2011. *African Smallholders: Food Crops, Markets and Policy.* Wallingford, UK: CABI Publishing.

Djurfeldt, G., and M. Jirström. 2005. "The Puzzle of the Policy Shift: The Early Green Revolution in India, Indonesia and the Philippines." In *The African Food Crisis: Lessons from the Asian Green Revolution*, edited by G. Djurfeldt, H. Holmén, M. Jirström, and R. Larsson, 43–63. Wallingford, UK: CABI Publishing.

Drabo, Z. 1976. "Rice Cultivation in Upper Volta." Unpublished, Rice Project Managers Meeting, Monrovia, Liberia, February 4, West Africa Rice Development Association (WARDA), Monrovia.

Dupraz, C. L., and A. Postolle. 2013. "Food Sovereignty and Agricultural Trade Policy Commitments: How Much Leeway Do West African Nations Have?" *Food Policy* 38: 115–125.

Durham, C., and R. J. Sexton. 1996. "Spatial Competition, Uniform Pricing, and Transportation Efficiency in the California Processing Tomato Industry." *American Journal of Agricultural Economics* 78: 115–125.

Ehigiamusoe, U. K. 2012. "A Comparative Analysis of Agricultural Performance between the Military and Civilian Regimes in Nigeria." *International Journal of Humanities and Social Science Invention* 1: 13–23.

Emodi, I. A., and M. C. Madukwe. 2008. "A Review of Policies, Acts and Initiatives in Rice Innovation System in Nigeria." *Journal of Agricultural Extension* 12 (2): 76–83.

Enplan Group. 2004. "Review of the Public Irrigation Sector in Nigeria." Draft Status Report, Rev. 1 of Project UTF/046/NIR/UTF, Lagos, Nigeria. Accessed on July 31, 2015. ftp://ftp.fao.org/agl/aglw/ROPISIN/ROPISINreport.pdf.

Erenstein, O., F. Lançon, S. O. Akande, S. O. Titilola, G. Akpokodje, and O. O. Ogundele. 2003. "Rice Production Systems in Nigeria: A Survey." WARDA, Abidjan, Côte d'Ivoire. Accessed on July 31, 2015. www.inter-reseaux.org/IMG/pdf_nigeria_rice_production_systems.pdf.

Erenstein, O., F. Lançon, O. Osiname, and M. Kebbeh. 2004. "Operationalizing the Strategic Framework for Rice Sector Revitalization in Nigeria." Project technical report prepared for The Nigerian Rice Economy in a Competitive World: Constraints, Opportunities and Strategic Choices, West Africa Rice Development Association (WARDA)/Africa Rice Center, Abidjan, Côte d'Ivoire. Accessed on July 31, 2015. http://pdf.usaid.gov/pdf_docs/Pnadb854.pdf.

Estudillo, J. P., and K. Otsuka. 2012. "Lessons from the Asian Green Revolution in Rice." In *An African Green Revolution: Finding Ways to Boost Productivity on Small Farms,* edited by K. Otsuka and D. Larson, 17–42. New York: Springer.

Evenson, R. E. 1977. "Comparative Evidence on Returns to Investment in National and International Research Institutions." In *Resource Allocation and Productivity in National and International Agricultural Research,* edited by T. Arndt, D. Dalrymple, and V. W. Ruttan, 237–258. Minneapolis: University of Minnesota.

———. 1982. "Government Policy and Technological Progress in U.S. Agriculture." In *Government Support of Technological Progress: A Cross Industry Analysis,* edited by R. Nelson. New York: Pergamon Press.

———. 2003. "GMOs: Prospects for Increased Crop Productivity in Developing Countries." Economic Growth Center Discussion Paper 878, Yale University, New Haven, CT, US. Available at SSRN: http://ssrn.com/abstract=487503.

Evenson, R. E., and P. M. Flores. 1978. "Social Returns to Rice Research." In *Economic Consequences of the New Rice Technology,* IRRI, 243–265. Los Baños, Philippines: International Rice Research Institute (IRRI).

Evenson, R. E., and D. Gollin. 1997. "Genetic Resources, International Organizations, and Improvement in Rice Varieties." *Economic Development and Cultural Change* 45 (3): 471–500.

Ezedinma, C. 2005. "Impact of Trade on Domestic Rice Production and the Challenge of Self-Suffiency in Nigeria." International Institute of Tropical Agriculture, Ibadan, Nigeria. Accessed on July 31, 2015. www.warda.cgiar.org/workshop/RicePolicy/Chuma.E/Chuma.E.Nigeria.Paper.pdf.

Fafchamps, M., and B. Minten. 1999. "Relationships and Traders in Madagascar." *Journal of Development Studies* 35 (6): 1–35.

Falola, T., and A. G. Adebayo. 2000. *Culture, Politics and Money among the Yoruba*. Piscataway, NJ, US: Transaction Publishers.

Fan, S., ed. 2008. *Public Expenditure, Growth and Poverty in Developing Countries: Issues, Methods and Findings*. Baltimore: Johns Hopkins University Press.

FAO (Food and Agriculture Organization of the United Nations). 2000. *Rivers of Africa*. Rome: FAO.

———. 2012. FAOSTAT online database. Food Balance Sheets. Rome: FAO. Accessed on June 10, 2015. www.fao.org/statistics/en/.

———. 2014. FAOSTAT online database. Accessed on June 10, 2015. www.fao.org/statistics/en/.

FAO/IIASA (International Institute for Applied Systems Analysis)/ISRIC (International Soil Reference and Information Centre)/ISSCAS (Institute of Soil Science, Chinese Academy of Sciences)/JRC (Joint Research Centre of the European Commission). 2012. Harmonized World Soil Database (version 1.2). Accessed on June 10, 2015. http://webarchive.iiasa.ac.at/Research/LUC/External-World-soil-database/HTML/.

Feder, G., R. E. Just, and D. Zilberman. 1985. "Adoption of Agricultural Innovation in Developing Countries: A Survey." *Economic Development and Cultural Change* 33 (2): 255–298.

Feeny, D., F. Berkes, B. McCay, and J. M. Anderson. 1990. "The Tragedy of the Commons: Twenty-two Years Later." *Human Ecology* 18 (1): 1–19.

Fischer, G., F. Nachtergaele, S. Prieler, E. Teixeira, G. Toth, H. van Velthuizen, L. Verelst, and D. Wiberg. 2012. "Global Agro-Ecological Zones (GAEZ v3.0)." IIASA, Laxenburg, Austria, and FAO, Rome: International Institute for Applied Systems Analysis.

Fisher, M., W. A. Masters, and M. Sidibé. 2001. "Technical Change in Senegal's Irrigated Rice Sector: Impact Assessment under Uncertainty." *Agricultural Economics* 24 (2): 179–197.

Flaherty, K., G. Ayoola, J. Ogbodo, and N. Beintema. 2010. "Nigeria: Recent Developments in Agricultural Research." Washington, DC: Agricultural Science and Technology Indicators (ASTI). Accessed July 30, 2015. http://cdm15738.contentdm.oclc.org/utils/getfile/collection/p15738coll2/id/7305/filename/7306.pdf.

FMARD (Federal Ministry of Agriculture and Rural Development). 2012. "Rice Transformation Implementation Plan." Abuja, Nigeria: Federal Ministry of Agriculture and Rural Development of Nigeria.

Forssell, S. 2009. "Rice Price Policy in Thailand: Policy Making and Recent Developments." MS thesis, University of Lund, Sweden. Accessed on July 30, 2015. https://liveatlund.lu.se/intranets/LUSEM/NEK/mfs/MFS/189.pdf.

Foster, V., and N. Pushak. 2011. *Nigeria's Infrastructure: A Continental Perspective.* Washington, DC: IBRD and World Bank.

Gboyega, A., T. Søreide, T. M. Le, and G. P. Shukla. 2011. "Political Economy of the Petroleum Sector in Nigeria." Policy Research Working Paper 5779, World Bank, Washington, DC. Accessed on July 31, 2015. www-wds.worldbank.org/external/default/WDSContentServer/IW3P/IB/2011/08/24/000158349_20110824134316/Rendered/PDF/WPS5779.pdf.

Gupta, S., K. Miranda, and I. Parry. 1995. "Public Expenditure Policy and the Environment: A Review and Synthesis." *World Development* 23 (3): 515–528.

Hardin, G. 1968. "The Tragedy of the Commons." *Science* 162 (3859): 1243–1248.

Harris, K. M., and R. J. Gagné. 1982. "Description of the African Rice Gall Midge, *Orseolia oryzivora* sp. n., with Comparative Notes on the Asian Rice Gall Midge *O. oryzae* (Wood-Mason) (Diptera: Cecidomyiidae)." *Bulletin of Entomological Research* 72: 467–472.

Harriss-White, B. 2005. "Commercialisation, Commodification and Gender Relations in Post-Harvest Systems for Rice in South Asia." *Economic and Political Weekly* 40 (25) (June 18): 2530–2542.

Harvest Choice. 2012. "Rice Production (mt) (2000)." International Food Policy Research Institute, Washington, DC, and University of Minnesota, St. Paul. Accessed on June 10, 2015. http://harvestchoice.org/node/4798.

Hashim, Y., and K. Meagher. 1999. "Cross-Border Trade and the Parallel Currency Market: Trade and Finance in the Context of Structural Adjustment: A Case Study from Kano, Nigeria." Research Report 113. Uppsala, Sweden: Nordiska Afrikainstitutet.

Hayami, Y., E. Brennan, and R. Barker. 1977. "Price Incentive versus Irrigation Investment to Achieve Food Self-Sufficiency in the Philippines." *American Journal of Agricultural Economics* 59 (4): 717–721.

Hayami, Y., and M. Kikuchi. 2000. *A Rice Village Saga: Three Decades of Green Revolution in the Philippines.* Los Baños, the Philippines: IRRI.

Hayami, Y., and V. Ruttan. 1970. "Korean Rice, Taiwan Rice, and Japanese Agricultural Stagnation: An Economic Consequence of Colonialism." *Quarterly Journal of Economics* 84 (4): 562–589.

Hayami, Y., and S. Yamada. 1998. "Agricultural Research Organization in Economic Development: A Review of the Japanese Experience." In *Agricultural Growth and Japanese Economic Development.* Vol. 7, edited by M. Smitka, 112–138. Abingdon, Oxford, UK: Taylor and Francis.

Headey, D., and S. Fan. 2008. "Anatomy of a Crisis: The Causes and Consequences of Surging Food Prices." *Agricultural Economics* 39: 375–391.

Heltberg, R., and F. Tarp. 2002. "Agricultural Supply Response and Poverty in Mozambique." *Food Policy* 27 (1): 103–124.

Herdt, R. W. 1991. "Research Priorities for Rice Biotechnology." In *Rice Biotechnology,* edited by G. S. Khush and G. T. Toenniessen, 19–54. Los Baños, Philippines: IRRI.

Herdt, R. W., and C. Capule. 1983. *Adoption, Spread, and Production Impact of Modern Rice Varieties in Asia.* Los Baños, Philippines: IRRI.

Holtzman, J. S. 2002. "Using Sub-Sector Analysis to Assess the Impact of Policy Reform on Commodity Sub-sectors." Unpublished, Abt Associates Inc., Bethesda, MD, US.

Hossain, M., D. Gollin, V. Camanilla, E. Cabrera, N. Johnson, G. S. Kush, and G. McLaren. 2003. "International Research and Genetic Improvement in Rice: Evidence from Asia and Latin America." In *Crop Variety Improvement and Its Effect on Productivity: The Impact of International Agricultural Research*, edited by R. Evenson and D. Gollin, 71–108. Wallingford, UK: CABI Publishing.

Houthakker, H. S., and L. D. Taylor. 1970. *Consumer Demand in the United States, 1929–1970: Analyses and Projections.* Cambridge, MA, US: Harvard University Press.

Hsu, S. K. 1997. "The Agroindustry: A Neglected Aspect of the Location Theory of Manufacturing." *Journal of Regional Science* 37: 259–274.

Ibitoye, S. J., J. S. Orebiyi, and U. M. Shaibu. 2012. "Economic Effect of Inorganic Pesticide Use on Fadama II Rice Farming in Ibaji Local Government Area, Kogi State, Nigeria." *International Journal of Agricultural and Rural Development* 15 (2): 1063–1070.

Imai, K. S., R. Gaiha, and G. Thapa. 2011. "Supply Response to Changes in Agricultural Commodity Prices in Asian Countries." *Journal of Asian Economics* 22 (1): 61–75.

Imeokparia, P. O. 1994. "Weed Control in Flooded Rice with Various Herbicide Combinations in the Southern Guinea Savanna Zone of Nigeria." *International Journal of Pest Management* 40 (1): 32–39.

Imeokparia, P. O., S. T. O. Lagoke, and B. A. Olunuga. 1992. "Evaluation of Postemergence Herbicides for Broad-Spectrum Weed Control in Three Cultivars of Flooded Rice in Nigeria." *Crop Protection* 11 (2): 165–173.

IMF (International Monetary Fund). 2011. *International Financial Statistics.* Washington, DC: International Monetary Fund.

———. 2012. IMF e-Library database. Washington, DC: International Monetary Fund.

Inocencio, A., M. Kikuchi, M. Tonosaki, A. Maruyama, D. Merrey, H. Sally, and I. de Jong. 2007. "Costs and Performance of Irrigation Projects: A Comparison of Sub-Saharan Africa and Other Developing Regions." Colombo, Sri Lanka: International Water Management Institute (IWMI) Research Report 109. Accessed on July 30, 2015. www.protos.be/sites/default/files/library_assets/W_LAN_E40_costs_performance.pdf.

IRRI (International Rice Research Institute). 1978. *Economic Consequences of the New Rice Technology.* Los Baños, Philippines: IRRI.

———. 1982. *Rice Research Strategies for the Future.* Los Baños, Philippines: IRRI.

———. 1984. *Terminology for Rice Growing Environments.* Los Baños, Philippines: IRRI.

———. 1986. *Small Farm Equipment for Developing Countries.* Los Baños, Philippines: IRRI.

———. 2014. *World Rice Statistics Online Query Facility.* Los Baños, Philippines: IRRI.

Islam, R., and S. Ahmed. 2007. "Market Report on Adamawa State Rice Value Chain." Communication and Marketing Research Group Limited (CMRGL) for PrOpCom (Promoting Pro-poor Opportunities in Service and Commodity Markets), Abuja, Nigeria. Accessed on July 30, 2015. www.propcommaikarfi.org/wp-content/uploads/2013/08/Adamawa-State-Rice-Value-Chain-Report.pdf.

Ismaila, U., A. C. Wada, E. Daniya, and A. U. Gbanguba. 2013. "Meeting the Local Rice Needs in Nigeria through Effective Weed Management." *Sustainable Agriculture Research* 2 (2): 37–48.

ISRIC. 2013. "Soil Property Maps of Africa at 1 Km." Accessed on June 10, 2015. www.isric.org.

Iwuchukwu, J. C., and E. M. Igbokwe. 2012. "Lessons from Agricultural Policies and Programmes in Nigeria." *Journal of Law, Policy and Globalization* 5: 11–22.

Jamala, Y. J., H. E. Shehu, and A. T. Garba. 2011. "Evaluation of Factors Influencing Farmers' Adoption of Irrigated Rice Production in Fadama Soil of North Eastern Nigeria." *Journal of Development and Agricultural Economics* 3 (2): 75–79.

Jaroensathapornkul, J. 2007. "The Economic Impact of Public Rice Research in Thailand." *Chulalongkon Journal of Economics* 19 (2): 111–134.

Jatileksono, T. 1998. "Impact of Rice Research and Technology Dissemination in Indonesia." In *Impact of Rice Research*, edited by P. Pingali and M. Hossain, 293–310. Los Baños, Laguna, Philippines: International Rice Research Institute.

Jirström, M. 2005. "The State and Green Revolution in East Asia." In *The African Food Crisis: Lessons from the Asian Green Revolution*, edited by G. Djurfeldt, H. Holmén, M. Jirström, and R. Larsson, 25–42. Wallingford, Oxfordshire, UK: CABI Publishing.

Johnson, M., P. Hazell, and A. Gulati. 2003. "The Role of Intermediate Factor Markets in Asia's Green Revolution: Lessons for Africa?" *American Journal of Agricultural Economics* 85 (5): 1211–1216.

Johnson, M., H. Takeshima, and K. Gyimah-Brempong. 2013. "Assessing the Potential and Policy Alternatives for Achieving Rice Competitiveness and Growth in Nigeria." IFPRI Discussion Paper 01301. Washington, DC: International Food Policy Research Institute.

Jones, L. L., and M. D. Woods. 2002. "Economic Location Theory and Practice." Unpublished, Texas A&M University, College Station, TX, US.

Judd, M. A., J. K. Boyce, and R. E. Evenson. 1986. "Investing in Agricultural Supply: The Determinants of Agricultural Research and Extension Investment." *Economic Development and Cultural Change* 35 (1): 77–113.

Kelly, V. A., A. Adesina, and A. Gordon. 2003. "Expanding Access to Agricultural Inputs in Africa: A Review of Recent Market Development Experience." *Food Policy* 28 (4): 379–404.

Kherallah, M., C. L. Delgado, E. Gabre-madhin, N. Minot, and M. Johnson. 2002. *Reforming Agricultural Markets in Africa*. Baltimore: Johns Hopkins University Press.

Khiem, N. T., and P. L. Pingali. 1995. "Supply Response of Rice and Three Food Crops in Vietnam." In *Vietnam and IRRI: A Partnership in Rice Research,* edited by G. L. Denning and V.-T. Xuan, 275–290. Manila: IRRI and the Ministry of Agriculture and Food Industry.

Kikuchi, M., and P. B. Aluwihare. 1990. "Fertilizer Response Functions of Rice in Sri Lanka: Estimation and Some Applications." International Water Management Institute, Colombo, Sri Lanka. Accessed on July 30, 2015. http://publications.iwmi.org/pdf/H_6935.pdf.

Kikuchi, M., G. Dozina Jr., and Y. Hayami. 1978. "Economics of Community Work Programs: A Communal Irrigation Project in the Philippines." *Economic Development and Cultural Change* 26 (2): 211–225.

Kikuchi, M., A. Maruyama, and Y Hayami. 2003. "Phases of Irrigation Development in Asian Tropics: A Case Study of the Philippines and Sri Lanka." *Journal of Development Studies* 39 (5): 109–138.

Kolawole, A., O. I. Oladele, and T. Wakatsuki. 2011. "Profitability of Different Sawah Rice Production Models within Lowlands in Nigeria." *Journal of Food, Agriculture and Environment* 9 (1): 268–274.

Kranjac-Berisavljevic, G., R. M. Blench, and R. Chapman. 2003. "Rice Production and Livelihoods in Ghana." Multi-Agency Partnerships (MAPS) for Technical Change in West African Agriculture. Overseas Development Institute, London. Accessed on July 31, 2015. www.odi.org/sites/odi.org.uk/files/odi-assets/publications-opinion-files/3990.pdf.

Kreinin, M. E., and M. G. Plummer. 2012. *The Oxford Handbook of International Commercial Policy.* Oxford, UK: Oxford University Press.

Krueger, A. O., M. Schiff, and A. Valdès. 1988. "Agricultural Incentives in Developing Countries: Measuring the Effects of Sectoral and Economywide Policies." *World Bank Economic Review* 2 (9): 255–271.

Kuijs, L. 1998. *Determinants of Inflation, Exchange Rate, and Output in Nigeria.* Washington, DC: International Monetary Fund.

Kuku-Shittu, O., A. Mathiassen, A. Wadhwa, L. Myles, and A. Ajibola. 2013. "Comprehensive Food Security and Vulnerability Analysis." IFPRI Discussion Paper. Washington, DC: International Food Policy Research Institute.

Lançon, F., O. Erenstein, S. Akande, S. Titilola, G. Akpokodje, and O. Ogundele. 2003a. "Imported Rice Retailing and Purchasing in Nigeria: A Survey." West Africa Rice Development Association (WARDA) and Nigerian Institute for Social and Economic Research (NISER), Cotonou, Benin. Accessed on July 31, 2015. www.inter-reseaux.org/IMG/pdf_imported_rice_consumption_and_retailing_in_nigeria.pdf.

———. 2003b. "Rice Processing in Nigeria: A Survey." West Africa Rice Development Association (WARDA), Cotonou, Benin. Accessed on July 31, 2015. www.hubrural.org/IMG/pdf/nigeria-rice-processing-survey.pdf.

Leaver, R. 2004. "Measuring the Supply Response Function of Tobacco in Zimbabwe." *Agrekon* 43 (1): 113–131.

Leurquin, P. 1967. "Rice in Colombia: A Case Study in Agricultural Development." Food Research Institute, Stanford University, Stanford, CA, US. Accessed on July 30, 2015. http://ageconsearch.umn.edu/bitstream/134934/2/fris-1967-07-02-350.pdf.

Levy, S. 2006. "Public Investment to Reverse Dutch Disease: The Case of Chad." Development Strategy and Governance Division Discussion Paper No. 35. Washington, DC: International Food Policy Research Institute.

Lin, Y. J. 1994. "The Nature and Impact of Hybrid Rice in China." In *Modern Rice Technology and Income Distribution in China,* edited by C. C. David and K. Otsuka, 375–408. Boulder, CO, US: Lynne Reiner.

Linares, O. F. 2002. "African Rice (Oryza glaberrima): History and Future Potential." *Proceedings of the National Academy of Sciences 99* (25): 16360–16365.

Lipton, M. 1988. "The Place of Agricultural Research in the Development of Sub-Saharan Africa." *World Development* 16 (10): 1231–1257.

Liverpool-Tasie, S., B. Olaniyan, S. Salau, and J. Sackey. 2010. "A Review of Fertilizer Policy Issues in Nigeria." IFPRI NSSP Working Paper 19. Washington, DC: International Food Policy Research Institute.

Liverpool-Tasie, S., and H. Takeshima. 2013. "Input Promotion within a Complex Subsector: Fertilizer in Nigeria." *Agricultural Economics* 44 (6): 581–594.

Lodestar International. 2010. "Rice Investment and Processing in Nigeria: A Review of Survey Findings." PowerPoint presentation, USAID/MARKETS project, Abuja, Nigeria.

Longtau, S. R. 2003a. "Nigeria Case Study Report on Rice Production." Working paper, Eco-Systems Development Organization (EDO) for Overseas Development Institute (ODI), Jos, Nigeria. Accessed on July 30, 2015. www.odi.org/sites/odi.org.uk/files/odi-assets/publications-opinion-files/3986.pdf.

———. 2003b. "A Review and Description of Rice Production Systems in Nigeria." Working paper, Eco-Systems Development Organization (EDO) for Overseas Development Institute (ODI), Jos, Nigeria. Accessed on July 30, 2015. www.odi.org/sites/odi.org.uk/files/odi-assets/publications-opinion-files/3984.pdf.

Lopez-Pereira, M. A., and M. P. Filippello. 1994. "Maize Seed Industries Revisited: Emerging Roles of the Public and Private Sectors." Part 1 of CIMMYT *1993/94 World Maize Facts and Trends.* Mexico, DF: Centro Internacional de Mejoramiento de Maiz y Trigo.

Lucas, M. T., and D. Chhajed. 2004. "Applications of Location Analysis in Agriculture: A Survey." *Journal of the Operational Research Society* 55 (6): 561–578.

Maji, A. T., and S. O. Fagade. 2002. "Utilization of Genetic Resources for Rice Production in Nigeria." In *Genetic Diversity in Rice Production: Case Studies from Brazil, India and Nigeria*, edited by A. T. Maji, S. O. Fagade, and V. N. Nguyen, 93–123. Rome: FAO.

Maneechansook, C. 2011. "Value Chain of Rice Exported from Thailand to Sweden." MBA thesis, University of Boras, Sweden. Accessed on July 30, 2015. http://bada.hb.se/bitstream/2320/8306/1/2011MF07.pdf.

Manyong, V. M., A. Ikpi, J. K. Olayemi, S. A. Yusuf, B. T. Omonona, V. Okoruwa, and F. S. Idachaba. 2005. "Agriculture in Nigeria: Identifying Opportunities for Increased Commercialization and Investment." International Institute of Tropical Agriculture (IITA), Ibadan, Nigeria. Accessed on July 30, 2015. www.iita.org/c/document_library/get_file?uuid=d8a9beb7-c284-4c39-af3c-6a8c85e09f98&groupId=25357.

Mccann, P., and S. Sheppard. 2003. "The Rise, Fall and Rise Again of Industrial Location Theory." *Regional Studies* 37 (6-7): 649–663.

McDonald, J. F., and R. Moffitt. 1980. "The Use of Tobit Analysis." *Review of Economics and Statistics* 62: 318–321.

Mellor, J. 1978. "Food Price Policy and Income Distribution in Low-Income Countries." *Economic Development and Cultural Change* 27 (1): 1–26.

Mikkelsen, K. W., and N. N. Langam. 1983. "Innovation in the Philippine Agricultural Machinery Industry." In *Consequences of Small-Farm Mechanization,* edited by International Rice Research Institute and Agricultural Development Council, 31–38. Los Baños, Philippines: International Rice Research Institute.

Minten, B., K. A. S. Murshid, and T. Reardon. 2011. "The Quiet Revolution in Agrifood Value Chains in Asia: The Case of Increasing Quality in Rice Markets in Bangladesh." IFPRI Discussion Paper 01141. Washington, DC: International Food Policy Research Institute.

———. 2013. "Food Quality Changes and Implications: Evidence from the Rice Value Chain of Bangladesh." *World Development* 42: 100–113.

Mogues T., M. Morris, L. Freinkman, A. Adubi, E. Simeon, C. Nwoko, O. Taiwo, C. Nege, P. Okonji, and L. Chete. 2008. "Agricultural Public Spending in Nigeria." IFPRI Discussion Paper 00789. Washington, DC: International Food Policy Research Institute.

Molua, E. 2010. "Rice Production Response to Trade Liberalization in Cameroon." *Research Journal of Agriculture and Biological Sciences* 6 (2): 118–129.

Moormann, F. R., and N. van Breemen. 1978. *Rice, Soil, Water, Land.* Los Baños, Philippines: International Rice Research Institute.

Morris, M., V. Kelly, R. Kopicki, and D. Byerlee. 2007. *Fertilizer Use in African Agriculture: Lessons Learned and Good Practice Guidelines.* Washington, DC: World Bank Publications.

Moseley, W. G., J. Carney, and L. Becker. 2010. "Neoliberal Policy, Rural Livelihoods, and Urban Food Security in West Africa: A Comparative Study of The Gambia, Côte d'Ivoire, and Mali." *Proceedings of the National Academy of Sciences of the United States of America* 107: 5774–5779.

Moses, L. N. 1958. "Location and the Theory of Production." *Quarterly Journal of Economics* 72 (2): 259–272.

Munshi, K. 2004. "Social Learning in a Heterogeneous Population: Technology Diffusion in the Indian Green Revolution." *Journal of Development Economics* 73 (1): 185–213.

Murshid, K. A. S. 2011. "Traditional Market Institutions and Complex Change: Exploring Transition and Change in the Bangladesh Rice Market." CPD-CMI Working Paper 1 (Centre for Policy Dialogue and Chr. Michelsen Institute), Dhaka, Bangladesh, and Bergen, Norway. Accessed on July 30, 2015. www.cpd.org.bd/pub_attach/CPD_CMI_WP1.pdf.

Nation. 2013. "Government Imposes 10% Duty on Rice." February 7.

Nayak, P. 1996. "Problems and Prospects of Rice Mill Modernization: A Case Study." *Journal of Assam University* 1 (1): 22–28.

NCRI (National Cereals Research Institute). 2008. "Rice Research Division Report in Research Review and Planning Workshop P80-96." Unpublished, NCRI, Badeggi, Nigeria.

———. 2012. "Training Manual for Training of Trainers on Rice Production, Processing and Marketing in Bayelsa State." Unpublished, NCRI, Badeggi, Nigeria.

———. 2013. "Report on Research Operations Department in the Annual Research Review and Planning Workshop." June 10–12. Unpublished, NCRI, Badeggi, Nigeria.

Ngambeki, D. S., and F. S. Idachaba. 1985. "Supply Response of Upland Rice in Ogun State of Nigeria: A Producer Panel Approach." *Journal of Agricultural Economics* 36 (2): 239–249.

Nigeria, NAERLS (National Agricultural Extension and Research Liaison Services). 2009. "Crops Data from Annual Agricultural Survey Report of the 2009 Wet Season." Abuja, Nigeria: Federal Ministry of Agriculture and Rural Development.

Nigeria, NBS (National Bureau of Statistics). 2012. "Farmgate Prices of Agricultural Products." Accessed on May 1, 2015. www.nigerianstat.gov.ng/pages/download/143.

———. 2013a. "Annual Abstract of Statistics, 2012." Accessed on May 1, 2015. www.nigerianstat.gov.ng/pages/download/253.

———. 2013b. National Bureau of Statistics unpublished reports. NBS, Abuja, Nigeria.

Nigeria, NBS and World Bank. 2011. "Living Standards Measurement Study–Integrated Surveys on Agriculture (LSMS–ISA)—Wave 1." Nigeria Post-planting Survey Data. Accessed on June 11, 2015. http://go.worldbank.org/X5W6ED95T0.

———. 2013. *Living Standards Measurement Study-Integrated Surveys on Agriculture (LSMS–ISA) – Wave 2*. Nigeria Post-planting Survey Data. Accessed on June 11, 2015. http://go.worldbank.org/GEW72B4QO0.

Nin-Pratt, A., M. Johnson, E. Magalhaes, L. You, X. Diao, and J. Chamberlin. 2011. "Yield Gaps and Potential Growth in Western and Central Africa." IFPRI Research Monograph. Washington, DC: International Food Policy Research Institute.

Niranjan, S. K. D. F. F., T. Jogaratnam, and J. Weerahewa. 2001. "The Effect of Trade Policies on Returns to Rice Research in Sri Lanka: An Analysis." *Tropical Agricultural Research* 13: 267–277.

Nkonya, E., D. Phillip, T. Mogues, J. Pender, M. K. Yahaya, G. Adebowale, T. Arokoyo, and E. Kato. 2008. "From the Ground Up: Impacts of a Pro-Poor Community-Driven Development Project in Nigeria." IFPRI Discussion Paper. Washington, DC: International Food Policy Research Institute.

Nnabuife, C. 2014. "Why FG Okayed New Tariff on Rice—Minister." *Nigerian Tribune*, April 15.

Nnanna, J. D. 2002. "Monetary and Financial Sector Policy Measures in the Year 2002 Budget." *Bullion* (publication of the Central Bank of Nigeria) 26 (2): 27–35.

Nwaobiala, C. U. 2010. "Economic Analysis of Swamp Rice Production in Ebonyi Southern Agricultural Zone of Ebonyi State, Nigeria." *Journal of Agriculture and Social Research* 10 (2): 135–141.

Ogah, E. O., J. A. Odebiyi, A. A. Omoloye, and F. E. Nwilene. 2012. "Evaluation of Some Rice Genotypes for Incidence of African Rice Gall Midge and Its Parasitoid (*P. diplosisae*)." *African Crop Science Journal* 20 (2): 137–147.

Ogolo, E. O. 2011. "Regional Trend Analysis of Pan Evaporation in Nigeria (1970 to 2000)." *Journal of Geography and Regional Planning* 4 (10): 566–577.

Ogunbiyi, M. O. 2011. "Assessment of Postharvest Handling and Quality Control Practices of Rice in North Central Nigeria: A Case Study of Lafia, Nasarawa State." *Journal of Developments in Sustainable Agriculture* 6: 143–163.

Oguntade, A. 2011. "Assessment of Protection and Comparative Advantage in Rice Processing in Nigeria." *African Journal of Food, Agriculture, Nutrition and Development* 11: 4632–4646.

Ojehomon, V. E. T., J. Wayagiri, S. A. Tiamiyu, A. Y. T. Maji, and S. M. Misari. 1999. "Impact of Upland and Lowland Rice Varieties in Nigeria." Unpublished, National Agricultural Research Project (NARP), Abuja, Nigeria.

Ojo, E. O., and P. F. Adebayo. 2012. "Food Security in Nigeria: An Overview." *European Journal of Sustainable Development* 1: 199–220.

Okeke, E. C., H. N. Eneobong, A. O. Uzuegbunam, A. O. Ozioko, and H. Kuhnlein. 2008. "Igbo Traditional Food System: Documentation, Uses and Research Needs." *Pakistan Journal of Nutrition* 7 (2): 365–376.

Oko, A. O., B. E. Ubi, and A. A. Efisue. 2012. "A Comparative Study on Local and Newly Introduced Rice Varieties in Ebonyi State of Nigeria Based on Selected Agronomic Characteristics." *International Journal of Agriculture and Forestry* 2 (1): 11–17. Accessed on June 11, 2015. http://dx.doi.org/10.5923/j.ijaf.20120201.03.

Oko, A. O., B. E. Ubi, A. A. Efisue, and N. Dambaba. 2012. "Comparative Analysis of the Chemical Nutrient Composition of Selected Local and Newly Introduced Rice Varieties Grown in Ebonyi State of Nigeria." *International Journal of Agriculture and Forestry* 2 (2): 16–23.

Olam. 2011. "Olam to Invest US$49.2 Million in a Greenfield Fully Integrated, Mechanised and Irrigated Rice Farming and Rice Milling Facility in Nigeria." Accessed December 2, 2015. http://olamgroup.com/news/olam-to-invest-us49-2-million-in-a-greenfield-fully-integrated-mechanised-and-irrigated-rice-farming-and-rice-milling-facility-in-nigeria-3/#sthash.IXADz3jF.vhVEtxjP.dpbs.

Olubode-Awosola, O. O., O. A. Oyewumi, and A. Jooste. 2006. "Vector Error Correction Modelling of Nigerian Agricultural Supply Response." *Agrekon* 45 (4): 421–436.

Omamo, S. W., X. Diao, S. Wood, J. Chamberlin, L. You, S. Benin, U. Wood-Sichra, and A. Tatwangire. 2007. "Strategic Priorities for Agricultural Development in Eastern and Central Africa." IFPRI Research Report 150. Washington, DC: International Food Policy Research Institute.

Omonona, B. T., J. O. Lawal, and I. D. Oyebiyi. 2012. "Profitability of Production and Resource-Use Efficiency among Ofada Rice (*Oryza sativa* japonica) Farmers in Southwest, Nigeria." *Comunicata Science* 3 (2): 104–107.

Omotayo, A. M., O. D. Chikwendu, and K. Adebayo. 2001. "Two Decades of World Bank Assisted Extension Services in Nigeria: Lessons and Challenges for the Future." *Journal of Agricultural Education and Extension* 7 (3): 143–152.

Omotayo, A. M., K. Elemo, and J. Oyedepo. 2007. "Mapping of Ofada Rice Production Area." PrOpCom (Promoting Pro-poor Opportunities in Service and Commodity Markets), Abuja, Nigeria.

Oramah, B. O. 1996. "The Direct Private Benefits of Participation in a Publicly Provided Surface Irrigation Scheme in the High Rainfall Area of Nigeria." *African Development Review* 8 (1): 146–172.

Oseni, G., and P. Winters. 2009. "Rural Nonfarm Activities and Agricultural Crop Production in Nigeria." *Agricultural Economics* 40 (2): 189–201.

Ostrom, E. 1990. *Governing the Commons: The Evolution of Institutions for Collective Action*. Cambridge, UK: Cambridge University Press.

———. 1992. *Crafting Institutions for Self-Governing Irrigation Systems.* San Francisco: Institute for Contemporary Studies Press.

Otsuka, K., and K. Kalirajan. 2006. "Rice Green Revolution in Asia and Its Transferability to Africa: An Introduction." *Developing Economies* 44 (2): 107–122.

Otsuka, K., and D. F. Larson, eds. 2012. *An African Green Revolution: Finding Ways to Boost Productivity on Small Farms*. New York: Springer.

Oyejide, T. A. 1986a. "Agricultural Marketing and Pricing Policy in Nigeria." Unpublished Working Paper, Development Research Department, World Bank, Washington, DC.

———. 1986b. The Effects of Trade and Exchange Rate Policies on Agriculture in Nigeria. IFPRI Research Report 55. Washington, DC: International Food Policy Research Institute.

Pagiola, S., R. Martin-Hurtado, P. Shyamsundar, M. Mani, and P. Silva. 2002. "Generating Public Sector Resources to Finance Sustainable Development: Revenue and Incentive Effects." World Bank, Washington, DC. Accessed on July 30, 2015. https://www.elaw.org/system/files/Generating+Public+Sector+Resources_0.pdf.

Pardey, P., J. Alston, C. Chan-Kang, E. Magalhaes, and S. Vosti. 2006. "International and Institutional RD Spillovers: Attribution of Benefits among Sources for Brazil's New Crop Varieties." *American Journal of Agricultural Economics* 88 (1): 104–123.

Pesaran, M. H., Y. Shin, and R. J. Smith. 2001. "Bounds Testing Approach to the Analysis of Level Relationships." *Journal of Applied Econometrics* 16 (3): 289–326.

Phillip, D., E. Nkonya, J. Pender, and O. A. Oni. 2009. "Constraints to Increasing Agricultural Productivity in Nigeria: A Review." NSSP working paper. Washington, DC: International Food Policy Research Institute.

Pingali, P. 2007. "Agricultural Mechanization: Adoption Patterns and Economic Impact." In *Handbook of Agricultural Economics,* edited by R. Evenson and P. Pingali, 2779–2805. Amsterdam: Elsevier.

Pinto, B. 1987. "Nigeria during and after the Oil Boom: A Policy Comparison with Indonesia." *World Bank Economic Review* 1 (31): 419–445.

Pradhan, P. 1993. "Research Priority of Irrigation Management in Nigeria." In *Irrigation Research Priorities for Nigeria: Proceedings of a National Seminar Held at the University of Ilorin, Nigeria from 20 to 23 April, 1993,* edited by E. U. Nwa and P. Pradhan, 19–32. Kano, Nigeria: International Irrigation Management Institute.

Pray, C. E. 1997. "Public Policy, Public Investment, and Private Investment in Seed Supply: Experiences in Turkey and India." In *Alternative Strategies for Smallholder Seed Supply: Proceedings of an International Conference on Options for Strengthening National and Regional Seed Systems in Africa and West Asia, 10–14 Mar. 1997, Harare, Zimbabwe,* edited by D. D. Rohrbach, Z. Bishaw, and A. J. G. van Gastel, 211. Andhra Pradesh, India: International Crops Research Institute for the Semi-arid Tropics.

PrOpCom (Promoting Pro-poor Opportunities in service and Commodity Markets). 2007a. "Demand and Supply Study on Domestic and Imported Rice in Kano Area." Monograph Series 22, July 2007, Optimum Agricultural Consultants, Kano, Nigeria.

———. 2007b. "Baseline Survey of Kano Rice Value Chain Final Report by Optimum Agricultural Consultants." Monograph Series 22, April 2007, Optimum Agricultural Consultants, Kano, Nigeria.

———. 2007c. "Ofada Rice Production, Processing and Marketing Cost Survey." PrOpCom Monograph Series 34, December 2007, Kano, Nigeria.

Rahji, M. A. Y., and M. Adewumi. 2008. "Market Supply Response and Demand for Local Rice in Nigeria: Implications for Self-Sufficiency Policy." *Journal of Central European Agriculture* 9: 567–574.

Rahji, M. A. Y., O. Ilemobayo, and S. B. Fakayode. 2008. "Rice Supply Response in Nigeria: An Application of the Nerlove Adjustment Model." *Agricultural Journal* 3 (3): 229–234.

Ramani, N. P. 2007. "Comparative Analysis of Performance Efficiency of Fertilizer Industry in India." PhD dissertation, Saurashtra University, Rajkot, India. Accessed on July 30, 2015. http://etheses.saurashtrauniversity.edu/106/1/(7)%20RAMNP07MBA01.pdf.

Rashid S., P. A. Dorosh, M. Malek, and S. Lenma. 2013. "Modern Input Promotion in Sub-Saharan Africa: Insights from Asian Green Revolution." *Agricultural Economics* 44 (6): 705–721.

Robertson, J. W. 1983. "An Analysis of Agricultural Trade and Subsidy Policies in Nigeria." Country Policy Department Discussion Paper No. 1983-11, World Bank, Washington, DC.

Robinson, J. A., R. Torvik, and T. Verdier. 2006. "Political Foundations of the Resource Curse." *Journal of Development Economics* 79: 447–468.

Rodenburg, J., and D. E. Johnson. 2009. "Weed Management in Rice-Based Cropping Systems in Africa." *Advances in Agronomy* 103: 149–218.

Roseboom, J., N. Beintema, P. G. Pardey, and E. O. Oyedipe. 1994. "Statistical Brief on the National Agricultural Research System of Nigeria." ISNAR Indicator series project: Phase II Statistical Brief 15. International Service for National Agricultural Research, Netherlands. Accessed on July 30, 2015. http://ebrary.ifpri.org/cdm/ref/collection/p15738coll11/id/219.

Rosegrant, M. W., F. Kasryno, and N. D. Perez. 1998. "Output Response to Prices and Public Investment in Agriculture: Indonesian Food Crops." *Journal of Development Economics* 55 (2): 333–352.

Rosegrant, M. W., and E. Pasandaran. 1993. "Determinants of Public Investment: Irrigation in Indonesia." Paper presented at the 37th annual conference, Australian Agricultural Economics Society, University of Sydney, February 9–11.

Rutger, J. N., and D. J. MacKill. 2001. "Application of Mendelian Genetics in Rice Breeding." In *Rice Genetics IV*, edited by G. S. Khush, D. S. Brar, and B. Hardy, 27–38. Los Baños, Philippines: International Rice Research Institute.

Sackey, J., S. Liverpool-Tassie, S. Salau, and T. Awoyemi. 2012. "Rural-Urban Transformation in Nigeria." *Journal of African Development* 14 (2): 131–168.

Sadoulet, E., A. de Janvry, and S. Fukui. 1997. "The Meaning of Kinship in Sharecropping Contracts." *American Journal of Agricultural Economics* 79 (2): 394–406.

Sahrawat, K. L. 2004. "Iron Toxicity in Wetland Rice and the Role of Other Nutrients." *Journal of Plant Nutrition* 27: 1471–1504.

Sahrawat, K. L., C. K. Mulbah, S. Diatta, R. D. DeLaune, W. H. Patrick Jr., B. N. Singh, and M. P. Jones. 1996. "The Role of Tolerant Genotypes and Plant Nutrients in the Management of Iron Toxicity in Wetland Rice." *Journal of Agricultural Science—Cambridge* 126: 143–149.

Saito, K., K. Azoma, and J. Rodenburg. 2010. "Plant Characteristics Associated with Weed Competitiveness of Rice under Upland and Lowland Conditions in West Africa." *Field Crops Research* 116 (3): 308–317.

Sala-i-Martin, X., and A. Subramanian. 2013. "Addressing the Natural Resource Curse: An Illustration from Nigeria." *Journal of African Economies* 22 (4): 570–615.

Sarkarung, S., B. Somrith, and S. Chitrakorn. 2000. "Aromatic Rices of Thailand." In *Aromatic Rices*, edited by R. K. Singh, U. S. Singh, and G. S. Khush, 180–183. New Delhi and Calcutta, India: Oxford and IBH Publishing Co. Pvt. Ltd.

Seck, P. A, A. Diagne, S. Mohanty, and M. C. S. Wopereis. 2012. "Crops That Feed the World 7: Rice." *Food Security* 4: 7–24.

Short, D. E., and J. C. Jackson. 1971. "The Origins of Irrigation Policy in Malaysia: A Review of Developments Prior to the Establishment of the Drainage and Irrigation Department." *Journal of the Malaysian Branch of the Royal Asiatic Society* 44: 78–103.

Shwetha, M. K. 2009. "Business Efficiency in Rice Mills: A Comparative Analysis of Conventional and Modern Rice Mills in Davangere District." MBA thesis, University of Agricultural Sciences, Dharwad, India.

Singh, B. N., C. T. Williams, M. N. Ukwungwu, and A. T. Maji. 2004. "Breeding for Resistance to African Rice Gall Midge, *Orseolia oryzivora* Harris and Gagné." In *New Approaches to Gall Midge Resistance in Rice: Proceedings of the International Workshop*, edited by J. Bennett, J. S. Bentur, I. C. Pasalu, and K. Krishnaiah, 121–130. Los Baños, Philippines: IRRI.

Singh, I., L. Squire, and J. Strauss. 1986. *Agricultural Household Models.* Baltimore: Johns Hopkins University Press.

Small, L., M. Adriano, and E. D. Martin. 1986. "Regional Study on Irrigation Service Fees: Final Report." International Irrigation Management Institute, Colombo, Sri Lanka. Accessed on July 30, 2015. http://pdf.usaid.gov/pdf_docs/PNAAW833.pdf.

Stone, R. 1954. "Linear Expenditure Systems and Demand Analysis: An Application to the Pattern of British Demand." *Economic Journal* 64 (255): 511–527.

Takayama, T., and G. Judge. 1964. "Equilibrium among Spatially Separated Markets: A Reformulation." *Econometrica* 32: 510–524.

Takeshima, H. 2014. "Importance of Rice R and D in Rice Seed Policies: Insights from Nigeria." IFPRI Discussion Paper 01343. Washington, DC: International Food Policy Research Institute.

Takeshima, H., and H. Edeh. 2013. "Typology of Farm Household and Irrigation Systems: Some Evidence from Nigeria." IFPRI Discussion Paper 01267. Washington, DC: International Food Policy Research Institute.

Takeshima, H., and L. S. Liverpool-Tasie. 2015. "Fertilizer Subsidy, Political Influence, and Local Food Prices in Sub-Saharan Africa: Evidence from Nigeria." *Food Policy* 54: 11–24.

Takeshima, H., and I. Masias. 2013. "Integrations of Local Rice, Imported Rice, and International Market." Unpublished, International Food Policy Research Institute, Washington, DC.

Takeshima, H., and L. Nagarajan. 2015. "Farmer Preferences on Seed Purchase Timing: Some Evidence from Nigeria." *Journal of Crop Improvement* 29 (1): 131–158.

Takeshima, H., A. Nin Pratt, and X. Diao. 2013. "Mechanization and Agricultural Technology Evolution, Agricultural Intensification in Sub-Saharan Africa: Typology of Agricultural Mechanization in Nigeria." *American Journal of Agricultural Economics* 95 (5, ASSA Proceedings Issue): 1230–1236.

Takeshima, H., and E. Nkonya. 2014. "Government Fertilizer Subsidy and Commercial Sector Fertilizer Demand: Evidence from the Federal Market Stabilization Program (FMSP) in Nigeria." *Food Policy* 47: 1–12.

Takeshima, H., A. Oyekale, S. Olatokun, and S. Salau. 2010. "Demand Characteristics for Rice, Cowpea and Maize Seeds in Nigeria: Policy Implications and Knowledge Gaps." IFPRI NSSP Background Paper 0016. Washington, DC: International Food Policy Research Institute.

Takeshima, H., and A. Winter-Nelson. 2012. "Sales Location among Semi-Subsistence Cassava Farmers in Benin: A Heteroskedastic Double Selection Model." *Agricultural Economics* 43 (6): 655–670.

Taylor, D. S., and T. P. Phillips. 1991. "Food-Pricing Policy in Developing Countries: Further Evidence on Cereal Producer Prices." *American Journal of Agricultural Economics* 73 (4): 1036–1043.

Taylor, L. D., and H. S. Houthakker. 2010. *Consumer Demand in the United States.* 3rd edition. New York: Springer.

Thakkar, H. 1999. "Assessment of Irrigation in India." Contributing paper prepared for Thematic Review IV.2 for the World Commission on Dams. Accessed on July 30, 2015. ftp://ftp.ecn.purdue.edu/vmerwade/class/GDT/Reports_Books/assessment%20of%20irrigation%20in%20India.pdf.

Thiele, R. 2000. "Estimating the Aggregate Agricultural Supply Response: A Survey of Techniques and Results for Developing Countries." Kiel Working Paper No. 1016, Kiel Institute of World Economics, Kiel, Germany. Accessed on July 30, 2015. https://www.ifw-kiel.de/ifw_members/global-images/kap1016.pdf.

———. 2003. "Price Incentives, Non-Price Factors and Agricultural Production in Sub-Saharan Africa: A Cointegration Analysis." *African Development Review* 15 (2): 425–438.

Thomas, D., and W. Adams. 1999. "Adapting to Dams: Agrarian Change Downstream of the Tiga Dam, Northern Nigeria." *World Development* 27 (6): 919–935.

Thongrattana, P. T. 2012. "An Analysis of the Uncertainty Factors Affecting the Sustainable Supply of Rice Production in Thailand." PhD dissertation, University of Wollongong, Australia. Accessed on July 30, 2015. http://ro.uow.edu.au/cgi/viewcontent.cgi?article=4552&context=theses.

Timmer, P. 1998. "Choice of Technique in Rice Milling on Java." In *International Agricultural Development,* 3rd ed., edited by C. Eicher and J. Staatz, 494–511. Baltimore: Johns Hopkins University Press.

Tinsley, R. 2009. "Increasing Rice Productivity for the Kpong Irrigation Project, Akusa-Asutsuare, Ghana." ACDI/VOCA Farmer-to-farmer program. Consultant Report. Agricultural Cooperative Development International and Volunteers in Overseas Cooperative Assistance, Washington, DC. Accessed December 2, 2015. http://webdoc.agsci.colostate.edu/smallholderagriculture/KIP-Report.pdf.

———. 2011. "Improved Quality of Rice Processing." Working paper, Winrock International, Arlington, VA, US. Accessed on July 30, 2015. http://webdoc.agsci.colostate.edu/smallholderagriculture/RLT-RiceProcessing-NIG225.pdf.

Titapiwatanakun, B. 2012. "The Rice Situation in Thailand." Working paper, Technical Assistance Consultant's Report, Asia Development Bank (ADB), Manila. Accessed on July 30, 2015. www.adb.org/sites/default/files/project-document/73082/43430-012-reg-tacr-03.pdf.

Tóth, G., B. Kozlowski, S. Prieler, and D. Wiberg. 2012. *Global Agro-Ecological Zones (GAEZ v3.0): User's Guide.* Laxenburg, Austria: International Institute for Applied Systems Analysis; Rome: Food and Agriculture Organization of the United Nations.

Totin, E., L. Stroosnijder, and E. Agbossou. 2013. "Mulching Upland Rice for Efficient Water Management: A Collaborative Approach in Benin." *Agricultural Water Management* 125: 71–80.

Townsend, R. M. 1995. "Financial Systems in Northern Thai Villages." *Quarterly Journal of Economics* 110 (4): 1011–1046.

Traxler, G., and D. Byerlee. 1993. "A Joint-Product Analysis of the Evolution and Adoption of Modern Cereal Varieties in Developing Countries." *American Journal of Agricultural Economics* 75 (4): 981–989.

Ukaoha, K., and E. Ngene. 2013. "A Review of the Proposed 2013 Agriculture Budget for Achieving the Transformation Agenda." National Association of Nigerian Traders (NANTS), Abuja, Nigeria. Accessed on July 30, 2015. www.inter-reseaux.org/IMG/pdf/NANTS_Review_of_2013_proposed_Agric_budget.pdf.

Ukwungwu, M. N., and M. E. Abo. 2004. "Nigeria Rice in the Science and Technology Vista." In *Nigeria Rice Memorabilia*, edited by M. E. Abo and A. S. Abdullahi, 49. Abuja, Nigeria: Project Synergy Limited.

Ukwungwu, M. N., and R. C. Joshi. 1992a. "Distribution of the African Rice Gall Midge, *Orseolia oryzivora* Harris and Gagné and Its Parasitoids in Nigeria." *Tropical Pest Management* 38: 241–244.

———. 1992b. "Evaluation of Asian Gall Midge Resistant Varieties for the African Rice Gall Midge." *International Rice Research Newsletter* 17: 12.

Ukwungwu, M. N., and S. M. Misari. 1997. "Management of African Rice Gall Midge, *Orseolia oryzivora* Harris and Gagné in Nigeria." *African Plant Protection* 7: 27–34.

Ukwungwu, M. N., M. D. Winslow, and V. T. John. 1989. "Severe Outbreak of Rice Gall Midge (GM) in the Savanna Zone, Nigeria." *International Rice Research Newsletter* 14 (4): 36–37.

UN Statistics Division. 2012. United Nations Commodity Trade Statistics database (COMTRADE). Accessed on June 11, 2015. comtrade.un.org.

UNEP (United Nations Environmental Program). 2005. *Integrated Assessment of the Impact of Trade Liberalization: A Country Study on the Nigerian Rice Sector.* Geneva, Switzerland: United Nations Environmental Program. www.unep.ch/etb/publications/intAssessment/Nigeria.pdf.

University of East Anglia. 2012. Historical Data on Global Precipitation from the Climate Research Unit. Accessed on June 11, 2015. www.cru.uea.ac.uk/data.

Unnevehr, L. J. 1986. "Consumer Demand for Rice Grain Quality and Returns to Research for Quality Improvement in Southeast Asia." *American Journal of Agricultural Economics* 68 (3): 634–641.

USAID (United States Agency for International Development). 2010. *Transport and Logistics Costs on the Tema-Ouagadougou Corridor.* West Africa Trade Hub Technical Report #25. Washington, DC: United States Agency for International Development.

USDA (United States Department of Agriculture). 2013. "Production, Supply and Distribution Online." Foreign Agricultural Services, USDA. Accessed on June 11, 2015. www.fas.usda.gov/psdonline/.

———. 2014. Agricultural Exchange Rate Data Set. Washington, DC. Computer disk.

USDA–ERS (Economic Research Service). Various dates. International Macroeconomic Data Set. Accessed on June 11, 2015. www.ers.usda.gov/data-products/international-macroeconomic-data-set.aspx.

USDA Foreign Agricultural Service. 2013. "THAILAND: Deficient Irrigation Supplies Reduce Dry Season Rice Production." Foreign Agricultural Service. Accessed on June 11, 2015. www.pecad .fas.usda.gov/highlights/2013/03/Thailand.

von Grebmer, K., D. Headey, T. Olofinbiyi, D. Wiesmann, H. Fritschel, S. Yin, and Y. Yohannes. 2013. *2013 Global Hunger Index: The Challenge of Hunger: Building Resilience to Achieve Food and Nutrition Security.* Washington, DC: International Food Policy Research Institute.

von Oppen, M., and J. T. Scott. 1976. "A Spatial Equilibrium Model for Plant Location and Interregional Trade." *American Journal of Agricultural Economics* 58: 437.

von Thunen, J. H. 1826. *Der Isolierte Staat in Beziehung auf Landwirtschaft und National'konomie* ("The Isolated States"). Hamburg: Perthes Verlag.

Wada, A. C., S. O. Bakare, M. Bashir, and M. N. Ukwungwu. 2012. "Integrated Management of Africa Rice Gall Midge (AfRGM) (*Orseolia oryzivora,* Harris and Gagné), Weeds and Diseases of Lowland Rice." *Agricultural Sciences* 3 (4): 486–490.

Walkenhost, P. 2009. "Nigeria." In *Distortions to Agricultural Incentives in Africa,* edited by K. Anderson and W. A. Masters, 441–462. Washington, DC: World Bank.

Wang, H., S. Pandey, O. Velarde, and B. Hardy. 2012. "Patterns of Varietal Adoption and Economics of Rice Production in Asia." Unpublished, International Rice Research Institute, Los Baños, Philippines.

Wattanutchariya, S. 1983. "Economic Analysis of Farm Machinery Industry and Tractor Contractor Business in Thailand." In *Consequences of Small-Farm Mechanization,* edited by International Rice Research Institute and Agricultural Development Council, 39–50. Los Baños, Philippines: International Rice Research Institute.

Weber, A. 1929. *Theory of the Location of Industries.* Translated by C. Friedrich. Chicago: University of Chicago Press.

Welsch, D. 1965. "Response to Economic Incentive by Abakaliki Rice Farmers in Eastern Nigeria." *American Journal of Agricultural Economics* 47 (4): 900–914.

Wijayaratna, C. M., and K. Hemakeerthi. 1992. "Production and Profitability of Rice Cultivation in Sri Lanka under Different Water Regimes: A Trend Analysis." In *Advancements in IIMI's Research, 1992: A Selection of Papers Presented at the Internal Program Review,* 277–308. International Irrigation Management Institute, Colombo, Sri Lanka.

Williams, C. T., O. Okhidievbie, K. M. Harris, and M. N. Ukwungwu. 1999a. "The Host Range, Annual Cycle and Parasitoids of the African Rice Gall Midge *Orseolia oryzivora* (Diptera: Cecidomyiidae) in Central and Southeast Nigeria." *Bulletin of Entomological Research* 89: 589–597.

Williams, C. T., M. N. Ukwungwu, B. N. Singh, O. Okhidievbie, and J. Nnabo. 1999b. "Farmer-Managed Trials in South-East Nigeria to Evaluate the Rice Variety Cisadane and Estimate Yield Losses Caused by the African Rice Gall Midge, *Orseolia oryzivora* Harris and Gagné. *International Journal of Pest Management* 45 (2): 117–124.

Windmeijer, P., and W. Andriesse. 1993. *Inland Valleys in West Africa: An Agro-Ecological Characterization of Rice-Growing Environments.* Publication 52. Wageningen, Netherlands: International Institute for Land Reclamation and Improvement.

Wopereis, M. C. S., C. Donovan, B. Nebié, D. Guindo, and M. K. N'Diaye. 1999. "Soil Fertility Management in Irrigated Rice Systems in the Sahel and Savanna Regions of West Africa. Part I. Agronomic Analysis." *Field Crops Research* 61 (2): 125–145.

World Bank. 1998. "India—Water Resources Management Sector Review: Irrigation Sector." New Delhi: World Bank.

———. 2007. "World Development Report 2008: Agriculture for Development." Washington, DC: World Bank.

———. 2012. World Development Indicators database. Accessed on June 11, 2015. http://data.worldbank.org/data-catalog/world-development-indicators.

———. 2014. World Development Indicators. Washington, DC: World Bank.

World Meteorological Organization. 2013. "Precipitation." Accessed on June 11, 2015. http://data.un.org/Explorer.aspx?d=CLINO .

Yahaya, M. 2002. "Development and Challenges of Bakolori Irrigation Project in Sokoto State, Nigeria." *Nordic Journal of African Studies* 11 (3): 411–430.

Yoshida, S. 1971. *Fundamentals of Rice Crop Science.* Los Baños, Philippines: International Rice Research Institute.

———. 1978. "Tropical Climate and Its Influence on Rice." IRRI Research Paper 20. International Rice Research Institute, Los Baños, Philippines. Accessed on July 30, 2015. http://pdf.usaid.gov/pdf_docs/PNAAG711.pdf.

Authors

Kwabena Gyimah-Brempong (kgyimah@usf.edu) is a professor and former department chair of the Department of Economics, University of South Florida, Tampa, US. He was formerly a senior research fellow in the Development Strategy and Governance Division of the International Food Policy Research Institute (IFPRI), Washington, DC, and the program leader of IFPRI's Nigeria office. His publications include "Remittances and Investment in Education: Evidence from Ghana," *Journal of International Trade and Economic Development* 24 (2), coauthored with Elizabeth Asiedu; "Higher Education and Economic Development," *African Development Review* 23 (2); "The Effects of College Education on the Male-Female Wage Differential," *Southern Economic Journal* 58 (3), coauthored with Rudy Fichtenbaum and Greg Willis; "Export Instability and Economic Growth in Sub-Saharan Africa," *Economic Development and Cultural Change* 39 (4); "Elasticity of Factor Substitution in Police Production: The Florida Experience," *Journal of Business and Economic Statistics* 5 (2); and "Scale Elasticities in Ghanaian Cocoa Production," *Applied Economics* 19 (10).

Michael Johnson (m.johnson@cgiar.org) is a senior research fellow in the Development Strategy and Governance Division of IFPRI, Washington, DC. His publications include "Exploring Strategic Priorities for Regional Agricultural R&D Investments in East and Central Africa," *Agricultural Economics* 41 (2), coauthored with Liangzhi You; "Toward a Green Revolution in Africa: What Would It Achieve, and What Would It

Require?," *Agricultural Economics* 39 (S1), coauthored with Xinshen Diao and Derek Headey; "Complementarity and Sequencing of Innovations: New Varieties and Mechanized Processing for Cassava in West Africa," *Economics of Innovation and New Technology* 13 (1), coauthored with William A. Masters; "The Role of Intermediate Factor Markets in Asia's Green Revolution: Lessons for Africa?" *American Journal of Agricultural Economics* 85 (5), coauthored with Peter Hazell and Ashok Gulati; and *Reforming Agricultural Markets in Africa: Achievements and Challenges*, coauthored with Mylene Kherallah, Christopher Delgado, Eleni Gabre-Madhin, and Nicholas Minot.

Hiroyuki Takeshima (h.takeshima@cgiar.org) is a research fellow in the Development Strategy and Governance Division of IFPRI, Washington, DC. His publications include "Sales Location among Semi-subsistence Cassava Farmers in Benin: A Heteroskedastic Double Selection Model," *Agricultural Economics* 43 (6), coauthored with Alex Winter-Nelson; "Risks and Farmers' Investment in Productive Assets in Nigeria," *Agricultural Economics* 43 (2), coauthored with Futoshi Yamauchi; "Government Fertilizer Subsidy and Commercial Sector Fertilizer Demand: Evidence from the Federal Market Stabilization Program (FMSP) in Nigeria," *Food Policy* 47, coauthored with Ephraim Nkonya; and "Fertilizer Subsidies, Political Influence and Local Food Prices in Sub-Saharan Africa: Evidence from Nigeria," *Food Policy* 54, coauthored with Lenis Saweda O. Liverpool-Tasie.

Oluyemisi Kuku-Shittu (O.Kuku@cgiar.org) was an associate research fellow in the Development Strategy and Governance Division of IFPRI, Washington, DC, and the Nigeria Strategy Support Program, Abuja, at the time she contributed to this book. Her publications include the *Comprehensive Food Security and Vulnerability Analysis: Nigeria*, IFPRI Discussion Paper 01275, coauthored with Astrid Mathiassen, Amit Wadwa, Lucy Myles, and Akeem Ajibola; and "Differences in Food Insecurity between Adults and Children in Zimbabwe," *Food Policy* 36 (2), coauthored with Craig Gundersen and Steve Garasky.

Paul A. Dorosh (P.Dorosh@cgiar.org) is the director of the Development Strategy and Governance Division of IFPRI, Washington, DC. His publications include "Can Cities or Towns Drive African Development? Economywide Analysis for Ethiopia and Uganda," *World Development* 63, coauthored with James Thurlow; *Food and Agriculture in Ethiopia: Progress and Policy*, coedited with Shahidur Rashid; and *Economic Policy and Poverty in Africa: Structural Adjustment Reconsidered* coauthored with David E. Sahn and Stephen D. Younger.

Mehrab Malek (M.Malek@cgiar.org) is a research analyst in the Development Strategy and Governance Division of IFPRI, Washington, DC. His publications include "Modern Input Promotion in Sub-Saharan Africa," *Agricultural Economics* 44 (6), coauthored with Shahidur Rashid, Paul Dorosh, and Solomon Lemma.

Xinshen Diao (X.diao@cgiar.org) is a senior research fellow and the deputy director of the Development Strategy and Governance Division of IFPRI, Washington, DC. Her publications include *Strategies and Priorities for African Agriculture: Economywide Perspectives from Country Studies*, coedited with James Thurlow, Sam Benin, and Shenggen Fan; *A New Era of Transformation in Ghana: Lessons from the Past and Scenarios for the Future*, coauthored with Clemens Breisinger, Shashidhara Kolavalli, Ramatu Al Hassan, and James Thurlow; and *Agricultural Growth and Investment Options for Poverty Reduction in Rwanda*, coauthored with Shenggen Fan, Sam Kanyarukiga and Bingxin Yu.

Bakare Samuel Oladele (oladelebakare@yahoo.co.uk) is an agronomist and head of the Research Outreach Department of the National Cereals Research Institute, Bida, Niger State, Nigeria. His publications include "Evaluation of Aratibiotech Organomineral Fertilizer on Lowland Rice Production," *Nigerian Journal of Soil Science* 23 (2), coauthored with Abdulrazak Aliyu, Joy E. Ekaette, and Mark N. Ukwungwu; "Field Evaluation of Biowish-Crop Organic Fertilizer in Lowland Rice Production, *Nigerian Journal of Soil Science* 23 (2), coauthored with Gbanguba Alhassan, Joy E. Ekaette, and Mark N. Ukwungwu; "Integrated Management of African Rice Gall Midge (AfRGM) (*Orseolia oryzivora*, Harris and Gagné), Weeds and Diseases of Lowland Rice," *Agricultural Sciences* 3 (4), coauthored with Akaama Clement Wada, Mohammed Bashir, and Mark Noye Ukwungwu.

Akeem Ajibola (aajibola@fews.net) is an assistant national technical manager of the Famine Early Warning Systems Network, Abuja, Nigeria (FEWS NET

– NIGERIA). Formerly he was a senior research assistant in the Development Strategy and Governance Division of IFPRI, Washington, DC, and the Nigeria Strategy Support Program, Abuja, Nigeria. His publications include "Rural Livelihood in South-West Nigeria: Strategies, Asset Ownerships and the Non-farm Income," *Research Journal of Applied Sciences* 6 (5), coauthored with Kemi Adenegan.

INDEX

Page numbers for entries occurring in boxes are followed by a *b;* those for entries in figures, by an *f;* those for entries in notes, by an *n;* and those for entries in tables, by a *t.*